AF754364

"The maximum proportion of the chapters in this mega book project comprise the ones which describe significant genetically modified crops from the earliest ones to the latest. This piece of work will certainly be impactful over the controversies and challenges that have been hovering over these various genetically-modified foods. Apart from readership point of view, it will prove out to be a great guiding tool for agronomists, agriculturalists, food biotechnologists, food scientists as well etc. Great effort indeed."

Dr. Shweta Kumari, *Women Scientist, Department of Biochemical Engineering and Biotechnology, Indian Institute of Technology*

"This book will be a blessing to academics and researchers alike, since it will provide a complete review of one of the most contentious subjects today: genetically modified crops. This work stands out among other books on the subject because it provides a comprehensive and balanced overview of key aspects of a number of well-known genetically modified crops, including genetic modification requirements and mechanisms, genes in agronomic trait development, and societal benefits. Each of the book's chapters will be unique in terms of providing detailed information on various genetically modified crops."

Dr. Saurabh Kumar Jha, Ph.D., *Assistant Professor, Department of Biotechnology, School of Engineering and Technology, Sharda University*

Genetically Modified Crops and Food Security

This book reviews a wide-range of genetically modified (GM) crops to understand how they are produced, their impacts on the agricultural industry, and their potential for improving food security.

The production of GM crops has now become an invaluable asset in the agricultural toolbox. With a significant portion of the world suffering from hunger and poverty, this book examines how food security can be achieved through GM crops. A wide variety of crops are examined, from the earliest developments of GM tomatoes and potatoes to more recent interest in the development of low-cost, high yielding biofuels, such as microalgae. Chapters also discuss the role of GM crops in pest management and the consequential reduction in the use of insecticides. Overall, this book provides an important synthesis of GM crops from their commercial value to the agricultural industry, as well as their potential for improving food security.

This book will be of great interest to students and scholars of agricultural engineering, crop science, food biotechnology, food security, and those interested in food and agriculture and sustainable development more broadly.

Jasmeet Kour is an Assistant Professor (academic arrangement), Department of Food Science and Technology, Padma Shri Padma Sachdev, Govt PG College for Women Gandhi Nagar, Jammu-180003 Jammu & Kashmir, India.

Vishal Sharma is presently working as Manager Industrial Estate at District Industries Centre, India.

Imtiyaz Khanday is an Assistant Professor of Plant Reproductive Biology and also Director of Research at the Seed Biotechnology Center of the Department of Plant Sciences at the University of California, Davis, USA.

Earthscan Food and Agriculture

The United Nations' Declaration on Peasants' Rights
Edited by Mariagrazia Alabrese, Adriana Bessa, Margherita Brunori and Pier Filippo Giuggioli

The Agricultural Dilemma
How Not to Feed the World
Glenn Davis Stone

Farming with the Environment
Thirty Years of Allerton Project Research
Chris Stoate

Net Zero, Food and Farming
Climate Change and the UK Agri-Food System
Neil Ward

Conservation Agriculture in India
A Paradigm Shift for Sustainable Production
Edited by A.R. Sharma

Agricultural Commercialization, Gender Equality and the Right to Food
Insights from Ghana and Cambodia
Edited by Joanna Bourke Martignoni, Christophe Gironde, Christophe Golay, Elisabeth Prügl and Dzodzi Tsikata

The Sociology of Farming
Concepts and Methods
Jan Douwe van der Ploeg

Genetically Modified Crops and Food Security
Commercial, Ethical and Health Considerations
Edited by Jasmeet Kour, Vishal Sharma and Imtiyaz Khanday

For more information about this series, please visit: www.routledge.com/books/series/ECEFA/

Genetically Modified Crops and Food Security

Commercial, Ethical and Health Considerations

Edited by
Jasmeet Kour, Vishal Sharma and Imtiyaz Khanday

Routledge
Taylor & Francis Group
LONDON AND NEW YORK

earthscan
from Routledge

First published 2023
by Routledge
4 Park Square, Milton Park, Abingdon, Oxon OX14 4RN

and by Routledge
605 Third Avenue, New York, NY 10158

Routledge is an imprint of the Taylor & Francis Group, an informa business

© 2023 selection and editorial matter, Jasmeet Kour, Vishal Sharma and Imtiyaz Khanday; individual chapters, the contributors

The right of Jasmeet Kour, Vishal Sharma and Imtiyaz Khanday to be identified as the authors of the editorial material, and of the authors for their individual chapters, has been asserted in accordance with sections 77 and 78 of the Copyright, Designs and Patents Act 1988.

All rights reserved. No part of this book may be reprinted or reproduced or utilised in any form or by any electronic, mechanical, or other means, now known or hereafter invented, including photocopying and recording, or in any information storage or retrieval system, without permission in writing from the publishers.

Trademark notice: Product or corporate names may be trademarks or registered trademarks, and are used only for identification and explanation without intent to infringe.

British Library Cataloguing-in-Publication Data
A catalogue record for this book is available from the British Library

Library of Congress Cataloging-in-Publication Data
Names: Kour, Jasmeet, editor. | Sharma, Vishal (Biotechnologist), editor. | Khanday, Imtiyaz, editor.
Title: Genetically modified crops and food security : commercial, ethical and health considerations / Jasmeet Kour, Vishal Sharma, Imtiyaz Khanday.
Description: New York, NY : Routledge, 2023. | Includes bibliographical references and index.
Identifiers: LCCN 2022016427 (print) | LCCN 2022016428 (ebook) | ISBN 9781032244297 (hardback) | ISBN 9781032244310 (paperback) | ISBN 9781003278566 (ebook)
Subjects: LCSH: Transgenic plants. | Genetically modified foods. | Food security.
Classification: LCC SB123.57 .G478123 2023 (print) | LCC SB123.57 (ebook) | DDC 631.5/233--dc23/eng/20220822
LC record available at https://lccn.loc.gov/2022016427
LC ebook record available at https://lccn.loc.gov/2022016428

ISBN: 978-1-032-24429-7 (hbk)
ISBN: 978-1-032-24431-0 (pbk)
ISBN: 978-1-003-27856-6 (ebk)

DOI: 10.4324/9781003278566

Typeset in Goudy
by Taylor & Francis Books

MIX
Paper | Supporting responsible forestry
FSC™ C013985

Printed in the United Kingdom
by Henry Ling Limited

Contents

Illustrations

Figures

Tables

Preface

Genetically Modified Crops and Food Security: Commercial, Ethical and Health Considerations provides a comprehensive overview of what is currently a very contentious issue – genetically modified crops. All the individual chapters covered in this book are distinctive and unique in terms of delivering detailed information on various genetically modified crops. As an introduction, the first chapter presents an interesting overview of genetically modified plants and animal foods. This is followed by a series of chapters describing significant genetically modified crops, including the earliest ones to be engineered such as GM tomato, cotton, canola, corn, potato, and soyabean. Genetically modified tomato or *Flavr Savr* was the first commercially grown genetically engineered food which was granted a license for human consumption. Bt cotton is a genetically modified cotton variety created through the incorporation of genes encoding toxin crystals in the Cry group of endotoxin to produce an insecticide to counter bollworm and thus make the crop resistant to pests. This has led to a noticeable increase in the yield of cotton as a result of the effective control of three types of bollworms: American, Spotted, and Pink. Other significant GM crops that are discussed and reviewed this book are genetically modified alfalfa, that possesses high herbicidal resistance, and rainbow papaya.

Apart from the genetically modified crops themselves, in recent years enormous emphasis has been placed on how to reduce the cost of biofuel, lower greenhouse emissions, and meet land and water resource needs. Owing to an attribute of fourth-generation microalgae to photosynthetically fix carbon dioxide, irrespective of special and costly culture media, in this publication there will also be a review of the potential avenues of genetic engineering can take in order to utilize these microalgae as a biofuel to produce biohydrogen, starch-derived alcohols, diesel fuel surrogates, and/or alkanes.

Another pivotal chapter highlighting the outcomes of a genetic engineering technique concerns sugar beet; a technique which has noticeably minimized the use of chemicals such as weedicide. In addition, a further chapter reports on how genetically engineered crops have played an important role in integrated pest management, eventually reducing broad-spectrum insecticides and enhancing yield.

As a whole, *Genetically Modified Crops and Food Security* compiles, organizes, and explains various eminent areas related to genetic modification and will be essential reading for understanding the controversies and challenges that have been hovering over these various genetically modified foods since their department. Apart from a readership perspective, this volume will prove to be a great guiding tool for agronomists, agriculturalists, food biotechnologists, and food scientists etc.

Contributors

Dr Jasmeet Kour completed her Master's Degree in Food Science and Technology from Government College for Women, Gandhi Nagar, Jammu University, Jammu and Kashmir, India. She was awarded a doctoral degree at Sant Longowal Institute of Engineering and Technology, Longowal, Sangrur, Punjab, India from Department of Food Engineering and Technology in 2019. She has been serving as an assistant professor (academic arrangement basis) at the Department of Food Science and Technology, Government College for Women, Gandhi Nagar, Jammu, Jammu and Kashmir since 2009. She has conducted her vast research in various prominent nutraceuticals derived from plant origin and her work has been published in reputed journals with high impact factors in eminent publishing houses in the field of food science. She has presented her research and review paper presentations in various national and international conferences. She has authored and co-authored numerous book chapters and scientific articles published in international books with prestigious publishing names such as Elsevier and Springer as well. She is also a part of various international projects. As her research work left a great imprint, she was also cordially invited to share her newest research findings at Tokyo University of Agriculture, Japan at the 6th International Conference on Agricultural and Biological Sciences (ABS 2020). She is currently working as an editorial board member and peer reviewer of various journals of repute. She is also credited with editorship of various book projects published by names such as Elsevier and CRC Press (Taylor and Francis group).

Vishal Sharma completed his Master's Degree in Biotechnology from Shri Mata Vaishno Devi University, Katra, Jammu and Kashmir, India in 2012. He was awarded a doctoral degree in Science by the Academy of Scientific and Innovative Research for research work carried out by him at the Microbial Biotechnology Division, CSIR-Indian Institute of Integrative Medicine, Canal Road, Jammu, India. He has qualified under various national examinations like GATE, DBT-JRF, ICMR-JRF, ICAR-NET and CSIR-NET. He is presently working as Block Investigator at District Industries Centre, Samba, Jammu and Kashmir. His research title during his PhD was "Effect of Epigenetic Modifiers on Secondary Metabolite Production in Endophytic Fungi of Datura innoxia Mill.". He has published various research and review papers in both national and international journals and has also contributed to various books published by Springer and Elsevier. He has twice been awarded the Microbial Culture Repository Award by CSIR-IIIM for the large number of microbial cultures deposited by him. During his research, he has remained a part of many national and international projects and presented numerous research papers at both national and international conferences. He has great experience of writing research, review papers, and book chapters as is evident from his publications.

Imtiyaz Khanday obtained his PhD from the Department of Microbiology and Cell Biology, Indian Institute of Science, Bangalore, India in 2014. During his graduate studies, he worked on the molecular genetics of flower development in rice and deciphered the role of *OsMADS1*, a master regulator transcription factor, in floral organ identity regulation and floral meristem termination. He then moved to the United States and started his postdoctoral research in the Department of Plant Biology at University of California, Davis. Dr. Khanday has more than 14 years of experience as a researcher and a scientific writer. His research work has been published in various journals of international repute and also highlighted by various news outlets around the world for its potential to ensure food security for the growing world population. Dr. Khanday has also presented his work as 'plenary' or 'invited speaker' at various international conferences like PAG2020, ISRFG2019, plantbiology2019 etc. Dr. Khanday is also the recipient of many fellowships, patents, and awards including the 'Outstanding Postdoc award' (2019) from Department of Plant Biology, UC Davis.

Nidhi Agarwal, Department of Food Technology, Vasantrao Naik Marathwada Agriculture University, Prabhani, Maharashtra

Abida Bhat, Department of Immunology and Molecular Medicine, Sher-e-kashmir Institute of Medical Sciences, Kashmir

Akanksha Bakshi, Department of Botanical and Environmental Sciences, Guru Nanak Dev University, Amritsar, Punjab 143005

Manish Bakshi, Department of Horticulture, School of Agriculture, Lovely Professional University, Jalandhar, Phagwara, Punjab 144411

Kiran Bala, Department of Biotechnology, GGM Science College, Canal Road, Jammu, Jammu and Kashmir 180001

Rosy Bansal, Department of Food Processing and Engineering, GSSDGS Khalsa College Patiala-147001, Punjab

Nancy Bhura, Department of Molecular Biology and Genetic Engineering, School of Bioengineering and Biosciences, Lovely Professional University, Jalandhar, Phagwara, Punjab 144411

Sayantan Chakraborty, Department of Food Technology, Mizoram University, Aizawl, Mizoram

Ramesh Chandra Mishra, Department of Food Technology, Mahakaushal University, Jabalpur, Madhya Pradesh 482003

Himjyoti Dutta, Department of Food Technology, Mizoram University, Aizawl, Mizoram

Isha Gupta, Department of Food Science and Technology, SKUAST-Jammu, Jammu and Kashmir 180009

Jeena Gupta, Department of Biochemistry, School of Bioengineering and Biosciences, Lovely Professional University, Jalandhar, Phagwara, Punjab 144411

Prerna Gupta, Department of Food Technology & Nutrition, School of Agriculture, Lovely Professional University, Jalandhar, Phagwara, Punjab 144411

Bhaskar Jyoti, School of Agriculture, Mahakaushal University, Jabalpur, Madhya Pradesh

Mahaldeep Kaur, Department of Microbial Biotechnology, Panjab University, Chandigarh 160014

Mandeep Kaur, Department of Agriculture, Mata Gujri College, Fatehgarh Sahib, Punjab, 140407

Rupinderpal Kaur, Department of Bioengineering and Biosciences, Lovely Professional University, Jalandhar, Phagwara, Punjab 144411

Naveet Kaushal, Department of Agriculture, Mata Gujri College, Fatehgarh Sahib, Punjab, 140407

Jasmeet Kour, Department of Food Science and Technology, Padma Shri Padma Sachdev, Govt PG College for Women Gandhi Nagar, Jammu-180003 Jammu & Kashmir

Kawaljeet Kour, Department of Food Science and Technology, Padma Shri Padma Sachdev, Govt PG College for Women Gandhi Nagar, Jammu-180003 Jammu & Kashmir

Amarjeet Kumar, Department of Home Science, Rohtas Mahila College, Sasaram, Veer Kunwar Singh University, Arrah, Bihar

Varun Kumar, Department of Home Science, Ramesh Jha Mahila College, Saharsa, Bhupendra Narayan Mandal University, Madhepura, Bihar

Bharti Mittu, National Institute of Pharmaceutical Education and Research (NIPER), S.A.S. Nagar, Mohali

Shivali Panjgotra, Department of Environment Studies, Panjab University, Chandigarh, 160014

Sangeeta, Guru Nanak College, Budhlada, Mansa, Punjab

Anu Sharma, General Zorawar Singh Memorial Degree College Reasi, Jammu and Kashmir 182311

Bharati Sharma, Department of Food Technology, Guru Nanak Dev University, Amritsar-143005, Punjab

Kanchan Sharma, Department of Home Science, University of Jammu, Jammu and Kashmir

Namrata Sharma, Department of Botany, University of Jammu, Jammu 180006

Nitika Sharma, Department of Botanical and Environmental Sciences, Guru Nanak Dev University, Amritsar, Punjab 143005

Priya Sharma, Department of Botany, University of Jammu, Jammu, 180006

Renu Sharma, Department of Chemistry, Akal Degree College Mastuana Sahib, Sangrur, Punjab 148001

Richa Sharma, Department of Plant Pathology, Sher-e-Kashmir University of Agricultural Sciences and Technology – Jammu

Vishal Sharma, District Industries Centre, Samba, Department of Industries and Commerce, Jammu and Kashmir

Ranjeet Singh Jamwal, Govt. SPMR College of Commerce, Canal Road, Jammu, Jammu and Kashmir 180002

Talwinder Singh, Department of Agriculture, Mata Gujri College, Fatehgarh Sahib, Punjab, 140407

Nitesh Sood, Department of Agriculture, Mata Gujri College, Fatehgarh Sahib, Punjab, 140407

Amrit Sudershan, Institute of Human Genetics, University of Jammu, 180006

Mansi Tomar, Department of Food Engineering and Technology, Sant Longowal Institute of Engineering and Technology, Longowal, Punjab 148106

1 Genetically modified plant and animal foods

An overview

Jasmeet Kour, Sangeeta, Himjyoti Dutta, Mansi Tomar, Kiran Bala and Ramesh Chandra Mishra

1.1 Introduction

Since time immemorial, humans have been constantly trying to secure reliable sources of food provision. With the rapid growth in global population, the challenges of a persistent search for food and the means to preserve it have been added to, with the need to maximize the output from these natural sources. Biotechnological techniques (ancient or modern) have made it possible to carry out extensive research on crops and how they grow which has led to the creation of genetically modified (GM) organisms. The year 1983 marks the production of the first GM plant in a greenhouse using an antibiotic resistance gene taken from bacteria (Lemaux, 2008). At present, the principal transgenic crops grown commercially are herbicide-resistant and insect-resistant. Herbicide resistance arises from genes present in these GM plants that inactivate the ingredients in herbicides (Schütte et al., 2017); likewise, genes from *Bacillus thuringiensis* (Bt) are inserted into plants to impart insect-resistant properties (Gatehouse et al., 2011). In fact, plants with stacked genes – e.g. with a combination of herbicide and insect resistance – have become an integral part of genetic engineering.

In 1994, *Flavr Savr* tomato got its approval for sale in the USA (Wong & Chan, 2016). Europe also accepted its first herbicide-resistant GM crop in that year – genetically engineered tobacco. Following this, Bt maize, cotton resistant to bromoxynil, Bt cotton, soybeans that can withstand glyphosate, virus-resistant squash, and tomatoes with delayed ripening were approved in 1995 and vitamin A enriched Golden Rice was developed in 2000 (Tang et al., 2009). However, this is just a sample of GM crops; other transgenic crops and foods include Bt brinjal (aubergine), genetically engineered potato, herbicide-tolerant alfalfa, 'Rainbow' papaya, GM canola, GM sugar beet, etc.

1.2 Production of GM foods: an overview

The alteration of the genetic makeup of animals, plants, or microorganisms with the use of gene technology is referred to as genetic modification. The technology is also known as modern biotechnology or recombinant DNA

DOI: 10.4324/9781003278566-1

technology. GM organisms are made by transferring the genes of one organism into another, usually from a different species. In all organisms, the DNA component is composed of amino acid sequences which relate to the genetic make-up of the organism. The alteration or movement of these amino acids from one organism to another organism can be carried out by means of recombinant technology and/or molecular cloning. The two enzymes that have a significant role in this process are restriction enzymes, which are used to cut DNA segments from one genome, and DNA ligases, which integrate these cut segments into another genome.

Assembling the desirable genes in one plant using traditional methods is intricate and time-consuming. Therefore, from the perspective of economics and agriculture, it is beneficial to grow crops by inserting genes that encode information for higher yield or improved quality, pest or disease resistance, and/or heat, cold, and drought tolerance. In the agricultural field, GM crops can provide insect resistance, herbicide tolerance, and virus resistance. In the medical realm, gene technology can be used in the treatment of diabetes mellitus, the production of human growth hormone, the production of blood clotting factors VIII and IX, and the treatment of cystic fibrosis, among others. Transgenic exchange can also be used for the treatment of contaminated land and water, and in the manufacturing of useful chemicals such as enzymes, etc.

Currently available GM foods stem mostly from plants, and the availability of food derived from GM microorganisms or GM animals is limited – though in the future these are likely to be seen in the market in bulk. In the future, genetic modifications will be targeted to alter the nutritional profile of food, diminishing the allergenic content as well as improving the food production systems. Yet, all GM foods should be subjected to assessment before being launched onto the market and, that being so, a risk analysis of GM food is carried out by following guidelines set by the FAO/WHO Codex.

1.2.1 *World production of GM food*

According to a statistical report generated by Statista (Shahbandeh, 2022), out of the total area under coverage by genetically engineered crops across the globe in 2019 (based on acreage), the largest proportion was in the United States at 71.5 million hectares followed by Brazil at 52.8 million hectares. This was followed by Argentina (24 million hectares), Canada (12.5 million hectares), and India (11.9 million hectares). Vietnam came at the bottom, with 0.1 million hectares used for GM food.

1.2.2 *GM crop market*

As of 2019, the most widespread distribution of GM crops includes soybean, corn, cotton, and canola. Cotton grown using gene technology alone comprises of 13.5 percent of the total cotton grown globally. This is similar to the trend exhibited by GM soybeans, which constitute 48.2 percent of all soybeans grown. The market involved in GM seeds and pesticides is the concern of just a few major companies.

1.3 Benefits of GM foods

GM food is a viable alternative to meet the increasing global food demand. GM crops can be grown in marginal soil, which reduces the strain on the resources available, and these crops are even resistant to drought, salt, herbicides, pesticides, frost, disease, and floods which results in better yields. A few GM techniques aim to include pivotal vitamins such as A, C, E as well as unsaturated fatty acids, alimentary cellulose, and probiotics, thus delivering substances having a high therapeutic and pro-health value (Schell & Van Montagu, 1977). In this way, nutritional goals can be met by designing foods for specific purposes. Moreover, crops such as rice, soybean, maize, and potatoes are being given due consideration for delivering vaccines against different infections (Nicolia et al., 2014).

However, with the advancement of GM crops, there is also the possibility of disruption to the food chain. As an example, the development of insect-resistant plants may lead to an increase in the number of minor pests resulting from the reduction of major pests controlled by the modification, which will then compete for the existing food. Antibiotic resistance is another major concern. Although the US regulatory agencies have accepted GM foods to be safe (Belson, 2000), followed by their acceptance in Canada (Smyth, 2014), Argentina (Lewi & Vicién, 2020), and Brazil (Hakim et al., 2020), Europe (Stein, 2015) and Japan (Komoto et al., 2016) have largely a negative regulatory and consumer reaction. Recently, there has been controversy surrounding GM foods and their effects on the environment and health. In the United States, only a fraction of consumers believe GM food to be less healthy in comparison with non-GM foods.

1.4 GM plants and animal food – health and ethical issues/risks

Genetic modification is a potent biotechnological approach whereby a special set of genes, or assemblies of specifically constructed genes, are transferred to a non-GM crop or organism by means of a variety of techniques, thus altering the genetic machinery of a related living subject such as organisms (animals), plants or microorganisms (Bawa & Anilakumar, 2013). The most common GM plant crops grown at the commercial level are corn, soybean, canola, and cotton which have insect-resistant and herbicide-resistant properties. Other commercial crops/foods produced by genetic modifications are virus-resistant (viruses are responsible for most of the African harvest destruction) sweet potato, iron and vitamin-enriched rice, plants with survival capabilities in extreme weather, and the production of human vaccines for hepatitis B (infectious diseases) from bananas, etc.

Yet, regardless of many benefits of GM foods, they have been perceived mostly as a risk/threat to human health as well as to the environment. Most importantly, technologies that are used for genetic modifications offer the greatest challenge because of the risks (both known as well as unknown) associated with them; as is the case for all new technologies. Among all the controversies, risks, and public concerns associated with GM foods (plants and animals), the most common issues are the safety of the environment and humans, ethical issues, consumer interest and labelling, the reduction of poverty, intellectual property rights, food security, and the conservation of the environment (Prakash et al., 2011). Some indirect effects on the environment related to GM organisms have also been reported (Dively et al., 2004; Pidgeon et al., 2007).

The main questions of concern about this technique are: What are the main threats of this gene modification approach of "interfering with Mother Nature", and its effect on the environment? What are the main health-related issues that a consumer should be made aware of? And, last but not the least, "Is the gene manipulation approach truly beneficial?" (Bawa & Anilakumar, 2013). In Section 1.5 we discuss the health and ethical issues/risks associated with GM foods (both plants and animals).

1.5 Risks associated with GM foods

The concerns linked to the consumption of food are inevitable. The potential health risks of consuming GM foods include toxicity, allergenicity, and genetic hazards. These arise from three potential sources: the inserted gene, secondary or pleiotropic effects of the products of gene expression, and the possible disruption of natural genes in the manipulated organism (Bawa & Anilakumar, 2013). StarLink maize, with its potential for allergenicity, highlights these issues as it provides an example of a food hazard caused directly by the expression of the inserted gene (Tabashnik, 1994). Also, the production of transgenic soybeans enriched with amino acids can result in allergic reactions.

The gene modification technique has the application of transferring genetic material from any species into other organisms or plants. In this gene introduction process, the overall expression pattern of the gene can be changed or altered by a single gene and different outcomes can result (Prakash et al., 2011). Different characteristics can be controlled by each gene in a single organism. The release of GM foods in the open environment could therefore have various predictable as well as unpredictable risks. Though the role of the isolated gene and its sequence may have a well-characterized function in the donor organism, it does not mean that the complete outcome can be predicted (Ellstrand, Prentice, & Hancock, 1999).

Some of the potential risks related to GM foods may be due to the rearrangement and deletion of a gene sequence, the additionally inserted fragment, or integration of copies of a gene etc., as these can cause the instability of the gene and/or interfere in the functionality of other genes (Prakash et al., 2011). Various environmental risks related to the use of GM organisms have been listed by the World Conservation Union (2004), following the report of the IUCN Law Centre. Risks can be, however, difficult to predict; for example, in the host the entire genome can be affected even by a single gene insertion, which can result in unforeseen side effects, which may not all be perceptible at the same time (Prakash et al., 2011).

Risk factors associated with GM organisms include:

1) *Gene interbreeding and contamination.* The interbreeding of GM organisms with non-modified native species may result in the disappearance of their novel traits, unless the advantage in the recipient is selective. The relationship of ecological behaviour and native species may also alter due to the development of tolerance abilities.
2) *Economic and ecological damage.* This can happen as the result of competition of GM organisms with native species. For example, the faster growth rate of GM organisms as compared to native organisms could make them invasive and allow them to spread to new habitats.
3) *Increased pressure of selection.* Species (targeted and non-targeted) are pressured to adapt as the changes introduced by genetic modification may result in the evolution of a population with distinct resistant characteristics.
4) *Impact on the ecosystem.* Impact on the ecosystem in a destructive and damaging way may result from the extended effect of changes in a single species.
5) *Elimination.* Once a problem arises as the result of the introduction of GM organisms, elimination of the problem is likely to be impossible. However, the risks associated with GM organisms can be similar or almost identical to those exhibited by introducing conventional or natural species. Yet, the benefits and safety of GM organisms are still not proven, and on top of that, they are even not scrutinized well regardless of the fact stated above.
6) *Risk of horizontal gene transfer.* This is a particular concern associated with GM organisms (Martin, 1999; Ochman, Lawrence, & Groisman, 2000). This is because the introduction of a foreign gene may transfer a novel attribute to another organism, which, in turn, could have a potentially harmful effect on human health as well as the environment. Some examples are, antibiotic-resistant genes which when transferred to a pathogen can potentially compromise the treatment of humans or animals (Bennett et al., 2004). In spite of the precautions taken, the chance of spreading recombinant genes to other bacteria should not be ignored because a horizontal gene transfer process has been observed in various bacteria genes and in several different environments. Recent data have confirmed that the spread of transgenic DNA in GM products (both plant and animal) can be possible by direct

> transfer by bacteria, viruses, animal cells, and plant cells (Keeling & Palmer, 2008). The data reported by Yoshida et al. (2010) show that during horizontal gene transfer, a gene from a nuclear monocot moved into the genome of *Striga hermonthica* (eudicot parasite witch-weed), and this in turn infected many African grass species.

Thus, horizontal gene transfer from GM foods can have various potential adverse effects on health and the environment (Keese, 2008), such as: the advent of a new weed/pest or disease and increased infection or pathogenicity; which in turn have unpropitious repercussions on species, ecosystems, and communities. Horizontal gene transfer from GM foods may also have effects which are unintended and unpredictable such as the introduction of a gene into potential organisms (pest or pathogen) which can result in alteration of the recipient in terms of its ecological potential or niche (Heuer & Smalla, 2007). Unanticipated structural or functional changes can also occur (Prescott et al., 2005). These may be due to gene insertion in the recipient at different sites during the gene transfer, in which not only the novel gene but a gene responsible for endogenous gene disruption could be introduced. Sometimes, the creation of a new GM organism by the introduction of new gene or genes to another species and its spread may have various adverse effects which can even result in the loss of management control measures as implemented by the regulatory permission or original license. Occasionally, severe long-term impacts have been reported (Nielsen & Townsend, 2004).

Besides the ethical issues, the application of GM products (organisms) in agriculture poses various risks for the environment and human health (Labra et al., 2001). Genetic modifications can, however, provide several opportunities and benefits such as reducing worldwide hunger, combatting malnutrition, and alleviating food insecurity problems, yet most of the global population worry about GM foods. This is because, in their minds, GM foods are 'unnatural' and therefore they avoid buying them. Some people also fear that the genes could be transmitted to them if they consume GM foods. So, the labelling of GM food is also an issue and some companies even avoid labelling their foods as genetically modified. In fact, most of the people in Europe are not interested in consuming GM foods due to their fear of the risks and also ethical issues. Testing for the toxicity of GM foods to humans and animals is therefore necessary before releasing them in the market.

Resistance to antibiotics, food safety, toxicity risk, eating foreign or exogenous DNA, and various environmentally related issues are risk factors of concern related to the use of GM foods. The ecosystem may be polluted unintentionally by the release of GM foods; i.e. from the instability of the transgene, biodiversity loss, transfer to weeds, mutations in the environment, new virus (live) generation, etc. (Prakash et al., 2011). Farmer exploitation is also a worry as small-scale farmers are sometimes exploited by large international companies. Terminator genes (used for making seeds sterile) and their use is one of the possible examples of this exploitation; this terminator technology has no agricultural or economic benefits with the exception of exploiting poor farmers. GM products are not easily available to farmers as they require a lot of investment and have high costs, which can result in the loss of traditional farming practices (collection, storage, and replanting of seeds).

1.6 GM crops versus GM animals

In comparison to GM plant food, results related to GM animal food are not as abundant; the former has approved market entry yet the latter has not had much success. Increased problems related to ethical issues of animal welfare have diminished consumer interest towards GM animal food. Acceptance of consumers towards transgenic plants has been greater than that of transgenic animals. Because of various restrictions related to religion and culture, an alteration to animal husbandry is much less acceptable and has a broad range of implications. This is the reason for the current prohibition of the consumption of GM food derived from animals. Despite this, public acceptance of pharmaceutical products based on transgenic animals is better. This is because pharmaceutical drugs based on transgenic animals have more benefits and these outweigh the associated risks

However, although all the health risks and ethical issues stated above are valid, there is also no evidence that GM food is dangerous. Yet, that said, there is not much documented proof on the long-term consequences of these foods on human health and the environment. Above all, human health and environmental safety should be a main concern and thus reviewed and dealt with completely before the release of GM foods.

1.7 GM foods and food security

The term 'food security' encompasses the sufficient accessibility of the population to nutritious and safe food. People's inherited food preferences and socio-economic status need to be considered when discussing the assurance of global food security. Historically, humankind has developed several techniques like crop selection, selective breeding, and interbreeding for hybridization to develop crops and animals with higher growth and reproduction rates. Genetic recombination technology is the latest technique and has propelled the GM scenario globally. Today, more than 90 percent of corn, soy, and sugar beets in the United States are genetically modified; a situation which can be considered a glimpse into how GM crops could take over the world food scenario (Wong & Chan, 2016). Importantly, GM crops have been developed with herbicide-tolerance and disease- and infestation-resistance traits that positively relate to higher yields and safety of the produce.

Modern agricultural practices use a range of herbicidal compounds such as glyphosate and glufosinate to control weed-related damage. To protect the crop from the applied herbicides and reduce their loads in the tissues, GM techniques have been used to impart enzymatic insensitivity in the crop for the specific herbicide and also to express enzymatic action for detoxifying the compound. Similarly, to control insect and *Bacillus thuringiensis* (Bt), codons for specific enzyme and degradation compound inhibitors have been introduced in crops by means of genetic engineering. These reduce the requirement for chemical insecticides while supporting the growth of other beneficial insects in the cropland ecosystems. Expression of phytoalexins, ribosomal inhibitors, and cell wall degradation enzymes have been carried out to inhibit bacterial and fungal damage. Additionally, GM crops have been designed to combat and alter soil pH, salinity, flood, and drought, resulting in improved productivity (Engel et al., 2002; Fischer et al., 2015).

In the post-harvest crop protection scenario, GM tomatoes with delayed ripening characteristics were first developed and made available in the US market. Their development primarily involved the suppression of genes responsible for the production of ripening precursors such as polygalacturonase. This suppression, however, also resulted in a marked flavor alteration in the tomato and, rather than its slower ripening, this was what attracted consumer attention (Bergougnoux, 2014). Since then, GM techniques have been developed to alter and enhance secondary flavor, color, and bioactive compounds in a number of crops. Beta carotene-enhanced 'Golden Rice' is a popular example of a GM crop targeted for providing a nutritional boost to vitamin A deficient populations. Rice has been also identified as the ideal candidate for iron enhancement through GM approaches such as the introduction of the ferritin gene to the genome and expression of a heat-tolerant phytase from other crops (Stein et al., 2006; Lucca et al., 2002).

GM techniques have been used to improve macro- and micronutrient composition in different crops. High amylose maize and waxy corn and potato have been developed in which either of the two fractions of the starch macromolecule, amylose and amylopectin, have been depleted from the macromolecule by altering the genetic sequences responsible for them (Hebelstrup et al., 2015). In the same vein, edible vegetable oils with higher levels of unsaturated fatty acids could be developed (Cahoon & Schmid, 2008). Suppression of genes for allergic proteins and the enhancement of amino acids with higher biological values have also been successfully carried out (Habben & Larkins, 1995). As nutraceutical products are attaining popularity, GM crop variants with an enhanced rate of functional compound synthesis have been developed and released onto the market.

Trials on animals for the production of more omega-3 fatty acids, leaner meat, and higher meat mass have been carried out (Uzogara, 2000). Genetic modification has been successfully implemented in cows to yield more milk; a product that is commercially sold in the US and some European countries. In the fish industry, GM salmon and trout species pioneered the currently growing GM fish research on high-yielding, rapidly growing, and high-quality oil-producing transgenic fishes.

Genetic technology entered the processed food industry when GM microorganisms began to be used in fermented food and beverage products during the last decade of the twentieth century. Currently, GM lactic acid bacteria and yeasts are in high demand across the globe to produce lactic and alcoholic fermented products with the desired characteristics. Food processing enzymes like rennet are largely extracted from microorganisms. A range of microorganisms have been genetically modified for synthesizing calf prochymosin to be used for cheese fermentation (Trivedi et al., 2016). Genetic modification has been also used by the single-cell protein industry to enhance yield and expand the substrate specificity of the species. The natural and nature-identical flavor industries use microorganisms extensively for compound synthesis and breakdown, indicating the great scope of genetic modification to be employed there. In fact, biotechnology has an immense role in the production of popular ranges of foods, beverages, and food ingredients. Genetic manipulation of microorganisms can be an effective way to enhance the efficiency of such processes.

Genetic materials naturally undergo changes over time; historically this process has been exploited to enhance edible yield and could in future play a significant role in enhancing food production. GM for food security has been endorsed by many; however, a number of short- and long-term health and environmental risks with ecological consequences have been identified by researchers. Major risks arising from natural cross-pollination with other crops are evident, which would necessitate isolated cultivation of the GM crops – not an ideal situation considering it is hoped to be a means of tackling global food insecurity involving marginalized farmer communities. With the huge capital involvement and intellectual property rights protection in the GM crop industries in recent decades, the encouragement of GM crops for a hunger-free and nutritionally secured world has faced ethical questions and doubts. There have been incidents wherein attempts to improve certain productive traits in a crop have also introduced some detrimental traits. For example, in an approach to enhance methionine in soybean by expressing a gene from Brazil nut, the unintended transfer of an allergenic protein also occurred (Nordlee et al., 1996). Though GMO regulations have been set by various nations, wide differences exist based on local and administrative issues and the governance systems in place. Hence, a constant state of debate on GM use in food industries can be observed. Nevertheless, considering the growing global population and shrinking arable land areas, the possibly of it emerging as a solution for food security cannot be denied.

Because of these issues, increasing consumer awareness and information via the labelling of GMO-based food products has been considered necessary.

1.8 Labelling of GM foods

GM food products are currently abundant in several countries where they have been largely accepted by the local consumers. To date, more than 60 countries, of which a good number fall under the EU, have GM labelling regulations for products containing or manufactured from GM materials. Specifically, products where GM ingredients make up 0.9 percent or more are required to be labelled (Dizon et al., 2015/6). The USA has been the highest producer and user of GM foods and are pioneers in the volunteer labelling of GMO/non-GM products. However, the statutory USDA regulations on the labelling of GMOs and products containing GM ingredients have only been put into place recently (since 2020). An alternative term 'bioengineered' has been used instead of genetically modified for those products to be packaged and stored in grocery stores but not for readily processed foods and takeaways. In addition, meat, poultry, and egg products are not covered by the labelling law. In India, since 2013 it has been mandatory to mark food packages that contain GM foods and ingredients as 'GM' (Bansal, 2013).

However, be it mandatory or voluntary, only labelling a product for its GM content does not serve the purpose of knowledge dissemination on GM crops and products. A large majority of consumers and producers, especially from those nations facing food insecurity, do not have the requisite knowledge on the processes, advantages, and risks associated with GM crops. Therefore, research-based knowledge generation and distribution through public channels before the implementation of GM policies and guidelines are necessary.

The policy options concerning the labelling of these GM crops can be voluntary or mandatory. The best voluntary labelling system includes the US labelling policy. In 1992 the US Food and Drug Administration (FDA) declared that the regulation of GM food and food products should be the same as for foods created by conventional means (Belson, 2000). In 1997 the European Union (EU) brought in the mandatory labelling of GM products wherein the presence of GM ingredients was detected as greater than 0.9 percent (Gruère & Rao, 2007). It is interesting to note that, with respect to these two tolerances concerning GM labelling, the first was set by a government regulatory agency and the latter by the commercial trade industry (Wilson & Dahl, 2005).

1.9 Toxicology of GM foods

With the ever-increasing prevalence of GM foods, consumers increasingly want to know how safe these are. Major concerns are based on the short- and long-term health issues that could be an outcome of this sophisticated technique; specifically the possibility of GM foods negatively affecting consumer health. The major causes of ill health rumoured to be due to the intake of GM foods include dietary changes, allergens, toxicity, and organ damage. To address these concerns, numerous papers have been published in peer-reviewed journals evaluating the pros and cons of GM foods and evaluating the impact of conventional foods against GM foods. In future, this information will need to be communicated to the general public.

Gene manipulation is a biological phenomenon inducing modifications in the genetic makeup of various living organisms. The World Health Organization (WHO) defines GM as "organisms such as plants, animals or microbes in which the genetic material is altered in a manner other than by natural process primarily either by mating and/or natural recombination" (Zhang et al., 2016). This definition distinguishes the direct manipulation of genetic material (the transference of genes from unrelated organisms via DNA recombinant technology) from the conventional technique of modifying the genetic stock of plants and animals by selective breeding. Similarly, the FAO (Food and Agriculture Organization of the United Nations) and the European Commission (EC) refer to a GMO as "a product that is not a result of natural processes such as mating and/or natural recombination" (FAO, 2016).

The use of GM foods and their potential impact on the health of human beings and environmental safety has been a topic of debate since their introduction. Being experimental, modified organisms has meant there has always been bias concerning their use, with many researchers reporting GM foods to be toxic for consumption and their long-term use to be harmful to consumers Yet, conversely, there is a body of scientists who claim that GM foods are totally safe for human consumption and many countries have approved their use. Nevertheless, although GM crops have certain advantages over conventional crops, their potentially adverse effects on both ecology and human health remain uncertain.

As of now, the prominent food safety risks associated with GM foods are mainly based on the potential allergens present in them, unexpected toxic components, a reduction in nutrition compared with the original food, along with other unintended effects. Concerns regarding the safe utilization of GM foods have been spiking up owing to their infiltration into our diet. Numerous toxicological studies conducted on various animals have revealed the potential for GM foods to be hazardous for a variety of organs and systems. The majority of research on GM foods suggests that they may produce several common toxic consequences such as hepatic, pancreatic, renal, or reproductive impacts, as well as modify the immune system. The most imminent hazards related to the consumption of GM foods for animals and the human population include the pleiotropic and insertional effects, adverse effects on animal and human health owing to an increase of anti-nutritional factors, viral DNA in plants, the transfer of antibiotic-resistant genes to bacteria in the gastrointestinal tract, as well as the effect of GM foods on allergic responses (Dona & Arvanitoyannis, 2009).

In addition, the three most significant health risks of GM foods are their strong correlation with toxicity, allergenicity, and genetic hazards arising from three potential sources: the inserted gene and its expressed protein, secondary or pleiotropic effects of the products of gene expression, and the degradation of natural genes in the manipulated organism (Bawa & Anilakumar, 2013). A recent study carried out by Shen et al. (2022) on the consumption of

GM foods has reported high mortality rates, onset of cancer, extremely low fertility rates, diminished learning and reaction abilities accompanied with organ abnormalities. The unexpected adverse effects of GM foods are reported in many similar studies such as the "Monarch Butterfly controversy" (Losey et al., 1999), the "Pusztai case" (Ewen & Pusztai, 1999), and the "Séralini case" (Séralini et al., 2014) etc., which has led to the ban of GM foods in many countries.

It has been more than two decades since genetically engineered crops were first produced and marketed on commercial basis, and the foods obtained from them have been consumed by millions of people and billions of livestock animals (Van Eenennaam & Young, 2014). In pre-market approval processes, the safety of those GM foods introduced into the market was assessed for their potential to have adverse effects on human and animal health or the environment. And, collectively, work conducted to date has identified no evidence of adverse health or nutritional effects from commercially available GE crops or the foods obtained from them (Delaney et al., 2018). In fact, numerous studies have proved that there is no detrimental effect of approved GM foods on animals and that the effect, if any, is low or non-existent. Many subchronic feeding studies have been documented for GM corn, soy, canola, rice, cottonseed meal, sweet potato, and tomato, and none of these studies has demonstrated evidence of adverse effects (Delaney et al., 2018). In addition to these subchronic studies, other types of toxicology studies such as on reproductive toxicology and studies of a year or greater in length have been conducted with whole grains and feed fractions from GE crops; not one reports any harmful effect of GM crops (Delaney et al., 2018; Tang et al., 2009; Steinberg et al., 2019).

While each new GM product is evaluated on a case-by-case basis, approved commercial products, such as those containing transgenes, have been subjected to rigorous scientific scrutiny through a risk assessment process that relies upon expertise from a variety of different fields. Yet, risks associated with GM crops have proven to be low to non-existent (Carzoli et al., 2018).

The debates over GM foods focus mostly on uncertainties concerning their *potential* adverse effects on human health and environmental safety. The anxiety among consumers can be attributed to four sources: the scientific community's difficulty in explaining concisely to the public the biological techniques involved; concerns about the improper dissemination of GM foods; the ethical principles inherent in traditional food processing; and misgivings with regards to the adequacy of evaluation of the GM foods (Schell & Van Montagu, 1977; Baulcombe et al., 2014; Gibson et al., 2013). Furthermore, negative reporting about GM technology by the media also significantly affects people's perception of GM foods (Huang & Peng, 2015).

Finally, although the benefits of this biotechnology to the environment and to human consumers have been well-documented, there remains a significant proportion of the population that is skeptical about the safety of GM crops/foods. In some cases, this fear has led to GM foods failing to get approval, such as *Golden Rice 2*, in places where their benefits could be both rapid and significant. In the future, individuals developing GE products should strongly consider more aggressive outreach to demonstrate the safety and benefits of their products. This outreach will likely need to go beyond the traditional audience of regulatory and scientific authorities and speak directly to consumers. Also, the GM food-related adverse events and the corresponding positive GM stories need to be validated by clinical trials and long-term cohort studies in human populations. Hence, it is better to prove the safety of GM foods before they are approved for consumption and it is essential that, as suggested by many researchers, they carry the appropriate labelling so that consumers can make their own choices.

1.10 Concluding remarks

Each new and innovative technology comes with its own set of advantages and disadvantages, and the condemnation of a beneficial technology is not on the basis of the few food hazards reported concerning them. One of the best solutions to this problem is to use such technology with great caution. GM foods have always been a topic of huge debate and discussion, with them being very prevalent in some parts of the world and banned in others. There are many challenges for us to overcome in terms of their safety testing, regulation, international policy, and labelling. Therefore, we must proceed with great caution and avoid unintended harm to human health and the environment as an outcome of our curiosity for GM foods.

References

Bansal, S. (2013). Labelling of genetically modified foods in India. *Economic and Political Weekly*, 48(34), 15–17.

Baulcombe, D., Dunwell, J., Jones, J., Pickett, J., & Puigdomenech, P. (2014). *GM Science Update: A Report to the Council for Science and Technology*. Council for Science and Technology. https://assets.publishing.service.gov.uk/government/uploads/system/uploads/attachment_data/file/292174/cst-14-634a-gm-science-update.pdf

Bawa, A. S., & Anilakumar, K. R. (2013). Genetically modified foods: Safety, risks and public concerns – a review. *Journal of Food Science and Technology*, 50(6), 1035–1046.

Belson, N.A. (2000). U.S. regulation of agricultural biotechnology: An overview. *AgBioForum*, 3(4), 268–280.

Bennett, P. M., Livesey, C. T., Nathwani, D., Reeves, D. S., Saunders, J. R., & Wise, R. (2004). An assessment of the risks associated with the use of antibiotic resistance genes in genetically modified plants: Report of the Working Party of the British Society for Antimicrobial Chemotherapy. *Journal of Antimicrobial Chemotherapy*, 53(3), 418–431.

Bergougnoux, V. (2014). The history of tomato: From domestication to biopharming. *Biotechnology Advances*, 32(1), 170–189.

Cahoon, E. B., & Schmid, K. M. (2008). Metabolic engineering of the content and fatty acid composition of vegetable oils. *Advances in Plant Biochemistry and Molecular Biology*, 1, 161–200.

Carzoli, A. K., Aboobucker, S. I., Sandall, L. L., Lübberstedt, T. T., & Suza, W.P. (2018). Risks and opportunities of GM crops: Bt maize example. *Global Food Security*, 19, 84–91.

Delaney, B., Goodman, R. E., & Ladics, G. S. (2018). Food and feed safety of genetically engineered food crops. *Toxicological Sciences*, 162(2), 361–371.

Dively, G. P., Rose, R., Sears, M. K., Hellmich, R. L., Stanley-Horn, D. E., Calvin, D. D., Russo, J. M., & Anderson, P. L. (2004). Effects on monarch butterfly larvae (*Lepidoptera: Danaidae*) after continuous exposure to Cry1Ab-expressing corn during anthesis. *Environmental Entomology*, 33(4), 1116–1125.

Dizon, F., Costa, S., Rock, C., Harris, A., Husk, C., & Mei, J. (2015/6). Genetically modified (GM) foods and ethical eating. *Journal of Food Science*, 81(2), R287–R291.

Dona, A., & Arvanitoyannis, I. S. (2009). Health risks of genetically modified foods. *Critical Reviews in Food Science and Nutrition*, 49(2), 164–175.

Ellstrand, N. C., Prentice, H. C., & Hancock, J. F. (1999). Gene flow and introgression from domesticated plants into their wild relatives. *Annual Review of Ecology and Systematics*, 30(1), 539–563.

Engel, K. H., Frenzel, T., & Miller, A. (2002). Current and future benefits from the use of GM technology in food production. *Toxicology Letters*, 127(1–3), 329–336.

Ewen, S. W., & Pusztai, A. (1999). Effect of diets containing genetically modified potatoes expressing Galanthus nivalis lectin on rat small intestine. *Lancet, 354*, 1353–1354.

Fischer, K., Ekener-Petersen, E., Rydhmer, L., & Björnberg, K. E. (2015). Social impacts of GM crops in agriculture: A systematic literature review. *Sustainability*, 7(7), 8598–8620.

Food and Agriculture Organization of the United Nations (2016). http://wwwfaoorg/docrep/005/y2772e/y2772e04htm

Gatehouse, A. M. R., Ferry, N., Edwards, M. G., & Bell, H. A. (2011). Insect-resistant biotech crops and their impacts on beneficial arthropods. *Philosophical Transactions of the Royal Society B: Biological Sciences*, 366(1569), 1438–1452.

Gibson, D. J., Gage, K. L., Matthews, J. L., Young, B. G., Owen, M. D., Wilson, R. G., ... Jordan, D. L. (2013). The effect of weed management systems and location on arable weed species communities in glyphosate-resistant cropping systems. *Applied Vegetation Science*, 16(4), 676–687.

Gruère, G. P., & Rao, S. R. (2007). A review of international labeling policies of genetically modified food to evaluate India's proposed rule. *AgBioForum*, 10(1), 51–64.

Habben, J. E., & Larkins, B. A. (1995). Genetic modification of seed proteins. *Current Opinion in Biotechnology*, 6(2), 171–174.

Hakim, M. P., Zanetta, L. D. A., de Oliveira, J. M., & da Cunha, D. T. (2020). The mandatory labeling of genetically modified foods in Brazil: Consumer's knowledge, trust, and risk perception. *Food Research International*, 132, 109053.

Hebelstrup, K. H., Sagnelli, D., & Blennow, A. (2015). The future of starch bioengineering: GM microorganisms or GM plants? *Frontiers in Plant Science*, 6, 247.

Heuer, H., & Smalla, K. (2007). Horizontal gene transfer between bacteria. *Environmental Biosafety Research*, 6(1–2), 3–13.

Huang, J. K., & Peng, B. W. (2015). Consumers' perceptions on GM food safety in urban China. *Journal of Integrative Agriculture*, 14(11), 2391–2400.

Keeling, P. J., & Palmer, J. D. (2008). Horizontal gene transfer in eukaryotic evolution. *Nature Reviews Genetics*, 9(8), 605–618.

Keese, P. (2008). Risks from GMOs due to horizontal gene transfer. *Environmental Biosafety Research*, 7(3), 123–149.

Komoto, K., Okamoto, S., Hamada, M., Obana, N., Samori, M., & Imamura, T. (2016). Japanese consumer perceptions of genetically modified food: Findings from an international comparative study. *Interactive Journal of Medical Research*, 5(3), e5850.

Labra, M., Savini, C., Bracale, M., Pelucchi, N., Colombo, L., Bardini, M., & Sala, F. (2001). Genomic changes in transgenic rice (*Oryza sativa* L.) plants produced by infecting calli with Agrobacterium tumefaciens. *Plant Cell Reports*, 20(4), 325–330.

Lemaux, P. G. (2008). Genetically engineered plants and foods: A scientist's analysis of the issues (Part I). *Annual Review of Plant Biology*, 59, 771–812.

Lewi, D. M., & Vicién, C. (2020). Argentina's local crop biotechnology developments: Why have they not reached the market yet? *Frontiers in Bioengineering and Biotechnology*, 8, 301.

Losey, J. E., Rayor, L. S., & Carter, M. E. (1999). Transgenic pollen harms monarch larvae. *Nature*, 399(6733), 214–214.

Lucca, P., Hurrell, R., & Potrykus, I. (2002). Fighting iron deficiency anemia with iron-rich rice. *Journal of the American College of Nutrition*, 21(sup 3), 184–190.

Martin, W. (1999). Mosaic bacterial chromosomes: A challenge en route to a tree of genomes. *Bioessays*, 21(2), 99–104.

Nicolia, A., Manzo, A., Veronesi, F., & Rosellini, D. (2014). An overview of the last 10 years of genetically engineered crop safety research. *Critical Reviews in Biotechnology*, 34(1), 77–88.

Nielsen, K. M., & Townsend, J. P. (2004). Monitoring and modeling horizontal gene transfer. *Nature Biotechnology*, 22(9), 1110–1114.

Nordlee, J. A., Taylor, S. L., Townsend, J. A., Thomas, L. A., & Bush, R. K. (1996). Identification of a Brazil-nut allergen in transgenic soybeans. *New England Journal of Medicine*, 334(11), 688–692.

Ochman, H., Lawrence, J. G., & Groisman, E. A. (2000). Lateral gene transfer and the nature of bacterial innovation. *Nature*, 405(6784), 299–304.

Pidgeon, J. D., May, M. J., Perry, J. N., & Poppy, G. M. (2007). Mitigation of indirect environmental effects of GM crops. *Proceedings of the Royal Society B: Biological Sciences*, 274(1617), 1475–1479.

Prakash, D., Verma, S., Bhatia, R., & Tiwary, B. N. (2011). Risks and precautions of genetically modified organisms. *International Scholarly Research Notices: Ecology*.

Prescott, V. E., Campbell, P. M., Moore, A., Mattes, J., Rothenberg, M. E., Foster, P. S., Higgins, T. J. V., & Hogan, S. P. (2005). Transgenic expression of bean α-amylase inhibitor in peas results in altered structure and immunogenicity. *Journal of Agricultural and Food Chemistry*, 53(23), 9023–9030.

Schell, J., & Van Montagu, M. (1977). The ti-plasmid of *Agrobacterium Tumefaciens*, a natural vector for the introduction of nif genes in plants? In *Genetic Engineering for Nitrogen Fixation*, edited by A. Hollaender et al. (New York: Plenum Press), 159–179.

Schütte, G., Eckerstorfer, M., Rastelli, V., Reichenbecher, W., Restrepo-Vassalli, S., Ruohonen-Lehto, M., … Mertens, M. (2017). Herbicide resistance and biodiversity: Agronomic and environmental aspects of genetically modified herbicide-resistant plants. *Environmental Sciences Europe*, 29(1), 1–12.

Séralini, G. E., Mesnage, R., Defarge, N., & Spiroux de Vendômois, J. (2014). Conflicts of interests, confidentiality and censorship in health risk assessment: The example of an herbicide and a GMO. *Environmental Sciences Europe*, 26(1), 1–6.

Shahbandeh, M. (2022). Global genetically modified crops by countries 2019 based on acreage. https://www.statista.com

Shen, C., Yin, X. C., Jiao, B. Y., Li, J., Jia, P., Zhang, X. W., … Liu, J. P. (2022). Evaluation of adverse effects/events of genetically modified food consumption: A systematic review of animal and human studies. *Environmental Sciences Europe*, 34(1), 1–33.

Smyth, S. J. (2014). The state of genetically modified crop regulation in Canada. *GM Crops & Food*, 5(3), 195–203.

Stein, A. J. (2015). *Acceptance of "GM food" in Europe: What People Say and Do.* ResearchGate Technical Report. doi:10.13140/RG.2.1.1052.9127.

Stein, A. J., Sachdev, H. P. S., & Qaim, M. (2006). Potential impact and cost-effectiveness of Golden Rice. *Nature Biotechnology*, 24(10), 1200–1201.

Steinberg, P., van der Voet, H., Goedhart, P. W., Kleter, G., Kok, E. J., Pla, M., … Wilhelm, R. (2019). Lack of adverse effects in subchronic and chronic toxicity/carcinogenicity studies on the glyphosate-resistant genetically modified maize NK603 in Wistar Han RCC rats. *Archives of Toxicology*, 93(4), 1095–1139.

Tabashnik, B. E. (1994). Evolution of resistance to *Bacillus thuringiensis. Annual Review of Entomology*, 39(1), 47–79.

Tang, G., Qin, J., Dolnikowski, G. G., Russell, R. M., & Grusak, M. A. (2009). Golden Rice is an effective source of vitamin A. *The American Journal of Clinical Nutrition*, 89(6), 1776–1783.

Trivedi, M., Singh, R., Shukla, M., &Tiwari, R. K. (2016). GMO and food security. In *Ecofriendly Pest Management for Food Security*, edited by Omkar (London: Academic Press), 703–726.

Uzogara, S. G. (2000). The impact of genetic modification of human foods in the 21st century: A review. *Biotechnology Advances*, 18(3), 179–206.

Van Eenennaam, A. L., & Young, A. E. (2014). Prevalence and impacts of genetically engineered feedstuffs on livestock populations. *Journal of Animal Science*, 92(10), 4255–4278.

Wilson, W. W., & Dahl, B. L. (2005). Costs and risks of testing and segregating genetically modified wheat. *Applied Economic Perspectives and Policy*, 27(2), 212–228.

Wong, A. Y. T., & Chan, A. W. K. (2016). Genetically modified foods in China and the United States: A primer of regulation and intellectual property protection. *Food Science and Human Wellness*, 5(3), 124–140.

World Health Organization (2016). https://www.who.int/news-room/questions-and-answers/item/food-genetically-modified

Yoshida, S., Maruyama, S., Nozaki, H., & Shirasu, K. (2010). Horizontal gene transfer by the parasitic plant *Striga hermonthica*. *Science*, 328(5982), 1128–1128.

Zhang, C., Wohlhueter, R., & Zhang, H. (2016). Genetically modified foods: A critical review of their promise and problems. *Food Science and Human Wellness*, 5(3), 116–123.

2 *Flavr Savr* and other GM tomatoes

Renu Sharma, Kawaljeet Kour, Jasmeet Kour and Rupinderpal Kaur

2.1 Introduction

Tomatoes are nutrient-dense, very popular, and extensively consumed fruits that belong to the family *Solanaceae*. It is a red-colored, savory edible fruit with a flavor that's balanced by the amount of sugars, acids, and volatile compounds present within it (Spricigo et al., 2021). Botanically, it is a fruit, but is considered to be a vegetable for culinary purposes. It is utilized all over the world in raw, cooked, and processed forms. Tomatoes come in a variety of sizes, shapes, and colors. It has a high water content and is a good source of vitamins and minerals such as vitamin C, pro-vitamin A, β-carotene, folate, potassium, calcium, phosphorous, and iron (Nardos, 2021). Other components found in tomatoes include lycopene, flavonoids, phytosterols, and polyphenols (Luthria et al., 2006).

Because tomatoes have good nutritive value and are rich in vitamins and minerals, they have a variety of health benefits of which a reduction in cholesterol and cardiovascular disease (Silaste et al., 2007), cancer protection (Tzonou et al., 1999), and vision improvement are just a few examples (Ganesan et al., 2012). It also exhibits strong antioxidant properties due to the presence of lycopene, which helps to protect against UV damage. Due to their versatility, tomatoes can be utilized to make a variety of products, but their sole drawback is their short shelf life. When a tomato is allowed to ripen naturally, it becomes soft and unable to withstand outer stresses (biotic and abiotic). To combat this, tomatoes have undergone a lot of genetic modifications (Gerszberg & Hnatuszko-Konka, 2017). A genetically modified (GM) tomato is one in which genes have been altered with the help of genetic engineering to obtain certain results. As tomatoes are highly perishable in nature and have a short shelf life, there was a need to develop a variety of tomatoes that would last longer without deteriorating.

Based on the concept of delayed ripening in tomatoes, the first GM tomato – "Flavr Savr" (or CGN-89564) – was introduced by the company Calgene in 1994 and it was granted a license for human ingestion by the US Food and Drug Administration (Gerszberg et al., 2015). However, it was not a success as the varieties that had been chosen for genetic modification did

DOI: 10.4324/9781003278566-2

not have flavor and were susceptible to various diseases. As a result, the production of GM tomatoes was stopped in 1997 (Halford, 2012). In China, Huazhong Agricultural University developed its first GM tomato with a longer shelf life (Huafan No 1), which was approved for commercialization in 1996. The USA, Japan, Mexico, and Canada have also approved other GM tomato varieties, including *1345–4* from DNA Plant Technology Corp (Oakland, USA), *351N* from Agritope Inc (Portland, USA), and *8,338* and *5,345* from Monsanto (St. Louis, USA) (Gerszberg et al., 2015). In addition, in 1996 the company Zeneca produced GM tomato paste much cheaper than purees from conventional tomatoes; however, sales of this were dropped in 1999 when unsubstantiated and misleading campaigning against GM foods began (Soil Association, 2005).

2.2 Why are genetic modifications required?

As mentioned earlier, genetic modification is the modification of genes in a particular species with the help of genetic engineering to achieve a desired outcome. The whole idea behind genetic modification is to produce pest-resistant, environmental stress-tolerant, nutritious, healthy, and cost-effective varieties. Tomatoes are utilized as a prototype to analyse and comprehend the mechanics of genetic engineering (Vitale et al., 2014). Fruit and vegetable cell structures are mainly made up of cellulose, hemicellulose, and pectin, i.e., polysaccharides, which soften or deteriorate during storage (Cosgrove, 2005). Naturally occurring tomatoes need to be harvested early (when they are still green in color) so that they can be kept for longer periods (Grierson et al., 1986). If a tomato is harvested after ripening, then its outer layer gets soft in a very short span of time. The deterioration of fruit is due to increased ethylene production during ripening process.

One of the greatest disadvantages of harvesting tomatoes early is that their natural flavor has not developed completely. After procuring unripe tomatoes from the field, the fruit is then treated with ethylene to achieve the desired color and ripeness before being shipped to different locations (Griffiths et al., 1999). Conversely, GM tomatoes are allowed to ripen on the vine, which aids in the development of their natural flavour and color (Gao et al., 2010). They are also less expensive to produce and process; the Zeneca tomato paste is an example of this (Soil Association, 2005). Redenbaugh (1992) reported that vitamin, protein, and mineral content, and also toxic glycoalkaloids, were not altered in the parent tomatoes when kanr genes were used in them via a GM antisense technique. Hence, the parent tomatoes and genetically modified tomatoes seemed to be "substantially equivalent" (Bawa & Anilakumar, 2013).

2.3 Genes involved in agronomic trait development

To figure out which genes are involved in agronomic trait development, we must first understand the normal process of tomato ripening. Ethylene is the main component in fruit ripening, which works in conjunction with transcription factors (TF) (Vrebalov et al., 2002). It brings about various biochemical and physiological processes in tomatoes: the color changes from green to red due to the degradation of chlorophyll and the synthesis of lycopene (Li et al., 2018) and the texture of the tomato transforms from hard to soft as the enzyme polygalacturonase (PG, pectin degrading enzyme) acts on its cell wall. Thus, in order to prevent the depletion or degradation of the cell wall, a gene having the capability of producing polygalacturonase enzymes was introduced in an antisense orientation. This reduces the production of enzymes in tomatoes and helps in maintaining their cell wall, thereby increasing their shelf life and flavor quality (Hamilton et al., 1995).

Literature studies show that amino acid sequences obtained from TOM clones indicate that TOM 6 is the gene encoding the PG enzyme (Grierson et al., 1986). With the use of genetic engineering, the PG gene was expressed in the tomato, which resulted in mRNA synthesis. It was then converted to cDNA using reverse transcriptase enzymes. This cDNA (complementary DNA) expresses itself and forms antisense mRNA (Krieger et al., 2008). The antisense mRNA was then inserted into normal tomatoes which are then known as "transgenic tomatoes." Upon reaching maturity, transgenic tomatoes become double stranded as they have their own mRNA and an antisense mRNA. At this stage, there will be no translation and no protein (polygalacturonase) formation as mRNA only translates when it is single stranded. In the absence of polygalacturonase, the cell wall pectin of the tomato remains intact for a longer duration, thereby increasing its shelf life (Engel et al., 1995).

DNA Plant Technology (DNAP), Agritope, and Monsanto have also developed tomatoes with the target of delayed ripening. All three of these were based on the concept of preventing fruit ripening by reducing ethylene formation. In DNAP's tomatoes, a diminutive version of ACC (1-aminocyclopropane-1-carboxylic acid synthase) was inserted into a normal tomato that blocked the endogenous ACC (Klee et al., 1991). This transgenic tomato was known as "*Endless Summer*." Agritope, in their GM tomato, lowered the amount of S-adenosylmethionine (which is a precursor to ACC) by inserting an S-adenosylmethionine hydrolase (SAM) obtained from *E. coli* (Good et al., 1994). Monsato used the ACC deaminase allele obtained from the *Pseudomonas chlororaphis* to develop its tomato, which reduced ethylene concentrations by disintegrating ACC (Klee et al., 1991).

Other examples include a premature tomato with the gene afa3 having an antifreezing capability developed from the winter flounder (*Pseudopleuronectes americanus*), with the purpose of enhancing frost tolerance (Lemaux, 2008), yet it was never commercialized. Another tomato was injected with a rice gene (Osmyb4) to improve resistance to cold and drought. Drought tolerance rose as a result, although cold tolerance did not appear to be affected (Vannini et al., 2007). Additionally, tomatoes have been infused with a gene called cysteine proteinase inhibitor, obtained from taro tubers, so that they could be made immune to the root-knot nematode (Chan et al., 2010). Therefore, as can be seen from the above examples, tomatoes are often genetically manipulated to help researchers better understand specific processes.

2.4 Mechanism of genetic modification

The recent introduction of genetically manipulated technology has greatly facilitated reliable and realistic crop genome modification. To some extent, modern editing approaches like Clustered Regularly Interspaced Short Palindromic Repeats (CRISPR), Zinc Finger Nucleases (ZFNs), and Transcription Activator-Like Effector Nucleases (TALENs) have addressed various issues related to traditional random mutation and transgenesis. These genome-editing techniques are helpful for the formation of superior varieties via interventions such as directed mutations, precise endogenic genetic manipulation, and location-specific inclusion of a characteristic gene. Despite biosecurity and ecosystem issues, recombinant techniques (altering genetic materials outside an organism) have become the preferred way for rapidly developing better crop plants and stacking several beneficial features (Kumar et al., 2020).

The basic principle of genome editing technologies is that specific engineered nucleases cause double-stranded breaks in DNA (dsDNA). Homology-dependent recombination (HDR) or non-homologous end joining (NHEJ) are then utilized to rebuild the dsDNA break. HDR allows for more accurate point mutations, omissions, or gene knock-in that are favorable to crop breeding, whereas NHEJ is prone to error and allows for small random insertions, deletions, and substitutions, leading to gene knockout (Veillet et al., 2019). CRISPR/Cas9 has been a very effective gene editing method, among others, which can be used to uncover genetic resources for crop enhancements in terms of stress resistance and nutrient content. CRISPR/Cas9, unlike ZFNs and TALENs, makes genome editing much more reliable and efficient for generating knockout mutants (Mohanta et al., 2017). A guide RNA (gRNA) is used to locate the allele and a Cas9 endonuclease – an RNA-dependent DNA endonuclease – is used to modify the gene in CRISPR/Cas9. Cas9 identifies a protospacer adjacent motif (PAM) with a consensus sequence of 5' NGG 3' at the 3' end of the target sequence when it is complexed with gRNA (Shah et al., 2013; Wang et al., 2015). Since its preliminary report in 2013, this method has been widely used to alter a variety of crop plants due to its efficiency, precision, and low cost. CRISPR/Cas9 genome engineering has been shown in studies to improve rice, maize, tobacco, cucumber, tomato, wheat, potato, and cassava variants in terms of handling biotic and abiotic stress. It was also found that their nutritive and other qualities improved too (Jaganathan et al., 2018). In comparison, TALENs and ZFNs use FokI endonuclease, which forms a complex consisting of two linked identical molecules (dimer), for improved precision in binding to specific DNA. However, developing active FokI nucleases is a time-consuming and expensive operation.

2.5 Impact on modern agriculture – trade of GM crops

Food products produced as a result of agricultural biotechnology are becoming one of the fastest growing agricultural commodities serving the livestock industry (Turnbull et al., 2021). It was in 1994 when the Calgene *Flavr Savr* tomato was approved as the first ever GM plant for human consumption by the US FDA. Created by Calgene (California) in 1992, these tomatoes are considered as one of the most successful GM crops. As mentioned earlier, these tomatoes were genetically engineered wherein an antisense gene was expressed which led to an appreciable extension in shelf life (Chheda & Vernekar, 2014).

The current germplasm of the tomato has fewer variations in genes with respect to wild relatives as the result of genetic constraints derived from its initial domestication followed by its migration to Europe in the sixteenth century and subsequent return to North America in the eighteenth century (Nesbitt & Tanksley, 2002). In traditional artificial selection, more emphasis is placed on increasing the size and producing variations in color, shape, and flavor in order to meet the growing needs of the market and tomato-processing industry in terms of growing habits, disease resistance, pest resistance, uniformity of ripening, and nutritional profile (D'Ambrosio et al., 2016). Due to its rapid consumption in many industrialized nations, tomatoes act as a food vehicle for essential nutrients to prevent deficiency diseases (D'Ambrosio et al., 2016). Tomatoes, when fortified with the combination of traditional breeding and standard agronomic practices in combination with sophisticated genetic engineering, could therefore be beneficial to developing as well as in industrialized countries (Martin, 2013).

2.6 Benefits to society

Presently, GM crops hold a significant place in biotechnological advancements. Recently developed GM methods have proven to be highly beneficial for improving the quality characteristics of commercial crops like tomato. The resultant GM food could therefore be a critical component in eradicating hunger, malnutrition, and poverty from the world. The various benefits of GM tomatoes are discussed below.

2.6.1 Hunger elimination

Among the various merits of developing GM products, elimination of world hunger is considered to be most significant (Thompson, 2011; Steier, 2018). The continual upsurge in worldwide population has engaged researchers to focus on the advantages of developing GM crops rather than the possible threats they represent (Herring, 2013). In recent years, tomato cultivation has gained tremendous popularity because of its specific nutritive significance. The presence of various chief nutrients such as β-carotene, vitamin C, lycopene, flavonoids, and derivatives of hydroxycinnamic acid make it highly protective food for human consumption (Wu et al., 2011; Raiola et al., 2014). Advancements in plant biotechnology and genetic engineering tools have unlocked unlimited opportunities for engineering tomato plants.

2.6.2 *Economic benefits*

Research studies have shown the net economic benefits of utilizing GM crop technology in agriculture at the farm level. Between 1996–2018, the utilization of GM crops elevated global gross farm revenue by $225.1 billion and by $18.9 billion in 2018 (Brookes & Barfoot, 2020). The United States, India, Argentina, and China have reaped the maximum economic benefits, with production expenses dropping drastically (Jones et al., 2017). In some studies the commercialization of GM crops has been found to be beneficial for both the economy and the environment. Their higher productivity provides the population with more food as these foods grow better than traditionally grown foods. Furthermore, GM food crops can also be grown in areas with unfavorable climates and where the soil is unsuitable for traditional agriculture. Though the seed for GM crops is fairly expensive, the production cost is reported to be lower in comparison to regular crops because of their natural insect and pest resistance. Hence, the cultivation of GM foods results in chemical-free and environmentally friendly foods as these crops eliminate the necessity of toxic herbicides and insecticides (Bawa & Anilakumar, 2013). Thus, agricultural scientists, growers, and environmentalists throughout the world have commended GM crops.

2.6.3 *Insect resistance*

The tomato is an important part of our everyday diet. Its significance in the food industry cannot be ignored as it is involved in the formulation of various foodstuffs. The crop is known to be highly attractive to several insects and pests that ultimately can cause various diseases. Insects of the order lepidoptera are reported to pose a serious risk and can overwhelm this valuable crop. In order to protect tomato crops from these insects, an Agrobacterium-mediated transformation method was adopted to transform modified Bt (*Bacillus thuringiensis*) genes (Cry1Ab) into a high-yield commercial tomato variety. This transformation has played a vital role reducing attack by lepidopteron insects (Ahuja & Fladung, 2014). In 2014, Xin et al. reported that the chloroplast localized tea gene, CsiHPL1, is also highly effectual in protecting the tomato crop. They showed that transformation of CsiHPL1 not only shelters the tomato crop against fungal attacks but also provides herbivore resistance.

2.6.4 *Improved nutrition and taste*

Tomatoes have been transformed in several ways to alter their nutritional composition and flavor. A bacterial gene encoding phytoene desaturase has been added to enhance the concentration of pro-vitamin A in tomatoes (Römer et al., 2000). Blue tomatoes have also been developed by scientists to promote the synthesis of anthocyanin – an antioxidant found in fruits and vegetables (Gonzali et al., 2009). In order to enhance anthocyanin production, various researchers have added a transcription factor from thale cress, *Arabidopsis thaliana* (Zuluaga et al., 2008) and snapdragon (*Antirrhinum*) (Tohge et al., 2015). The use of snapdragon genes has led to the production of tomatoes with anthocyanin concentrations comparable with blueberries and blackberries. The taste and smell of the transgenic tomatoes was found to be improved when geraniol synthase from lemon basil (*Ocimum basilicum*) was expressed in tomatoes under a fruit-specific promoter (Bartoszewski et al., 2003). Additionally, researchers have attempted to boost isoflavone levels in tomatoes by including soybean isoflavone synthase, which has been linked to cancer prevention (Shih et al., 2008). In 2021, Sanatech Seed of Japan released the *Sicilian Rouge High GABA* tomato variety. This variety contains elevated levels of gamma-amino butyric acid (GABA), an amino acid believed to aid relaxation and help lower blood pressure (Tilahun et al., 2021).

2.6.5 Environmental stress tolerance

Abiotic stresses like frost, drought, and higher salinity in the surrounding environment are limiting factors to the growth of tomato crops (Foolad, 2007). Cold stress is one of the environmental stress factors that significantly affects crop yield, growth, and quality. The intolerance of any crop towards cold temperatures may be due to the accumulation of saturated lipids. Literature studies have revealed that the accumulation of glycine betaine desaturates the lipid molecules by increasing the expression of naturally desaturating genes. The mode of action of glycine betaine is not properly understood but its role is well-defined. Results show that plant crops with a greater accumulation of glycine betaine exhibited higher resistance to cold temperatures without disturbing other parameters (Karabudak et al., 2014). Hence, the genetic modification of plants has led to the formation of cold-resistant varieties.

In tomato plants, cold temperatures cause cell membrane shrinkage due to the formation of ice crystals in the extracellular matrix and dehydration in the cytoplasm. A frost-tolerant tomato was established by the transformation of an antifreeze protein gene from carrot. Compared with non-transgenic plants, transgenic plants have proven to be more resistant against frost damage (Kumar et al., 2014).

2.7 Trade in the global market

In the current era, agriculture delivers clothes, fuels, and food to a large population with limited land availability, fewer farmers, and additional modern challenges. Humans have always found a way to build on previous knowledge for improving agricultural capabilities and these improvements have led to increased production and availability. The introduction of biotechnology in modern agricultural practices is considered to be the major contributor to agricultural success (Turnbull et al., 2020). In 2018, globally about 184 million hectares were planted with crops having GM traits and the production of GM crops has experienced more than a 100-fold increase in the last 25 years (Brookes & Barfoot, 2013; Mathur et al., 2018).

GM crops have proven to be mostly successful in addressing various agricultural challenges while providing numerous benefits to growers all over the world. They contributed a profit of $117.6 billion in global agriculture income over a time period of 17 years from 1996 to 2013 (Chen & Lin, 2013; Brookes & Barfoot, 2014). In 2010–2012, the annual net income was amplified by 34.3 percent worldwide. Furthermore, while boosting global yield by 22 percent, GM crops also lowered the use of pesticides by 37 percent and environmental impact (herbicide and insecticide use) by 18 percent (Klümper & Qaim, 2014).

Since the introduction of GM techniques in the mid-1990s, they have contributed significantly to improving global production of the staple crops including soybeans and maize to 278 million tonnes and 498 million tonnes, respectively. Farmers have benefited by an average US$3.75 in extra income for every extra dollar invested in GM crop seeds (relative to the conventional seed cost). The average return was $4.41 in developing countries while in developed countries, it was $3.24 for each extra dollar invested in GM crop seeds (Brookes & Barfoot, 2020).

2.8 Health and ethical issues/risks

Before considering GM foods, it is very important to understand their benefits and drawbacks, particularly in terms of their safety. It is believed that the intake of GM foods might result in the development of antibiotic-resistant diseases. Furthermore, as these food products are comparatively new, their long-term impacts on human health are not known (Bawa & Anilakumar, 2013). Because the allergenic potential of GM products might be more intense compared with conventional products, they may cause allergenic effects in individuals who are prone to allergies or other adverse health impacts (Verma et al., 2011). This is due to the fact that the introduction of a non-allergenic protein expressing gene does not guarantee that the product will be free of allergenic action (Ntona & Arvanitogiannis, 2009; Arjó et al., 2013). Other literature studies have reported the occurrence of necrosis and stomach erosion in rats fed with GM *Flavr Savr* tomatoes, while GM potatoes expressing *Galanthus nivalis* (GNA; snowdrop) lectin encouraged proliferative growth in rat stomachs. Such glomelular stomach erosions can lead to life-threatening haemorrhage, especially in the aged and patients taking nonsteroidal anti-inflammatory drugs (Pusztai et al., 2003; Dona & Arvanitoyannis, 2009).

In addition, scientifically possible practices may not be ethically right (Preston, 2007). For example, if genetic modification of an organism is found to be risky for human health, it could be excluded on an ethical basis. Yet, any new advancement in science always challenges traditional practices. On a utilitarian dimension, logical thinking and assigning a value to the various merits and demerits of GM crops should be the deciding factor. However, evaluating its impact on organisms, ecology, and the environment as a whole is a very tough and cumbersome task (Thompson, 2000). Various decisions regarding the research, production, use, trade, and release of GM crops should always be undertaken by keeping in view the rights and welfare of all sections of society. Therefore, a thorough study of the science, economy, and morals governing usage of genetic engineering is a pre-requisite *before* the decision-making process.

2.9 Current status and future prospects

Recently, economists have revealed that consumer acceptability of GM food crops is closely related to the features incorporated into the various crops such as food labels on retail foodstuffs and the information made available. For instance, the marketing of GM foods has been exposed to a lot of contradictions that have made it hard for producers and consumers to make informed decisions (Rousu et al., 2007).

Advancements in genome sequencing technologies have enabled scientists to utilize genome editing tools for characterizing various genes that are beneficial for improving crops in terms of yield, quality, nutritional composition, disease resistance, and abiotic stress tolerance. Such tools, including insertional mutagenesis, RNA interference, and Clustered Regularly Interspaced Short Palindromic Repeats (CRISPR/Cas9) have been successfully employed for manipulating the genes exhibiting resistance to abiotic stresses including salinity, drought, cold, and heat. Among the various gene-modifying techniques, CRISPR/Cas9 has gained much attention as this technology makes the gene-editing procedure easier and more accurate (Salava et al., 2021).

Genetic engineering tools have also been employed to target phytoene synthase 1 (PSY1), MYB transcription factor 12 (MYB12), and anthocyanin 2 (ANT2) for the production of yellow (Hayut et al., 2017), pink (Deng et al., 2018), and purple (Čermák et al., 2015) tomatoes, respectively. The usage of CRISPR technology for deactivating ripening inhibitor or DNA demethylase 2 (DML2) resulted in fruits with longer shelf life. Yet, in this case, partially ripened fruit with a bland flavor and little nutritional value were obtained (Lang et al., 2017; Ito et al., 2015). On the other hand, however, fruits with a longer shelf life without compromising tomato organoleptic or nutritional quality are possible by silencing pectate lyase (PL) and alcobaca (ALC), thus implying that the CRISPR system could be useful in the future (Uluisik et al., 2016; Wang et al., 2019).

As tomato is a highly sensitive crop and cold stress can easily damage its fruit quality. To increase their cold and drought resistance, a gene from rice

(Osmyb4), reported for exhibiting tolerance to drought and cold in transgenic *Arabidopsis thaliana* plant species, was introduced into the tomato (Vannini et al., 2007). However, the inserted gene, though resulting in drought tolerance did not affect cold tolerance. Similarly, transgenic tomatoes with choline oxidase gene (codA) were reported to show higher resistance under salt-stressed environments (Wei et al., 2017). Viral-resistant tomatoes are engineered with Agrobacterium-mediated transformation by integrating coat protein (CP) genes into their genetic makeup. CP genes are found to provide more than 90 percent resistance against infection by tomato mosaic virus (ToMV) and tobacco mosaic virus (TMV) and thus prevent the crop yield loss in tomato plants (Ali et al., 2014). With the application of CRISPR/Cas9 technology, crucial genes such as Tomato Dicer-like 2 (DCL2), which are involved in resistance pathways, have been knocked out to assess their immunity against viruses. On infecting with potato virus X, tobacco mosaic virus, or tomato mosaic virus, these DCL2 mutants showed viral symptoms, implying the involvement of DCL2 in the defensive system against RNA viruses (Wang et al., 2018).

2.10 Conclusion

GM plants have the potential to be one of the most significant inventions in agricultural history. GM methods have greatly improved the tolerance of crops for insects and various diseases and thus reduced the cost of food production. Newly introduced genetic engineering techniques could dramatically enhance beneficial food attributes such as vitamins, antioxidants, flavor, and shelf life. These improvements in plant quality are only possible through genetic modification, not by conventional breeding. More research is required to improve essential traits including tolerance to pathogens and abiotic stress, crop yield and dietary composition such as lycopene-enriched tomatoes. With the lately introduced CRISPR/Cas9 system, plant genetics could expand the opportunity of producing various crops with increased flavor, nutrition and stress resistance.

References

Ahuja, M. R., & Fladung, M. (2014). Integration and inheritance of transgenes in crop plants and trees. *Tree Genetics & Genomes*, 10(4), 779–790.

Ali, A., Muzaffar, A., Awan, M. F., Din, S., Nasir, I. A., & Husnain, T. (2014). Genetically modified foods: Engineered tomato with extra advantages. *Advancements in Life Sciences*, 1(3), 139–152.

Arjó, G., Portero, M., Piñol, C., Viñas, J., Matias-Guiu, X., Capell, T., ... Christou, P. (2013). Plurality of opinion, scientific discourse and pseudoscience: An in depth analysis of the Séralini et al. study claiming that Roundup^{TM} Ready corn or the herbicide Roundup^{TM} cause cancer in rats. *Transgenic Research*, 22(2), 255–267.

Bartoszewski, G., Niedziela, A., Szwacka, M., & Niemirowicz-Szczytt, K. (2003). Modification of tomato taste in transgenic plants carrying a thaumatin gene from *Thaumatococcus daniellii Benth. Plant Breeding*, 122(4), 347–351.

Bawa, A. S., & Anilakumar, K. R. (2013). Genetically modified foods: Safety, risks and public concerns—a review. *Journal of Food Science and Technology*, 50(6), 1035–1046.

Brookes, G., & Barfoot, P. (2013). The global income and production effects of genetically modified (GM) crops 1996–2011. *GM Crops & Food*, 4(1), 74–83.

Brookes, G., & Barfoot, P. (2014). Economic impact of GM crops. *GM Crops & Food: Biotechnology in Agriculture and the Food Chain*, 5, 65–75.

Brookes, G., & Barfoot, P. (2020). GM crop technology use 1996–2018: Farm income and production impacts. *GM Crops & Food*, 11(4), 242–261.

Čermák, T., Baltes, N. J., Čegan, R., Zhang, Y., & Voytas, D. F. (2015). High-frequency, precise modification of the tomato genome. *Genome Biology*, 16(1), 1–15.

Chan, Y. L., Yang, A. H., Chen, J. T., Yeh, K. W., & Chan, M. T. (2010). Heterologous expression of taro cystatin protects transgenic tomato against Meloidogyne incognita infection by means of interfering sex determination and suppressing gall formation. *Plant Cell Reports*, 29(3), 231–238.

Chen, H., & Lin, Y. (2013). Promise and issues of genetically modified crops. *Current Opinion in Plant Biology*, 16(2), 255–260.

Chheda, A. H., & Vernekar, M. R. (2015). A natural preservative ε-poly-L-lysine: Fermentative production and applications in food industry. *International Food Research Journal*, 22(1), 23–30.

Cosgrove, D. J. (2005). Growth of the plant cell wall. *Nature Reviews Molecular Cell Biology*, 6(11), 850–861.

Deng, L., Wang, H., Sun, C., Li, Q., Jiang, H., Du, M., … Li, C. (2018). Efficient generation of pink-fruited tomatoes using CRISPR/Cas9 system. *Journal of Genetics and Genomics [Yi chuan xue bao]*, 45(1), 51–54.

Dona, A., & Arvanitoyannis, I. S. (2009). Health risks of genetically modified foods. *Critical Reviews in Food Science and Nutrition*, 49(2), 164–175.

D'Ambrosio, C., Stigliani, A. L., & Giorio, G. (2016). Food from genetically engineered plants: Tomato with increased β-carotene, lutein, and xanthophylls contents. In *Genetically Modified Organisms in Food. Production, Safety, Regulation and Public Health*, edited by R. R. Watson & V. R. Preedy (San Diego, CA: Academic Press), 361–380.

Engel, K. H., Takeoka, G. R., & Teranishi, R. (1995). *Genetically Modified Foods* (Washington, DC: American Chemical Society).

Foolad, M. R. (2007). Current status of breeding tomatoes for salt and drought tolerance. In *Advances in Molecular Breeding Toward Drought and Salt Tolerant Crops*, edited by M. A. Jenks, P. M. Hasegawa, & S. M. Jain (Dordrecht: Springer), 669–700.

Ganesan, M., Rajesh, M., Solairaj, P., & Senthilkumar, T. (2012). Tomato as a pioneer in health management. *International Journal of Pharmaceutical, Biochemical and Medical Sciences*, 2(3), 210–217.

Gao, G., Bergefurd, B., & Precheur, B. (2010). Growing tomatoes in the home garden. *Ohio State University Extension Bulletin*, 1624.

Gerszberg, A., & Hnatuszko-Konka, K. (2017). Tomato tolerance to abiotic stress: A review of most often engineered target sequences. *Plant Growth Regulation*, 83(2), 175–198.

Gerszberg, A., Hnatuszko-Konka, K., Kowalczyk, T., & Kononowicz, A. K. (2015). Tomato (*Solanum lycopersicum* L.) in the service of biotechnology. *Plant Cell, Tissue and Organ Culture (PCTOC)*, 120(3), 881–902.

Gonzali, S., Mazzucato, A., & Perata, P. (2009). Purple as a tomato: Towards high anthocyanin tomatoes. *Trends in Plant Science*, 14(5), 237–241.

Good, X., Kellogg, J. A., Wagoner, W., Langhoff, D., Matsumura, W., & Bestwick, R. K. (1994). Reduced ethylene synthesis by transgenic tomatoes expressing S-adenosylmethionine hydrolase. *Plant Molecular Biology*, 26(3), 781–790.

Grierson, D., Maunders, M. J., Slater, A., Ray, J., Bird, C. R., Schuch, W., ... Knapp, J. E. (1986). Gene expression during tomato ripening. *Philosophical Transactions of the Royal Society of London. B, Biological Sciences*, 314(1166), 399–410.

Griffiths, A., Barry, C., Alpuche-Solis, A. G., & Grierson, D. (1999). Ethylene and developmental signals regulate expression of lipoxygenase genes during tomato fruit ripening. *Journal of Experimental Botany*, 50(335), 793–798.

Halford, N. G. (2012). Toward two decades of plant biotechnology: Successes, failures, and prospects. *Food and Energy Security*, 1(1), 9–28.

Hamilton, A. J., Fray, R. G., & Grierson, D. (1995). Sense and antisense inactivation of fruit ripening genes in tomato. In *Gene Silencing in Higher Plants and Related Phenomena in Other Eukaryotes*, edited by P. Meyer (Berlin: Springer-Verlag), 77–89.

Hayut, S. F., Melamed Bessudo, C., & Levy, A. A. (2017). Targeted recombination between homologous chromosomes for precise breeding in tomato. *Nature Communications*, 8(1), 1–9.

Herring, R. J. (ed.). (2013). *Transgenics and the Poor: Biotechnology in Development Studies* (New York: Routledge).

Huffman, W. E. (2010). Consumer acceptance of genetically modified foods: Traits, labels and diverse information. Working Paper 10029, Department of Economics, Iowa State University, Ames, IA.

Ito, Y., Nishizawa-Yokoi, A., Endo, M., Mikami, M., & Toki, S. (2015). CRISPR/Cas9-mediated mutagenesis of the RIN locus that regulates tomato fruit ripening. *Biochemical and Biophysical Research Communications*, 467(1), 76–82.

Jaganathan, D., Ramasamy, K., Sellamuthu, G., Jayabalan, S., & Venkataraman, G. (2018). CRISPR for crop improvement: An update review. *Frontiers in Plant Science*, 9, 985.

Jones, P. J., McFarlane, I. D., Park, J. R., & Tranter, R. B. (2017). Assessing the potential economic benefits to farmers from various GM crops becoming available in the European Union by 2025: Results from an expert survey. *Agricultural Systems*, 155, 158–167.

Karabudak, T., Bor, M., Özdemir, F., & Türkan, İ. (2014). Glycine betaine protects tomato (*Solanum lycopersicum*) plants at low temperature by inducing fatty acid desaturase7 and lipoxygenase gene expression. *Molecular Biology Reports*, 41(3), 1401–1410.

Klee, H. J., Hayford, M. B., Kretzmer, K. A., Barry, G. F., & Kishore, G. M. (1991). Control of ethylene synthesis by expression of a bacterial enzyme in transgenic tomato plants. *The Plant Cell*, 3(11), 1187–1193.

Klümper, W., & Qaim, M. (2014). A meta-analysis of the impacts of genetically modified crops. *PLoS One*, 9(11), e111629.

Krieger, E. K., Allen, E., Gilbertson, L. A., Roberts, J. K., Hiatt, W., & Sanders, R. A. (2008). The Flavr Savr tomato, an early example of RNAi technology. *HortScience*, 43(3), 962–964.

Kumar, K., Gambhir, G., Dass, A., Tripathi, A. K., Singh, A., Jha, A. K., ... Rakshit, S. (2020). Genetically modified crops: Current status and future prospects. *Planta*, 251(4), 1–27.

Kumar, S., Kiruba, R., Balamurugan, S., Cardoso, H. G., Birgit, A. S., Zakwan, A., & Sathishkumar, R. (2014). Carrot antifreeze protein enhances chilling tolerance in transgenic tomato. *Acta Physiologiae Plantarum*, 36(1), 21–27.

Lang, Z., Wang, Y., Tang, K., Tang, D., Datsenka, T., Cheng, J., ... Zhu, J. K. (2017). Critical roles of DNA demethylation in the activation of ripening-induced genes and inhibition of ripening-repressed genes in tomato fruit. *Proceedings of the National Academy of Sciences*, 114(22), E4511–E4519.

Lemaux, P. G. (2008). Genetically engineered plants and foods: A scientist's analysis of the issues (Part I). *Annual Review of Plant Biology*, 59, 771–812.

Li, S., Xu, H., Ju, Z., Cao, D., Zhu, H., Fu, D., ... Zhu, B. (2018). The RIN-MC fusion of MADS-box transcription factors has transcriptional activity and modulates expression of many ripening genes. *Plant Physiology*, 176(1), 891–909.

Luthria, D. L., Mukhopadhyay, S., & Krizek, D. T. (2006). Content of total phenolics and phenolic acids in tomato (*Lycopersicon esculentum Mill.*) fruits as influenced by cultivar and solar UV radiation. *Journal of Food Composition and Analysis*, 19(8), 771–777.

Martin, C. (2013). The interface between plant metabolic engineering and human health. *Current Opinion in Biotechnology*, 24, 344–353.

Mathur, V., Javid, L., Kulshrestha, S., Mandal, A., & Reddy, A. A. (2018). World cultivation of genetically modified crops: Opportunities and risks. In *Sustainable Agriculture Reviews*Volume 25, edited by E. Lichtfouse (Cham: Springer), 45–87.

Mohanta, T. K., Bashir, T., Hashem, A., Abd Allah, E. F., & Bae, H. (2017). Genome editing tools in plants. *Genes*, 8(12), 399.

Nesbitt, T. C., & Tanksley, S. D. (2002). Comparative sequencing in the genus *Lycopersicon*: implications for the evolution of fruit size in the domestication of cultivated tomatoes. *Genetics*, 162, 365–379.

Ntona, A. A., & Arvanitogiannis, I. S. (2009). Genetically modified food and health impact, review [in Greek]. *Database of Greek Medicine*, 26, 727–740.

Preston, R. (2007). An error in the code: What can a rare disorder tell us about human behavior? *New Yorker*, August 6, p. 30.

Pusztai, A., Bardocz, S., & Ewen, S. W. B. (2003). 16 genetically modified foods: Potential human health effects. In *Food Safety: Contaminants and Toxins*, edited by J. P. F. D'Mello (Wallingford, UK: CAB International), 347–372.

Raiola, A., Rigano, M. M., Calafiore, R., Frusciante, L., & Barone, A. (2014). Enhancing the human-promoting effects of tomato fruit for biofortified food. *Corporation Mediators of Inflammation*. doi:10.1155/2014/139873.

Redenbaugh, K. (1992). *Safety Assessment of Genetically Engineered Fruits and Vegetables: A Case Study of the Flavr Savr Tomato* (Boca Raton, FL: CRC Press).

Römer, S., Fraser, P. D., Kiano, J. W., Shipton, C. A., Misawa, N., Schuch, W., & Bramley, P. M. (2000). Elevation of the provitamin A content of transgenic tomato plants. *Nature Biotechnology*, 18(6), 666–669.

Rousu, M., Huffman, W. E., Shogren, J. F., & Tegene, A. (2007). Effects and value of verifiable information in a controversial market: Evidence from lab auctions of genetically modified food. *Economic Inquiry*, 45(3), 409–432.

Salava, H., Thula, S., Mohan, V., Kumar, R., & Maghuly, F. (2021). Application of genome editing in tomato breeding: Mechanisms, advances, and prospects. *International Journal of Molecular Sciences*, 22(2), 682.

Shah, S. A., Erdmann, S., Mojica, F. J., & Garrett, R. A. (2013). Protospacer recognition motifs: Mixed identities and functional diversity. *RNA biology*, 10(5), 891–899.

Shih, C. H., Chen, Y., Wang, M., Chu, I. K., & Lo, C. (2008). Accumulation of isoflavone genistin in transgenic tomato plants overexpressing a soybean isoflavone synthase gene. *Journal of Agricultural and Food Chemistry*, 56(14), 5655–5661.

Silaste, M. L., Alfthan, G., Aro, A., Kesäniemi, Y. A., & Hörkkö, S. (2007). Tomato juice decreases LDL cholesterol levels and increases LDL resistance to oxidation. *British Journal of Nutrition*, 98(6), 1251–1258.

Soil Association (2005). Flavr-Savr tomato & GM tomato puree: The failure of the first GM foods. Briefing Paper 29.

Spricigo, P. C., Freitas, T. P., Purgatto, E., Ferreira, M. D., Correa, D. S., Bai, J., & Brecht, J. K. (2021). Visually imperceptible mechanical damage of harvested tomatoes changes ethylene production, color, enzyme activity, and volatile compounds profile. *Postharvest Biology and Technology*, 176, 111503.

Steier, G. (2018). *Advancing Food Integrity: GMO Regulation, Agroecology, and Urban Agriculture* (Boca Raton, FL: CRC Press).

Thompson, P. B. (2000). Discourse ethics for agricultural biotechnology: Its limits and its inevitability – a response to Jamieson. *Science and Engineering Ethics*, 6(2), 275–278.

Thompson, R. P. (2011). *Agro-technology: A Philosophical Introduction* (Cambridge: Cambridge University Press).

Tilahun, S., Choi, H. R., Baek, M. W., Cheol, L. H., Kwak, K. W., Park, D. S., … Jeong, C. S. (2021). Antioxidant properties, γ-aminobutyric acid (GABA) content, and physicochemical characteristics of tomato cultivars. *Agronomy*, 11(6), 1204.

Tohge, T., Zhang Y., Peterek S., Matros, S., Rallapalli, G., Tandron, Y. A., … Fernie A. R. (2015). Ectopic expression of snapdragon transcription factors facilitates the identification of genes encoding enzymes of anthocyanin decoration in tomato. *The Plant Journal*, 83, 686–704. doi:10.1111/tpj.12920.

Turnbull, C., Lillemo, M., & Hvoslef-Eide, T. A. (2021). Global regulation of genetically modified crops amid the gene edited crop boom – a review. *Frontiers in Plant Science*, 12, 258.

Tzonou, A., Signorello, L. B., Lagiou, P., Wuu, J., Trichopoulos, D., & Trichopoulou, A. (1999). Diet and cancer of the prostate: A case-control study in Greece. *International Journal of Cancer*, 80(5), 704–708.

Uluisik, S., Chapman, N. H., Smith, R., Poole, M., Adams, G., Gillis, R. B., … Seymour, G. B. (2016). Genetic improvement of tomato by targeted control of fruit softening. *Nature Biotechnology*, 34(9), 950–952.

Vannini, C., Campa, M., Iriti, M., Genga, A., Faoro, F., Carravieri, S., … Bracale, M. (2007). Evaluation of transgenic tomato plants ectopically expressing the rice Osmyb4 gene. *Plant Science*, 173(2), 231–239.

Veillet, F., Perrot, L., Chauvin, L., Kermarrec, M. P., Guyon-Debast, A., Chauvin, J. E., … Mazier, M. (2019). Transgene-free genome editing in tomato and potato plants using agrobacterium-mediated delivery of a CRISPR/Cas9 cytidine base editor. *International Journal of Molecular Sciences*, 20(2), 402.

Verma, C., Nanda, S., K Singh, R., B Singh, R., & Mishra, S. (2011). A review on impacts of genetically modified food on human health. *The Open Nutraceuticals Journal*, 4(1), 3–11.

Vitale, A., Rocco, M., Arena, S., Giuffrida, F., Cassaniti, C., Scaloni, A., … Leonardi, C. (2014). Tomato susceptibility to Fusarium crown and root rot: Effect of grafting combination and proteomic analysis of tolerance expression in the rootstock. *Plant Physiology and Biochemistry*, 83, 207–216.

Vrebalov, J., Ruezinsky, D., Padmanabhan, V., White, R., Medrano, D., Drake, R., … Giovannoni, J. (2002). A MADS-box gene necessary for fruit ripening at the tomato ripening-inhibitor (rin) locus. *Science*, 296(5566), 343–346.

Wang, J., Li, J., Zhao, H., Sheng, G., Wang, M., Yin, M., & Wang, Y. (2015). Structural and mechanistic basis of PAM-dependent spacer acquisition in CRISPR-Cas systems. *Cell*, 163(4), 840–853.

Wang, T., Zhang, H., & Zhu, H. (2019). CRISPR technology is revolutionizing the improvement of tomato and other fruit crops. *Horticulture Research*, 6.

Wang, Z., Hardcastle, T. J., Pastor, A. C., Yip, W. H., Tang, S., & Baulcombe, D. C. (2018). A novel DCL2-dependent miRNA pathway in tomato affects susceptibility to RNA viruses. *Genes & Development*, 32(17–18), 1155–1160.

Wei, D., Zhang, W., Wang, C., Meng, Q., Li, G., Chen, T. H., & Yang, X. (2017). Genetic engineering of the biosynthesis of glycine betaine leads to alleviate salt-induced potassium efflux and enhances salt tolerance in tomato plants. *Plant Science*, 257, 74–83.

Wu, Z., Sun, S., Wang, F., & Guo, D. (2011). Establishment of regeneration and transformation system of *Lycopersicon esculentum* MicroTom. *British Biotechnology Journal*, 1(3), 53–60.

Xin, Z., Zhang, L., Zhang, Z., Chen, Z., & Sun, X. (2014). A tea hydroperoxide lyase gene, CsiHPL1, regulates tomato defense response against Prodenia Litura (Fabricius) and Alternaria Alternata f. sp. Lycopersici by modulating green leaf volatiles (GLVs) release and jasmonic acid (JA) gene expression. *Plant Molecular Biology Reporter*, 32(1), 62–69.

Zuluaga, D. L., Gonzali, S., Loreti, E., Pucciariello, C., Degl'Innocenti, E., Guidi L., … Perata, P. (2008). Arabidopsis thaliana MYB75/PAP1 transcription factor induces anthocyanin production in transgenic tomato plants. *Functional Plant Biology*, 35, 606–618.

3 Bt cotton

Shivali Panjgotra, Amrit Sudershan, Vishal Sharma and Priya Sharma

3.1 Introduction

Modern biotechnology has allowed us to identify and separate genes, transfer the functions performed by these genes to another organism, and as a result improve the quality of our food. The various developments in biological sciences have been widely used for improving the productivity of plants. They have also been used for improving the resistance of plants to pests and diseases (Gandhi & Namboodiri, 2006).

Cotton is a crop with global significance that is cultivated in over 70 countries. It is largely grown in the sub-tropical and tropical regions of the world. In total, global cotton cultivation has reached 31.8 million hectares. In 2017, India's cotton cultivation reached over 12 million hectares and, with a productivity of 524 kg lint/ha, India has become the first country in the world to become self-sufficient in cotton. In 2017–18 India's cotton production surpassed that of China, with 6.2 million tonnes of cotton (lint) produced. Worldwide, India grows the greatest percentage of cotton, accounting for 25 percent of global cotton production (Kumar et al., 2019).

Cotton is cultivated in three agro-ecological regions of India. These zones are the Northern zone (Rajasthan, Haryana, and Punjab), the Central zone (Gujarat, Odisha, Madhya Pradesh, and Maharashtra), and the Southern zone (Tamil Nadu, Karnataka, Telangana, and Andhra Pradesh). During the time of independence, India produced mainly short and medium staple cottons. Today, it is one of the biggest producers of cottons globally and has wide variety of cottons from 6 to 120s counts, from non-spinnable, coarse to medium, extra elongated and superfine cotton. Cotton productivity has increased significantly in the past decade. This crop supports more than 60 million people through its use in agriculture, processing, and textiles (Kumar et al., 2019). It accounts for approximately 29.8 percent of the country's gross domestic productivity.

But farmers face various challenges due to the insect pests and diseases that can damage their crops. In terms of global production, cotton accounts for 16 percent of insecticide use; in India, it accounts for around 40 percent of the total pesticide use (Venugopal, 2004). In most regions of India, such as

DOI: 10.4324/9781003278566-3

Central and Southern India, cotton growers use around 12–15 sprays per season. Pesticide use is very heavy in irrigated cotton areas. Despite the heavy use of chemicals, many of them are ineffective against certain pests and diseases as these have developed resistance to the use of certain chemicals (Birthal et al., 2000). In recent times, Bt cotton technology has been successfully used to control bollworms, which are the most common cotton pest.

Cotton is cultivated in many states in India. According to the Directorate of Cotton Development (2019), during 2017–18 the most cotton was cultivated in Maharashtra (4.2 M ha), followed by Gujarat (2.6 M ha), Telangana (1.9 M ha), and Andhra Pradesh (0.7 M ha). Although, Maharashtra has the biggest area under cultivation of cotton crop, it only harvested 52 percent of its total area (Ministry of Agriculture, 2019). In terms of productivity, India was ranked 34th among the top cotton producing countries in the world. Among the top 10 cotton producing nations, Turkey had the highest cotton yield at 1884 kg/ha. The other countries that have high cotton production are Australia (1814 kg/ha), Brazil (1675 kg/ha), China (1787 kg/ha), and the USA (955 kg/ha) (Kumar et al., 2019). India became one of the largest producer, exporter, and consumer of cotton globally with total production of around 460kg/ha in 2021–2022 (Cotton Corporation of India, 2022).

3.1.1 Cotton cultivation in India

Cotton, which is also known as the White Gold, is an essential cash crop in India and has a crucial role in increasing the country's wealth. There is a lot of evidence of cotton cultivation in the Indus Valley during the time of the Indus Valley civilization (3300–1300 BCE) and, since then, India has been a major exporter. Cotton is the important raw material for India's textile industry, contributing about 75 percent of the total raw material employed (Barwale et al., 2004). India is leading globally in terms of the area of cotton cultivation although most of it is rainfed and has low productivity. Yet, all four species of *Gossypium*, viz. *barbadense*, *herbaceum*, *hirsutum*, and *arboreum*, can be cultivated commercially in India and this can explain the overall high production status.

India has the distinction of growing all the cultivated cotton species and their inter- and intra-specific hybrids. Hybrid cotton was first introduced in India in 1970; it was developed in Gujarat state and later commercialized in Maharashtra. Hybrid cotton cultivation is a major milestone achievement for the Indian cotton industry. It represents 55 percent of the total area given over to cotton and 45 percent of its total production. However, the introduction of Bt cotton in 2002 has changed the dynamics of the cotton value chain in India. The acceptance and use of Bt cotton by Indian farmers has led to a considerable enhancement in the production of the crop and to a decrease in the expenditure on chemicals (James, 2003; Sahai & Rahman, 2003). The total area under cultivation of cotton, its production, and its yield during recent years is given in Table 3.1.

Table 3.1 Area, production, and yield of cotton in India

Year	*Area under cotton cultivation* (in lakh hectares)*	*Area under Bt. cotton cultivation** (in lakh hectares)*	*Production (in lakh bales)*	*Yield (in kgs per hectare)*
2002–03	86.24	0.29	86.21	191
2003–04	75.98	0.92	137.28	307
2004–05	87.87	4.85	164.29	318
2005–06	86.78	12.34	184.99	362
2006–07	91.44	33.53	226.32	421
2007–08	94.14	54.72	258.84	467
2008–09	90.07	66.69	222.76	403
2009–10	101.32	85.52	240.22	403
2010–11	111.23	96.32	330.00	499
2011–12	121.78	107.58	352.00	491
2012–13	119.77	105.43	342.20	486
2013–14	119.60	110.35	359.02	510
2014–15	128.19	119.40	348.05	462
2015–16	122.92	106.83	300.05	415
2016–17	108.28	89.43	325.77	511
2017–18	124.29	110.76	328.05	477
2018–19	126.58	117.81	287.08	386
2019–20***	125.84	117.47	322.67	436

Source*: Directorate of Economics and Statistics.
Source**: Directorate of cotton Development, Nagpur.
Source***: First Advancement Estimate (Directorate of Economics and Statistics).

3.1.2 Bt cotton in India

Through the introduction of Bt cotton, there has been a considerable technological change in cotton cultivation globally. Bt cotton is a pest-resistant plant that was developed to provide resistance against bollworms. Bt cotton was formed by changing the DNA of cotton with a microbial protein which is produced by an aerobic soil bacterium i.e. *Bacillus thuringiensis.* It is characterized by its capability to generate crystalline inclusions during sporulation. The introduction of the bacterial gene into cotton seeds has protected the plants from bollworm parasites. The plant's transgene produces toxins that are different from the ones produced by the normal plant. When ingested by a certain organisms, these crystals can kill the organism. *Bacillus thuringiensis* was first discovered in Japan in 1901 by Japanese bacteriologists. The crystal protein was isolated by a German scientist in 1945. The insecticide *Bacillus thuringiensis* was first registered in the USA in 1961 as a pest control agent (Gandhi & Namboodiri, 2006).

Bt cotton was the first transgenic crop to be introduced in India to address the bollworm problem. The introduction of this crop took six years of experimentation and multiple trials. The greater yield and the reduced application of pesticides were the main factors that contributed to its superior performance (Choudhary & Gaur, 2010). Transgenic technology is suitable for Indian farmers even though they have small farm holdings. Although the per hectare yield of cotton in India is lower as compared to the world average cotton yield of about 762 kg/ha, the changes in the realm of cotton cultivation in India are giving it the ability to take its current yield near to the world average in the near future (Cotton Corporation of India, 2022).

Prior to the arrival of the British, many different varieties of cotton were cultivated in various parts of India, each of which suitable to the regional soil and climatic conditions. The Bourbon cotton variety was introduced in 1797 by the British to suit the needs of the Lancashire cottonmills. However, it was prone to bollworm and other insects and the problem was getting worse. To overcome this issue, various pesticides and insecticides were increasingly used. This was the reason why, in 2002, the government introduced a new type of cotton called Bt cotton to control the pests that affected the cotton industry (Shah, 2012).

Bt cotton was first approved in the US in 1993 for field trials and in 1995 it was first permitted for US market use. In 1997, it was also permitted for trade in China. In 2002, Bt cotton was introduced to India by Monsanto and Mahyco under the name of Bollgard (BG); Bollgard II was introduced in 2006. Bollgard has simply one gene, i.e., the Cry 1Ac, and Bollgard II has two genes, i.e., CryAb and Cry1Ac. Bollgard provides defence against cotton bollworms but Bollgard II offers protection against cotton leafworm (*Spodoptera litura*) in addition to bollworm. Bt cotton was initially approved by the Genetic Engineering Appraisal Committee (GEAC) of the Ministry of Environment, Forests and Climate Change for commercial cultivation in the Central (Gujarat, Maharashtra, and Madhya Pradesh) and South zone states (Tamil Nadu, Andhra Pradesh, and Karnataka); approval for the North zone states was given in 2005–06. In addition to Bollgard and Bollgard II technologies, four other Bt technologies were approved in India, although these have yet to be used commercially. Currently, only Bollgard and Bollgard II technologies are used out of which about 95 percent of the area under cultivation in India is planted with double-gene Bt cotton – i.e. Bollgard II (Najork et al., 2022).

From 2002 to 2018, cotton acreage increased significantly from 7.67 M ha to 12.1 M ha and productivity grew at a fast pace, i.e. from 191 kg/ha to 520 kg/ha of lint (Kumar et al., 2019), making India the largest producer of GM cotton globally. By 2014, 95 percent of cotton grown in the US was genetic modified, and 96 percent of that grown in India. Bt cotton is presently cultivated in a wide range of countries such as Argentina, Australia, China, Indonesia, India, South Africa, and the United States. Bt cotton has gained positive feedback from many countries, including China, Australia, and USA. This is mainly due to its economic benefits and the reduction of harmful chemicals in its cultivation. The year of Bt cotton's adoption in main cotton growing nations is given in Table 3.2. Table 3.3 shows the chronology of expansion of Bt cotton in India.

Table 3.2 Year of Bt cotton's adoption in main cotton growing nations

Country	*Year of adoption of Bt cotton*
Australia	1996
Argentina	1998
Brazil	2005
Burkina Faso	2008
China	1997
Colombia	2002
India	2002
Indonesia	2002
Kenya	2020
Mexico	1996
Pakistan	2010
South Africa	1997
USA	1996

Table 3.3 Chronology of expansion of Bt cotton in India up to its approval

Year	*Description*
1994	The development of Institutional Biosafety Committees (IBSC) and proposals for trade of genetically modified Bt cotton seed were some of the steps taken by the government to ensure the safety of the products.
1995	Import of the Coker variety of Bt cotton seed (having Cry1Ac gene) from Monsanto, USA was allowed by the Department of Biotechnology.
1996	Bt cotton seed was imported and green house trials were carried out including evaluation of pollen escape. Back crossing breeding for Bt gene transfer was done in green house.
1997–1998	Field trials were carried out at locations to evaluate the pollen escape. Various toxicological studies on Ruminant goat model and Allergenicity BNR model were done. Permission for small trials of Bt cotton was given by Monsanto.
1998–1999	On 15–25 locations, multi-centric trials were carried out to assess the efficacy of Bt genes in Indian elite germplasm. The results of these trials were shared with the Committee on Genetic Manipulation.
2000–2001	(a) Extensive trials were conducted to evaluate the effectiveness of the Bt gene in the germplasm of Indian elite; (b) production of hybrid seed was increased to 150 ha; (c) different studies on biosafety were made; and (d) six location trials were set up by ICAR (Indian Council of Agricultural Research).
2001–2002	(a) Extensive trials on about 100 ha of area were done to estimate the performance of Bt hybrids and the effectiveness of Bt gene in Indian elite germplasm; (b) production of hybrid seed increased to 300 ha; (c) biosafety studies were made; and (d) ICAR increased the trial locations from six to eleven.

Source: Authors.

3.2 Why are genetic modifications required?

A successful genetic alteration is the practice which changes the genome of an organism. Indirectly, this process has been happening for thousands of years by controlling or selectively breeding plants and animals. Modern biotechnology, through genetic engineering, has made it possible to target a certain gene for a more specific change in the genome of the recipient organism.

For decades, scientists have been using traditional breeding techniques to produce organisms with specific traits by selecting the genomes of plants and animals. These artificial selection processes have led to the development of a variety of organisms with specific traits, but these are limited to naturally occurring variations. Through the advancement in genetic engineering scientists have been able to control the exact changes in the genes that organisms carry. We can now introduce new genes from different species into one organism. This allows us to improve the productivity and efficiency of various crops. Plants, soil bacteria, and farm animals are some of the notable examples of organisms that have been successfully transformed through genetic modification.

There are many types of insects that are beneficial to crops. Unfortunately, some of them are harmful pests. Pests can damage a crop or kill it completely. In rare cases, they can even destroy an entire field of produce. Farmers have various ways to combat these pests, one of which is by using chemicals. These chemicals – pesticides – kill pests and prevent them from damaging the plants. They also form a toxic barrier around the plant to keep further pests away. However, the use of these pesticides can negatively affect the environment, of which one such example is their ability to kill beneficial organisms. Through the development of GM crops, scientists have been able to create pest-resistant plants that are more effective at killing pests, which eliminates the need for chemicals on many crops.

When it comes to crops, farmers often worry about three factors that could affect their harvests: insects, weeds, and the weather. Fortunately, most of the GM crops that are grown today have been developed to address issues caused by weeds or insects. When it comes to insects these GM plants can repel particular kinds of insects that feed on them, so avoiding the use of harmful chemicals. Additionally, some plants with GM seeds have been developed to resist certain herbicides. This allows farmers to control weeds much more easily. Those who grow genetically engineered crops often see the biggest benefits, which usually go unnoticed by the general public. These include: lower costs, better soil erosion, and less pesticide use. Apart from plants, GMOs are also used to make various medicines and vaccines. These products benefit from the use of genetically engineered crops which can be produced without using any antibiotics and/or other unsafe substances.

Genetic engineering in agriculture can offer various advantages to the world's population. These include increased crop yields, better quality food, the reduced need for chemicals and fertilizers, the reduced need for antibiotics, and greater resistance to pests and disease. As already mentioned, GM technology has helped minimize the losses caused by pests and insects. The use of GM crops also means a reduction in the use of chemicals as they do not need any kind of insecticides or pesticides. It also helps farmers to reduce their labor force and increase their productivity.

Also, although adding desirable traits to living organisms can be done by crossbreeding, this process can often lead to unwanted properties. And, although these unwanted properties can subsequently be removed through interbreeding, this process can take decades. By contrast, the development of new variants through genetic modifications can transfer only the known and characterized genes and also in much shorter timeframes than conventional techniques. With the world's population expected to more than double by 2050, the demand for food crops is expected to increase significantly and the world might soon be hungry for new varieties of vegetables (Tabashnik, 1994). The UN Food and Agriculture Organization (FAO) has approximated that global agricultural production must increase by at least 60 percent to reach the levels needed to sustain the developing population of the world (McGuire, 2013).

Although the productivity of agriculture has improved significantly over the years, it is starting to wane as food demand increases. Rice and wheat production are expected to grow significantly less than they did in the past due to the slowing down of global economic growth. If the trend continues, production may not be enough to meet demand unless we start using significantly more fertilizer, land, and water. Climate change is already causing problems for farmers, with higher temperatures and more frequent storms expected to make matters worse. These conditions could spread diseases and insect pests across new regions. With the help of genetic engineering, new varieties can be created almost immediately to adapt to the changes brought about by climate change.

Genetic resistance is a strategy used to manage the pest populations of cotton. It is available in the cotton gene pool and is widely used to protect the plant from tiny sap sucking insects such as snow flies or whitefly and jassid. This technique has been used to develop resistant/tolerant varieties and hybrids in India. However, this technique did not work against bollworm. An alternative approach was proposed to overcome this issue. This method involved cloning and transferring the genes of the toxic protein (i.e. endo toxin protein) from the bacteria *Bacillus thuringiensis*. Bt cotton which, as mentioned before, is a type of genetically engineered cotton developed by Monsanto that has been successfully developed in the USA to suppress the growth of bollworms during the early phase of crop development.

Worldwide area under Bt cotton has increased significantly over the years. In total, around 12 percent of the cotton that is cultivated globally is now being planted with GM crops. According to the ICAC (International Cotton Advisory Committee), the area under the cultivation of Bt cotton is estimated to rise to 50 percent in the next 5–7 years (ICAC, 2021). GM crops also reduce the need for chemicals that harm the environment. However, these GM crops can cause insects to develop resistance to the proteins that they consume. This could be a warning sign for the future as some of the technologies that worked well in the past may become less effective in the future due to the evolution of certain insects. However, the future is very promising for the development of GM technologies that can meet the growing global demand for food in a sustainable manner.

3.3 Mechanism of genetic modification leading to desired trait

The commercialization of Bt cotton was initiated in the year 1981 by the advancement of new biotechnology techniques to enable its transformation and its characterization. Characterization and cloning of the Bt genes in *Escherichia coli* revealed that the protein maintained its insecticidal activity. The high activity against insects that was observed in plants that had undergone transformation suggested that the plants did not require high protein expression to achieve high efficiency (Purcell et al., 2004). Two different techniques were employed to attain the traits of Bt cotton varieties that can trigger the genes encoding δ-endotoxin or exotoxin from *Bacillus thuringiensis*. Firstly, the development of two commercial varieties, Ingard_R and Bollgard_R, was carried out by using *Agrobacterium tumefaciens* (Firoozabady et al., 1987). The process of direct DNA transfer, which is known as the biolistic method or microprojectile bombardment plant transformation, was used for the development of Bollgard II (McCabe & Martinell, 1993; Doherdty et al., 2000).

The steps of development of transgenic cotton through *Agrobacterium tumefaciens* and the biolistic method are shown in Figure 3.1. The ability to transfer a piece of Ti-DNA (Ti stands for tumor-inducing) from one cell to another (host plants) was the key component that allowed the use of *Agrobacterium*. The genes introduced through Ti plasmids into the Ti-DNA region are co transferred and incorporated into the plant genome. *Agrobacterium tumefaciens* is a type of bacteria which develops naturally in wounded tissues of dicotyledonous plants. This bacterium causes the growth of asymmetrical tissue called crown gall tumors in plants that produces opines which are a metabolite and used by the bacterium as a source of food (Tempe & Petit, 1982). As for the genes that cause tumor induction and production of opines from the Ti plasmids, *A. tumefaciens* can replace them with genes of interest by transferring them to the plant tissue. The commercial varieties of Bollgard R have been developed through the *Agrobacterium* transformation involve the transformation of cotton (*Gossypium hirsutum* L. cv Coker 312) with the plasmid PV–GHBKO4 (Perlak et al., 1990; Zambriski, 1992). Although there are many varieties of cotton, most of the genetic transformation is restricted to the Coker 312 variety because of the difficulties in regenerating cotton from undifferentiated callus tissues. Due to the lack of resistance character in Coker 312, it is necessary to transfer this trait to elite varieties through back crossing (Perlak et al., 1990). The PV–GHBKO4 plasmid vector has two plant gene expression

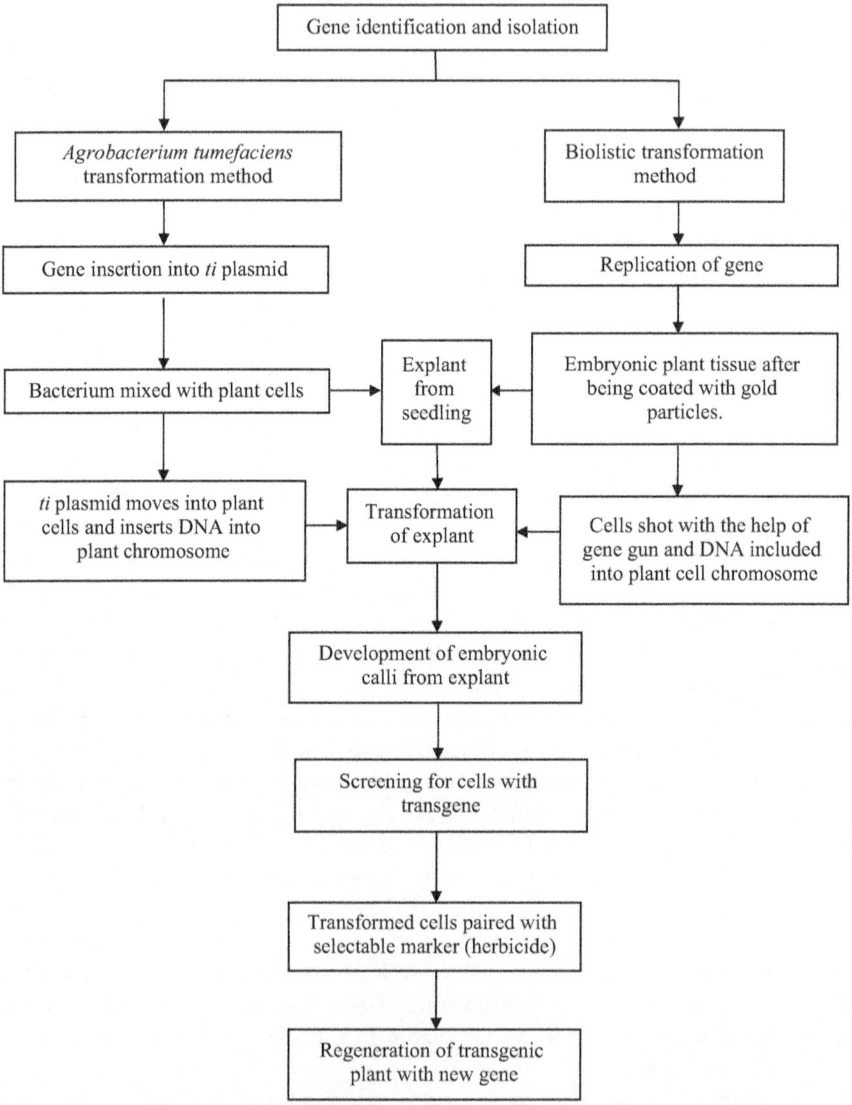

Figure 3.1 Steps of development of transgenic cotton through *Agrobacterium tumefaciens* and the biolistic method.
Source: Authors.

cassettes. The first cassette contains the insecticidal protein Cry1Ac *Bt*, while the second cassette contains the selectable marker (nptII). This vector can also overcome the disease-causing capability of *A. tumefaciens* as it contains the bacterial antibiotic resistant genes such as nptII and aad. The aad gene provides resistance to spectinomycin and streptomycin. The nptII gene is also known to provide resistance to neomycin and kanamycin. nptII and aad genes are actually related to the bacterial promoter. These two antibiotic resistance genes also do not possess the regulatory plant sequences, and the protein is not expressed in transformed cotton plants. The transformation of cotton from a native *Bt* coding gene sequence to an antibiotics-resistant phenotype was initially not expected (Vaeck et al., 1987).

A chimeric transgene in plants showed that the coding region of protein genes can inhibit the efficient expression of plants. Since typical plant genes have high G+C content and the cry protein genes are rich in A+T content, together they eliminate the potential polyadenylation signals and ATTTA sequences generated by these genes. They also replace the bacterial codons with the preferred plant codons. This gene modification increases the plant's expression 1,000-fold (Perlak et al., 1991; Purcell et al., 2004). The optimal expression of the Cry1Ab protein in Bollgard was also achieved through the modification of the Cry1Ac1 gene. The Bt protein is a hybrid protein that's made up of the amino acids 1–466 of the Cry1Ab protein extracted from *B. thuringiensis* subspecies kurstaki (Btk) strain HD-1 and 467–1178 of the Cry1Ac protein extracted from Btk HD-73. Consequently, the expression of the Cry1Ac protein in Bollgard is slightly different from that of native Cry1Ac. This gene is controlled by the CaMV promoter 35S, with a nontranslated region of the soybean alpha subunit of the beta-conglycinin gene. Bollgard_R II provides control against lepidopteran and advanced resistance to pest management.

3.4 Genetics of Bt cotton

The *Bacillus thuringiensis* bacterium is known to produce two types of toxins. The first is a Cry toxin which is encoded by different cry genes and the second is a Cytolytic toxin (Greenplate et al., 2003). This bacterium is known to have more than 50 genes with encoded Cry toxin and are sequenced for different studies (Agbios, 2002). Different genes along with their properties are detailed in Table 3.4.

Table 3.4 Different genes and their properties

Gene	*Cry I A(a), A (b), A(c), B, C, D, E, F, G*	*Cry II (Sub group)*	*Cry III (Sub group)*	*Cry IV (Sub group)* A, B, C, D	*Cry V–IX*
Crystal Shape	Bipyramidal	Cuboidal	Flat irregular	Bipyramidal	Various
Protein Size (kDa)	130–138	69–71	73–74	73–134	35–129
Insect	Lepidopteran larva	Lepidoptera, dipteral	Coleoptera	Diptera	Various

Source: Authors.

3.5 Mode of action of the toxin

In plants, the pH level is less than 5 and at this pH the toxin is in its insoluble non-toxic form. Also at this pH, the protein size is of around 130–138 kDa. This means it is non-toxic to humans and other higher organisms. The insect's mid gut is known to have a pH level of 9 which is ideal for dissolution of the toxic substances and changing the protein size to 60 kDa. When this toxin is ingested by insects, dissolution of cry proteins takes place in the alkaline juices found in the mid gut lumen. The proteases present in the insect's gut hydrolyze the protein and release the toxic fragments which are bound to a specific high-affinity receptor present on the brush border of mid gut epithelial cells. Because of this the brush border mid gut membrane builds up pores which are non-specific in nature. This leads to inflammation in the mid gut of larva which then makes the insect stop eating. Due to the lowered pH, bacterial spores begin germinating and cause the death of the larva via septicaemia (Sheikh et al., 2017) (Figure 3.2).

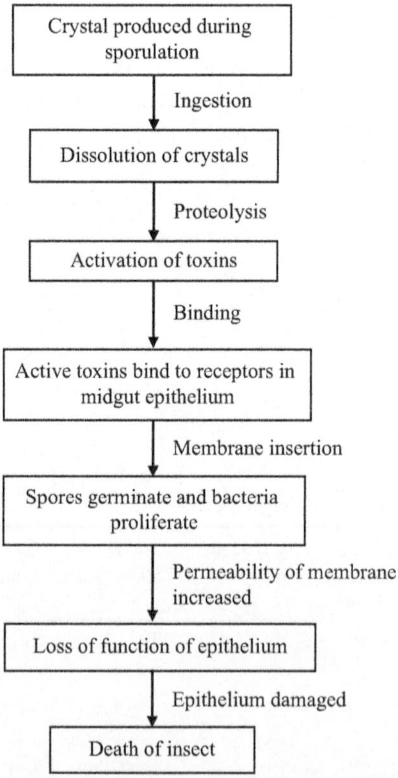

Figure 3.2 Mechanism of mode of action.
Source: Authors.

3.6 Benefits to society

With the introduction of Bt cotton in the production process, cotton producing countries have witnessed a momentous increase in their cotton acreage. For example, in 2014–15 India's cotton production grew by almost 20 percent to reach a record high of 39 million bales from just 14 million bales in 2014–15. India's total cotton production in 2020 amounted to approximately 36.5 million bales. India's cotton production is projected to increase to 43 million bales by 2030 and will contribute to as much as 40 percent global increase in cotton during the outlook period (Abdullayev, 2022). Its export has also grown at a fast pace. In 2015–16 India became the biggest cotton producer in the world overtaking the prior leader China. Bt cotton has been widely used by cotton producers for the management of bollworms in cotton. Various advantages of this technology include reduced use of chemicals, improved crop management, reduced production costs, lesser risk to cotton farming, feasible cotton growth even in regions of severe pest infestation, improved yield, and better profitability. The various indirect advantages of this technology include the reduced pesticide runoff, better growth of necessary insects and fauna in the cotton fields, reduced health hazard for farm workers, improved soil fertility, and curbs to air pollution, fossil fuel use, and labor cost and time.

The most profound benefit from biotechnology cotton is the decline in the use of pesticides for many crops. Several studies performed across China, the United States, Mexico, Australia, and Spain have shown a reduction in the usage of harmful pesticides after the application of Bt cotton with a significant reduction in the number of sprays per crop season from 7.7 to 1.0. In 2000–2001, large scale DBT (Department of Biotechnology) testing of MAHYCO hybrids successfully proved that pesticidal usage in India was reduced by 3.6 sprays.

In 2000, different scientific groups from around the globe, i.e. the Brazilian Academy of Sciences, the Chinese Academy of Sciences, the Indian National Science Academy, the Mexican Academy of Sciences, the Royal Society of London, Third World Academy of Sciences and the US National Academy of Sciences, released a report which called for the use of biotechnology in agriculture to improve the situation of poverty and hunger throughout the world. In reality, the use of the best scientific evidence will help governments to make informed decisions when it comes to the development of biotechnology. It was pointed out in the report that to prevent the possible harmful effects of introducing GM plants, public health regulation systems needed to be in place in every country. Concerns about the environment also needed to be addressed when it came to the use of agricultural technologies. It was suggested that establishing a formal risk-assessment process could help to minimize the uncertainties associated with the use of new crop plants. Many countries had already established protocols for the endorsement of new crops. In 2000, the US National Academy of Sciences published a report which clearly stated that there was no evidence that introducing recombinant DNA techniques or genes from unrelated organisms into the environment represented special risks. The report separately mentioned that the risk assessment should also analyse the environment where the organisms are introduced.

The primary purpose of commercial GE cotton crops is to combat pests. They also lessen the possible collateral harm to non-target species, including humans. A reduction in the cost of pesticide and labor use and an increase in seed cost and yield were seen in India during the period between 1990 and 2012 (Kranthi & Stone, 2020; Qaim & De Janvry, 2003; Qiao, 2015). Use of controlled pesticides and reduced crop loss are believed to have increase yields (Qaim & De Janvry, 2003). The decrease in the time spent on pesticide spraying can be attributed to the decreased labor use; nevertheless an overall increase in seed cost was observed because the Bt cotton is more expensive than non-Bt cotton (Qiao, 2015). This one apparent negative is, however, offset by the benefits of Bt cotton as discussed in detail below.

1 Bt cotton is genetically resistive to bollworm. This resistance is controlled by a single dominant gene and is extremely potent at minimising the losses caused by these pests without causing any damage to the cotton.
2 Growing Bt cotton allows for the use of fewer chemicals. This saves the environment and lowers the cost of production.
3 Cotton cultivation becomes feasible even in areas where the bollworm infestation is severe.
4 It helps to raise crop yields, thus significantly increasing the profitability for farmers.
5 It is an eco-friendly cultivation of cotton as it promotes the growth of beneficial insects i.e. predators and parasites of bollworms (Fitt et al., 1994; Luttrell & Herzog, 1994).
6 Bt cotton has significantly lowered the number of sprays per crop season in India by 3.6 sprays as compared to non Bt cotton. Consequently, it curbs various health risks related to the usage of insecticides.

3.7 Trade in the global market

The USA is now the leading exporter of Bt cotton. After it started using GE technology for cotton production, it has been estimated that the country's farmers have gained an additional US$150 billion in economic benefits (Carlson, 2016). Despite the disagreements over whether Bt and other technologies are advantageous or not overall, it is still believed that they will be beneficial in the long run (Kathage & Qaim, 2012; Mandal et al., 2020). Bt cotton has provided China's farmers with a 10-fold increase in their products since 1996 (ISAAA, 2020). Exports of cotton from India improved from 0.05 million bales (2002) to 8.5 million bales (2007) and the income of the country's cotton producers increased from US$10.4 million (2002–2003) to US $2.2 billion (2007–2008). The quantity of cotton imported also decreased by

Table 3.5 Country-wise export of cotton from India

Country	*Export of cotton from India in 2020–2021 (in lakhs bales)*
Bangladesh	22.09
People's Republic of China	21.97
Vietnam	6.40
Indonesia	2.70
Thailand	0.39
Oman	0.31
Turkey	0.27
Italy	0.14
Mauritius	0.12
Others	0.44
Total	54.83

Source: Ministry of Textiles, Government of India.

almost 2 million bales during the same period. Exports of Indian cotton textile products grew from US$3.4 billion (2002) to US$4.7 billion (2007) (Cotton Corporation of India, 2009). Even though the increase in cotton export was largely due to higher yields, it was also influenced by changes in international trade regulations.

Out of total export of 54.83 lakh bales of cotton, 21.97 lakh bales of cotton were imported to China in 2020–21. After Bangladesh, China was the second largest importer of cotton from India in 2020–21. Table 3.5 shows the country-wise data on the export of cotton from India in 2020–2021.

3.8 Health and ethical issues/risks

All genetically engineered food products have to meet the same safety and quality standards as traditional crops. Food and food products from genetically engineered plants have been approved by the US FDA through a voluntary consultation process before being sold on the market. The US FDA did not finalize its decision on the safety of the food produced by the plants until the questions raised by the US FDS about the safety of these foods and food products had been clarified.

The introduction of Bt cotton and its effects have been widely debated in India and extensively studied in the past few years. It is widely agreed that the introduction of Bt cotton has greatly improved the productivity of cotton, trade, farmers' livelihoods, and the environment. A study conducted by scientists in India showed that introducing GM crops has led to a significant drop in the use of chemical insecticides (37 percent) which has subsequently brought about an increase in the yield (22 percent) and profits for farmers (68 percent) (Gosh et al., 2022).

3.8.1 Risks and potential impacts of Bt cotton on human health

In the US, the Food and Drug Administration (FDA) has already approved the utilization of Bt cottons for human consumption (its oil, and meat from animals fed on it) after investigating the influence of Bt cottons on human health in 2019 (Krippel, 2020). In order to promote voluntary labelling, the FDA has issued a review that is mandatory prior to use. Australia has robust regulations for the approval of new drugs and technology and the Gene Technology Regulator carries out estimations related to the safety and effectiveness of the technology. Other countries and international organizations also conduct similar evaluations prior to approval.

Safety assessment of Bt cotton is a risk-based process that aims to evaluate the health and welfare of animals and humans and is concentrated on the following points:

1 A comprehensive knowledge of cotton's biology, its uses, and its various products.
2 A biochemical evaluation of the inserted protein, appraisal of concentrations of the protein, and an evaluation of the safety of the introduced protein are conducted to identify its usefulness. The safety assessment for various proteins includes:

   a a history of safe consumption of the proteins by humans or animals;
   b animal toxicity tests of the proteins before consumption;
   c results from the field and lab safety studies to assess the allergic effect, toxicity, and digestibility of the expressed proteins; and
   d assessment of the dietary consumption of the proteins by humans and animals of cotton products.

3 A determination of any unintended effects on the quality traits of the crop as a result of the insertion of the genetic material or the resulting protein expression. The concept is termed as 'Substantial Equivalence'. Various trials were conducted in cotton to test the various agronomic characteristics of the plant, including its plant morphology and oil quality, fibre quality, and nutritional components of its seed oil. These nutritional studies include its amino acid spectrum, fatty acid spectrum, gossypol carbohydrates, fats, calories.
4 Feeding studies have been carried out on rats and other animals to investigate the effects of cotton seed meal.
5 Cotton products used in medical and personal hygiene products and food were reviewed and tested.

Changes at the molecular level can also be employed to produce compounds that can trigger actions that are exceptionally important for the cotton industry. For instance, a system has been developed by researchers from various sectors, both private and public, that would allow the plant to only produce infertile seed – the Technology Protection System (TPS). The TPS means that farmers usually have to buy seeds for their crops every year. Although the concept of the TPS is mainly related to cotton, it can also be utilized for other crops. However, plans to commercialize the technology were withdrawn due to public objections. This shows that GE could be used to develop traits that could be objectionable to some farmers as they go against traditional practices.

Even before the introduction of Bt cotton, concerns were raised about the potential for insects to resist the toxin produced by the Bt gene. Currently it is globally accepted that insects *will* develop resistance to the toxins, and precautionary measures have been taken to prevent this. The reengineering of cotton with new genes which will produce toxins with different modes of action can help to minimize the development of resistance in the insects that feed on it.

Only two types of GE cottons that involve three different genes have been commercialized so far. They do not show deleterious effects from the interaction with other genetic material in the plant. However, interactions with other genes in the plant are not impossible. According to a review of safety information, Bt cotton does not pose any risk to humans or animals. Although the EPA has not yet approved the use of certain proteins in Bt cotton, it has been shown that they do not require a tolerance level, which means that these proteins are considered safe for human or animal consumption (Zia et al., 2015). The tolerances set by the EPA for the use of pesticides in food i.e. cottonseed oil and feed – cottonseed, cottonseed hulls, and cottonseed meal – are intended to prevent the accidental use of harmful chemicals. More approvals have been granted following scientific review for the use of Bt cotton in food and feed products in various countries such as Australia, Argentina, Canada, China, Mexico, Japan, and South Africa. It is also used as an ingredient in cosmetic and medical products.

3.8.2 Impacts on the environment

The US Department of Agriculture (USDA) has carried out field tests for the safety and efficacy of biotechnology crops. The USDA assessed if a new technology could threaten the health or welfare of plants or animals. The US Environmental Protection Agency (EPA) is the authority which enforces the regulations regarding the use and marketing of crops having pesticidal properties such as Bt cotton. The US EPA also enforces the regulations concerning the environmental exposure of croplands to certain organisms, such as non-target insects and animals. The agency plans to update its regulations to clarify how it regulates the use of biological control agents for crop protection. Other countries and international organizations also conduct reviews before they allow the use of Bt cottons.

The various components of Bt cotton have been studied to identify its environmental impacts (Dhillon et al., 2011; Sood et al., 2021). Field observations are typically conducted to evaluate the agronomic performance of new cotton varieties through various parameters such as yield, morphology, the plant's susceptibility to pests and diseases, characteristics of plant growth, and lint quality. The traits of the above-mentioned parameters were not affected by the introduction of genetic material, but the effects of the material on these factors were still significant. No harmful impact on the environment has been reported by the consumption of Bt cotton since it was introduced in 1996 in any country. In fact, the use of Bt cotton has mostly had a considerable beneficial impact on the environment such as the reduced use of harmful pesticides (Dhillon et al., 2011; Sood et al., 2021).

The effects of Bt cotton on the environment, animal health, and human health are given in detail below (Pongpaiboon, 2022; Rana et al., 2022):

1 No harmful effects have been observed on non-target beneficial insects using Bt cotton.
2 The use of Bt cotton and its products such as its seed and cake do not affect the digestion of animals. Also, no noxious or allergic effects have been reported with the use of Bt cotton seed and meal.
3 There is no chance of cross pollination between Bt cotton and other species of *Gossypium* because of the presence of the Bt gene.
4 Feeding on Bt cotton seed by animals does not have detrimental effects on them.
5 Oil extracted from the Bt cotton seeds has been known to have no harmful effects on humans.
6 Although the Bt gene can be inserted into upland cotton, it cannot be used to cross-breed with the *Malvaceae* families.
7 It cannot be crossed with other tetraploid species such as *G. Tomentosum* that are located in extremely isolated regions or gardens that are regularly maintained by research institutes.
8 No adverse effects of Bt cotton have been reported from any of the countries where it is grown commercially.

3.9 Current status and future prospects

During the 1990s, some scientists doubted the safety of introducing GM crops into the agriculture industry. It was mostly expected that the adoption of GM technology would be difficult especially for developing nations. Despite the various controversies, the use of GM crops has gained popularity because of their benefits such as food security, environment and health issues, and benefits to the world's economy. For these reasons, the number of cultivated fields especially in the small or resource-poor regions has increased (James, 2010). With the rise in world population, the area suitable for agricultural production has decreased over the years (Zang et al., 2016). Various global issues, such as climate change and food shortages, have led to the development of new GM technologies that can improve the production of agriculture (Fedoroff, 2010).

Sustainability is the key to productivity in cotton production; it can meet the increasing environmental demands while producing high-quality cotton. Getting the most out of your crop requires improving cultivation techniques and genetics, and this will differ with different environments and regions. The modern techniques involve the development of hybrid cotton varieties that combine desirable traits with novel genes. These varieties can help in improving soil fertility, facilitating cultivation practices, and managing pest control as well as meeting both market and environmental demands. The use of Bt and herbicide-resistant cotton has significantly decreased the energy consumption and operational costs associated with processing cotton. Yet, the process of revenue maximization does not always involve high yielding plants; it can still be achieved by optimizing the fibre quality to get the best possible price per unit. Water-efficient and exceptional drought-resistant varieties are being used in biotechnology laboratories all around the world. Transgenic cotton is a versatile crop that can tolerate both drought and soil salinity. It can be used for new regions and soils that were previously unsuitable for growing.

Water availability is a vital issue for the expansion of agriculture in diverse parts of the world and will have a huge impact on agriculture in the future. Cotton is one of the crops that requires less water. Improvement in cotton's tolerance to drought and production with variable rainfall will help its success. With the increasing popularity of drought-tolerant crops, biotechnology companies are expecting robust growth rates of up to 10 percent (Ranjha et al., 2022). Similarly, the development of physiological mechanisms that can increase the uptake of nitrogen will help minimize the use of fertilizer. The widespread adoption of genetically engineered technologies in a variety of organisms expressing other genes than those already marketed will help minimize the threat of diseases and insects. As stated above, various modes of action are necessary to maintain the insects' susceptibility to disease and pests when they encounter the proteins produced by plants. These new stacked genes or proteins such as protease inhibitors, which are taken from other sources, will contribute to management of insect resistance and insect control.

Similarly new cotton varieties that can tolerate various herbicides will also provide options for weed-resistance management. Due to the availability of certain herbicide-tolerant varieties, the cultivation methods used for these crops have been refined on a scale that was not available previously. In addition to improving the quality of the cotton, recent varieties – with improved traits produced by various biotechnology companies – will also allow producers to expand their operations better suited to their field environment and improve their production. Cotton transformation regarding the enhancement of traits for its better quality and better yield will be the objective of scientists and cotton breeders in the future.

3.10 Conclusion

Bt crops have been beneficial to the environment by reducing the amount of chemicals used and thus the cost of production. With genetically engineered crops increasingly being used for the control of insects or pests, the number of strains with multiple Bt toxins could increase. Modified Bt toxins can be used to kill insects that are genetically resistant to standard toxins and, therefore, with the continuing development of genetically engineered crops, the use of these modified Bt-toxin crops could become widespread in developing nations. Further studies on the patterns of evolved resistance in different crops could help to develop future resistance-management techniques that can minimize the negative effects of introducing GM crops and maximize the benefits of current and future generations of transgenic crops.

References

Abdullayev, R. (2022). Trends in the world cotton market condition. *Berlin Studies – Transnational Journal of Science and Humanities*, 2 (1.1 Economical sciences).

Agbios (2002). Biotech crop database. http://www.agbios.com/main.php

Birthal, P., Sharma, O. P., & Kumar, S. (2000). Economics of integrated pest management: Evidences and issues. *Indian Journal of Agricultural Economics*, 55(4), 644–659.

Barwale, R. B., Gadwal, V. R., Zehr, U., & Zher, B. (2004) Prospects of Bt cotton technology in India. *The Journal of Agrobiotechnology Management and Economics*, 7 (1–2), 23–26.

Carlson, R. (2016). Estimating the biotech sector's contribution to the US economy. *Nature Biotechnology*, 34(3), 247–255.

Choudhary, B., & Gaur, K. (2010). *Bt Cotton in India: A Country Profile*. ISAAA Series of Biotech Crop Profiles (Ithaca, NY: ISAAA).

Cotton Corporation of India (2022). National cotton scenario. Ministry of Textiles, Government of India. https://cotcorp.org.in/national_cotton.aspx

Directorate of Economics and Statistics, Department of Agriculture and Farmers Welfare, Ministry of Agriculture and Farmers Welfare, Government of India (2021). Agricultural statistics at a glance. https://eands.dacnet.nic.in/

Dhillon, M. K., Gujar, G. T., & Kalia, V. (2011). Impact of Bt cotton on insect biodiversity in cotton ecosystem in India. *Pakistan Entomologist*, 33(2), 161–165.

Doherdty, S. C., Lirette, R. P., & Hamilton, K. A. (2000). *Molecular Report of the Stability of Cotton Event 15985*. Monsanto Company Laboratory project ID study 00-01-36-09, MSL-16749.

Fedoroff, N. V. (2010). The past, present and future of crop genetic modification. *New Biotechnology*, 27(5), 461–465.

Firoozabady, E., DeBoer, D. L., Merlo, D. J., Halk, E. L., Amerson, L. N., Rashka, K. E., & Murray, E. E. (1987). Transformation of cotton (*Gossypium hirsutum* L.) by Agrobacterium tumefaciens and regeneration of transgenic plants. *Plant Molecular Biology*, 10(2), 105–116.

Fitt, G. P., Mares, C. L., & Llewellyn, D. J. (1994). Field evaluation and potential ecological impact of transgenic cottons (*Gossypium hirsutum*) in Australia. *Biocontrol Science and Technology*, 4(4), 535–548.

Gandhi, V. P., & Namboodiri, N. V. (2006). *The Adoption and Economics of Bt Cotton in India: Preliminary Results from a Study*. Indian Institute of Management, Ahmedabad, India.

Ghosh, T., & Adhikari, D. (2022). Transgenic plants: A curtain raiser implying risks vis-a-vis benefits on present environment and society. *International Journal of Advanced Research in Biological Sciences*, 9(1), 71–95.

Greenplate, J. T., Mullins, J. W., Penn, S. R., Dahm, A., Reich, B. J., Osborn, J. A., … Shappley, Z. W. (2003). Partial characterization of cotton plants expressing two toxin proteins from Bacillus thuringiensis: Relative toxin contribution, toxin interaction, and resistance management. *Journal of Applied Entomology*, 127(6), 340–347.

Government of India (2019). *Status Paper of Indian Cotton*. Directorate of Cotton Development, Government of India. Ministry of Agriculture & Farmers Welfare, Department of Agriculture, Cooperation and Farmers Welfare, Nagpur.

ICAC (2021). Data portal. International Cotton Advisory Committee. https://www.icac.org/DataPortal/DataPortal?Units=Exports&Year=2021/22%20est

ISAAA (2020). International service for the acquisition of agri-biotech applications. Bt-cotton increased farmers' produce by 10-fold in China. Published on September 23, 2020. http://www.isaaa.org/kc/cropbiotechupdate/article/default.asp?ID=18344

James, C. (2003). *Global Status of Commercialized Transgenic Crops: 2003*. ISAAA Briefs No. 30, 1–38.

James, C. (2010). *Executive Summary of Global Status of Commercialized Biotech/GM Crops: 2010*. ISAAA Brief No. 42.

Kathage, J., & Qaim, M. (2012). Economic impacts and impact dynamics of Bt (*Bacillus thuringiensis*) cotton in India. *Proceedings of the National Academy of Sciences*, 109(29), 11652–11656.

Kranthi, K. R., & Stone, G. D. (2020). Long-term impacts of Bt cotton in India. *Nature Plants*, 6(3), 188–196.

Kranthi, K. (2015). Cotton in the climate trap. *Cotton Statistics and News*, 10, 6–9.

Kripal, K. (2020). Student Series: FDA approves genetically modified cotton for human consumption. Farm Bill Law Enterprise. http://www.farmbilllaw.org/2020/01/21/student-series-fda-approves-genetically-modified-cotton-for-human-consumption/#:~:text=In%20early%20October%20of%202019,source%20for%20humans%20and%20animals

Kumar, R., Kumar, S., Sivaramane, N., Dhandapani, A., & Meena, P. C. (2019). *Adoption of Insect Resistance Management Practices in Bt-Cotton Cultivation in India*. ICAR National Academy of Agricultural Research Management, Hyderabad, India.

Luttrell, R. G., & Herzog, G. A. (1994). Potential effect of transgenic cotton expressing BT on cotton IPM programs. In *Proceedings of the Beltwide Cotton Production and Research Conference*, edited by D. J. Herber (Memphis, TN: National Cotton Council of America), 806–809.

Mandal, A., Sarkar, B., Owens, G., Thakur, J. K., Manna, M. C., Niazi, N. K., … Patra, A. K. (2020). Impact of genetically modified crops on rhizosphere microorganisms and processes: A review focusing on Bt cotton. *Applied Soil Ecology*, 148, 103492.

McCabe, D. E., & Martinell, B. J. (1993). Transformation of elite cotton cultivars via particle bombardment of meristems. *Bio/technology*, 11(5), 596–598.

McGuire, S. (2013). WHO, World Food Programme, and International Fund for Agricultural Development. The State of Food Insecurity in the World 2012. Economic growth is necessary but not sufficient to accelerate reduction of hunger and malnutrition. Rome, FAO. *Advances in Nutrition*, 4(1), 126–127.

Ministry of Textiles, Government of India (2021, July). Press release. https://pib.gov.in/PressReleaseIframePage.aspx?PRID=1740344

Perlak, F. J., Deaton, R. W., Armstrong, T. A., Fuchs, R. L., Sims, S. R., Greenplate, J. T., & Fischhoff, D. A. (1990). Insect resistant cotton plants. *Bio/technology*, 8(10), 939–943.

Purcell, J. P., Oppenhuizen, M., Wofford, T., Reed, A. J., & Perlak, F. J. (2004). The story of Bollgard® cotton. In *Handbook of Plant Biotechnology*, edited by P. Christou & H. Klee (Hoboken, NJ: John Wiley and Sons).

Pongpaiboon, S. (2022). Understanding the ethical and environmental implications of genetically modified crops through the Bt insecticidal gene. *International Journal of Research in Engineering, Science and Management*, 5(3), 26–33.

Qaim, M., & De Janvry, A. (2003). Genetically modified crops, corporate pricing strategies, and farmers' adoption: The case of Bt cotton in Argentina. *American Journal of Agricultural Economics*, 85(4), 814–828.

Qiao, F. (2015). Fifteen years of Bt cotton in China: The economic impact and its dynamics. *World Development*, 70, 177–185.

Rana, I. A., Bakhsh, A., Tillaboeva, S., Haihong, S., & Azhar, M. T. (2022). Genetically modified cotton: Boom or dust. In *Cotton Breeding and Biotechnology*, edited by Z. Khan, Z. Ali, & A. A. Khan (Boca Raton, FL: CRC Press), 271–289.

Ranjha, M. M. A. N., Shafique, B., Khalid, W., Nadeem, H. R., Mueen-ud-Din, G., & Khalid, M. Z. (2022). Applications of biotechnology in food and agriculture: A mini-review. *Proceedings of the National Academy of Sciences, India Section B: Biological Sciences*, 1–5.

Sahai, S., & Rahman, S. (2003). Performance of Bt cotton: Data from first commercial crop. *Economic and Political Weekly*, 38(30), 3139–3141.

Shah, D. K. (2012). Bt cotton in India: A review of adoption, government interventions and investment initiatives. *Indian Journal of Agricultural Economics*, 67(3).

Sheikh, A. A., Wani, M. A., Bano, P., Un, S., Nabi, T. A. B., Bhat, M. A., & Dar, M. S. (2017). An overview on resistance of insect pests against Bt crops. *Journal of Entomology and Zoology Studies*, 5, 941–948.

Sood, N., Gupta, S., Rath, R., & Saxena, R. (2021). Genetically modified crops – need of the hour: An Indian perspective. *Journal of Thematic Analysis*, 2(2), 2582–7391.

Tabashnik, B. E. (1994). Evolution of resistance to *Bacillus thuringiensis*. *Annual Review of Entomology*, 39(1), 47–79.

Tempe, J., & Petit, A. (1982). Opine utilization by Agrobacterium. In *Molecular Biology of Plant Tumors*, edited by G. Kahl & J. S. Schell (New York: Academic Press), 451–459.

Vaeck, M., Reynaerts, A., Höfte, H., Jansens, S., De Beuckeleer, M., Dean, C., … Leemans, J. (1987). Transgenic plants protected from insect attack. *Nature*, 328 (6125), 33–37.

Venugopal, P. (2004). *Input Management*, Volume 8 (New Delhi: Academic Foundation).

Zambryski, P. C. (1992). Chronicles from the Agrobacterium-plant cell DNA transfer story. *Annual Review of Plant Biology*, 43(1), 465–490.

Zhang, C., Wohlhueter, R., & Zhang, H. (2016). Genetically modified foods: A critical review of their promise and problems. *Food Science and Human Wellness*, 5(3), 116–123.

Zia, M. A., Jan, S. A., Shinwari, Z. K., Shah, S. H., & Khalil, A. T. (2015). Impact of Bt cotton on animal health: A review. *Global Veterinaria*, 14(3), 377–381.

4 Transgenic soybean

Varun Kumar, Amarjeet Kumar and Nidhi Agarwal

4.1 Introduction

The practice of sequence technology in food production has become a fascinating subject because of increased food desires as well as the wish for improved quality. With the application of gene technology to plants and animals, these higher goals are often achieved more quickly than by traditional methods/selection. Sequence technology will transfer isolated genes between unrelated organisms – something that traditional selection cannot do (Jany, 1999a). Soybean (*Glycine max (L.) Merr.*) is a self-pollinating crop that pollinates itself before its flowers open. Soybean seeds contain around 40 percent protein, 20 percent oil, and 22–23 percent carbs by dry weight. Fibres and minerals make up about 3–6 percent of the total (Wang, 1997). FAO (Food and Agriculture Organization of the United Nations) statistics show that more than 80 countries grow this crop, with globally about 142 million metric tonnes produced in 1997 (Chee & Hu, 2000). For decades, traditional crop breeding, which need species be sexually attuned, has improved the genetics of soybeans and other crops. Yet, now, with breakthroughs in biotechnology techniques, scientists can now transfer genes from one organism to another, regardless of species or sexual suitability.

In the USA, the adaptation of planting of transgenic soybean crops rapidly increased by 72.29 million acres with 93 percent of acres in the year 2013 (Fernandez-Cornejo et al., 2014). In year of 2020, it reached 94 percent planting acres (USDA, 2020). Transgenic soybeans are being planted in the United States because of the weed control profit of utilizing a pesticide that affects normal soybeans. This herbicide, glyphosate is an economic solution to various weed issues that soybean producers in the United States confront each year (Chee & Hu, 2000). The purpose of this chapter is to describe the nature of soybean transformation, as well as to provide general accounts of various successful soybean transformation procedures, benefits, and trading in the global market, and to put forward ~~compute~~ the weed control benefits provided on transgenic soybean acreage.

DOI: 10.4324/9781003278566-4

4.2 Why are genetic modifications required?

Soybean is considered a major crop worldwide and is in high demand for its various applications such as edible soybean oil for human consumption and soybean meal as a source of protein in animal feed. Soybean oil is widely used edible oil for human consumption around the world. Soybean meal, on the other hand, is a well-known product that is a major source of protein for animal diets. Furthermore, soybeans are widely employed in industry, such as the creation of biodiesel. As a result of the increasing demand for soybeans, genetic modification for soybean crops is being pursued to close the supply–demand imbalance. Though GM crops are still controversial, considering the positive side of the debate, the technology has provided a mechanism to add new beneficial qualities to existing crops and has the potential to address present and future agricultural difficulties. Traditional soybean reproduction schemes are sometimes hampered through the limited variety of germplasm. In Brazil, biotech soybean seed adoption rate reached 96 percent in the 2018/19 crop season. Biotech soybeans had the most planted area, with 34.86 million hectares planted, up from 33.7 million hectares in 2017, representing a 3.4 percent increase (ISAAA, 2018).

4.3 Gene involved in agronomic trait development

Herbicide-tolerant soybean plants are Roundup Ready (RR) vegetation which is resistant to the herbicide glyphosate (N-[phosphonomethyl] glycine). Glyphosate resistance was achieved by expressing the 5-enol-pyruvyl-shikimate-3-phosphate synthase (EPSPS) gene from *Agrobacterium* spp. strain CP4. The inclusion of the glyphosate-tolerant EPSPS enzyme covers the plant's demand for aromatic amino acids and other metabolites essential for plant developmental processes in the presence of glyphosate (Padgette et al., 1995). Even though both transgenic actions – the binding of glyphosate to the indigenous EPSPS enzyme and the prevention of shikimate-3-phosphate production – generate identical CP4 EPSPS proteins, the second-technology RR soybean, which was developed by introduction of CP4 EPSPS gene cassette into the Asgrow soybean variety A3244, outperforms the former in terms of superior agronomic characteristics such as yield (Lundry et al., 2008; Levy-Booth et al., 2009). Soybean CV127 ("Cultivance"), a novel class of genetically engineered soybeans that is herbicide-resistant to imidazolinone. Scientists from Embrapa, and subsequently the German company BASF, used a transformed csr1–2gene from Arabidopsis thaliana to genetically engineer the CV127 soybean (Rech et al., 2008).

The entomopathogenic bacterium Bt (*Bacillus thuringiensis*) is frequently utilized as a bio-pesticide to regulate insect pests. B. thuringiensis generates proteins which form crystals on spores and are highly toxic to dipteran, lepidopteran, and coleopteran larvae (Peferöen, 1997; Hongyu et al., 2000). Transgenic Bt protein synthesis is believed to be very successful at regulating insect pests in a range of key crop plants, mainly maize and cotton (James, 2011). Various Bt soybean strains have also been developed, and stacking soybean (Bt/RR) was permitted for commercialization in Brazil in 2011. According to Parrott et al. (1989), the presence of the cry1Ab gene hindered *Anticarsia gemmatalis* larval feeding. A modified line expressing high amounts of a synthetic cry1Ac gene caused total larvae mortality of *A. gemmatalis* and dramatically lowered *Pseudoplusia includens* and *Helicoverpa zea* larval survival and feeding (Stewart Jr et al., 1996, Walker et al., 2000). Homrich et al. (2008a) reported on how a modified soybean generating a fake cry1Ac gene boosted production. In in vitro and in vivo bioassays, modified flowers were revealed to be severely toxic to *Anticarsia gemmatalis* and in another study it was found that the cry1Ac transgene had no effect on overall agronomic performance or yield (Homrich et al., 2008b). Further research is detailed in Table 4.1. that illustrates over-expression, gene silencing, and other methods for transgenic soyabean.

GM soybean plants up-regulation of Arabidopsis 1-pyrroline-5-carboxylate synthase (P5CR) gene demonstrated enhanced resistance to drought and heat stress due to its improved free proline content (Kocsy et al., 2005; Yorinori et al., 2005; De Ronde et al., 2004a, b; Manavalan et al., 2009). Stress tolerance may also be accomplished by modifying the appearance of strain-precipitated transcriptional regulators (TR), which would then affect the expression of a large number of down-regulating genes (Agarwal et al., 2006). Focusing on value-added attributes, such as greater nutritional content, is the greatest solution from a client's perspective. Changes in protein quality and level, key amino acids, lipids, and fatty acids, beneficial secondary metabolites, and mineral availability are examples of nutritionally enhanced transgenic soybean plants, according to McGloughlin (2008). The quality and composition of soybean oil have recently received a lot of consideration, and attempts have been made to enhance its oxidative strength, decorate-three fatty acid constituted materials, as well as raise the actual quantity of oil in soybean seeds (Clemente & Cahoon, 2009). The commercialization of a soybean GMO with high amounts of oleic acid has previously been approved by the FDA (CERA, 2010).

Additionally, Buhr et al. (2002) provided an overview of altered soybean plants wherein genetic variation in seeds was simultaneously down-regulated, resulting in soybean oil with substantially reduced palmitic acid and a higher oleic acid content. A few of these cases were looked into as a possible biodiesel source (Graef et al., 2009). The extruded oil displayed improved quick and painless glide, as well as oxidative stability.

4.4 Molecular mechanism of genetic modification leading to the desired traits

The initial proposal for transgenic soybean conversion was made in 1988 (Hinchee et al., 1988; Christou et al., 1988). Sustained soybean conversion, even after nearly three decades, cannot be called typical since it is based on

Table 4.1 Investigations of soybean genes include over-expression, silencing, transposon-based mutagenesis, protein sub-cellular localization, and/or promoter characterization

	Gene Explanation	*Method*	*References*
1	Expression of myo-inositol-1-phosphate (**GmMIPS1**) and seed expansion are linked.	Gene silencing RNAi	Nunes et al. (2006)
2	**Thioredoxin gene** in nodule growth and symbiotic relationship maintenance	Gene silencing RNAi	Lee et al. (2005)
3	The significance of the **GS52 ecto-apyrase gene** in the process of modulation	Gene silencing RNAi	Govindarajulu et al. (2009)
4	The **actin gene** was temporarily silenced, resulting in a reduction in the frequency of SMV infection foci.	Gene silencing VIGS	Zhang et al. (2009a)
5	The ability of 140 genes to compromise resistance to *Phakopsora pachyrhizi* was tested.	Gene silencing VIGS	Pandey et al. (2011)
6	Maize **dissociation (Ds) transposon**: a powerful tool for soybean mutagenesis.	Gene silencing transposon-based mutagenesis	Mathieu et al. (2009)
7	Over-expression of **soybean genes** in a heterologous context.	Over-expression	Zhou et al. (2010)
8	Transcription factor effect on defence genes for production of ethylene-responsive element-binding protein **(GmEREBP1)** encoding gene.	Over-expression	Mazarei et al. (2007)
9	**GmDREB3** over-expression increases cold, drought, and salt tolerance.	Over-expression	Chen et al. (2009)
10	Aldo/keto reductase 1 gene (GmAKR1) regulates nodule formation	Over-expression	Hur et al. (2009)
11	Heterologous systems used for subcellular soybean gene products.	Subcellular localization	Li and Dhaubhadel (2011)
12	**Adh gene** promoter used for transgenic hairy roots.	Promoter characterization	Preiszner et al. (2001)
13	The effect of plant/nitrogen-fixing bacteria interaction on **soybean isoflavone synthase promoters.**	Promoter characterization	Subramanian et al. (2004)
14	Study of the soybean polyubiquitin gene promoter (**GmUbi**).	Promoter characterization	Hernandez-Garcia et al. (2009)
15	**GmDREB3** promoter responsible for cold stress.	Promoter characterization	Chen et al. (2009)

	Gene Explanation	*Method*	*References*
16	A soybean heat shock protein 90-like was studied (**GmHSP90L**).	Promoter characterization	Chiera et al. (2007)
17	**BCSP666**, a seed-specific promoter fragment, was isolated and analyzed.	Promoter characterization	Qinggele et al. (2007)
18	**Soybean lectin gene promoter** applicable for foreign gene expression to change embryo-specific traits.	Promoter characterization	Cho et al. (1995)

Source: Authors.

the capacity to integrate economic and regeneration approaches. Two fundamental techniques have been proven to be effective in providing completely and reliably changed plants: particle bombardment and the *agrobacterium tumefaciens* technique. Both techniques are essential in both agriculture and fundamental science applications.

4.4.1 Target tissues and particle bombardment

Genetic modification through particle bombardment is the expedited delivery of deoxyribonucleic acid (DNA) into tissues and cells by means of tiny particles moving at an elevated rate (Sanford, 1988). These particles are capable of the crossing cell membranes as well as nuclear membranes. Exogenous DNA fragments are released in the nucleus and may be incorporated into body DNA by homologous recombination processes involving the cytoplasm component of the cell (Sanford, 1990; Kohli et al., 2003). One of the key benefits of particle bombardment is that it allows genes transfer to another cell or tissue type, independent of genotype and without the need for host–bacterium compatibility. The method, on the other hand involves the accumulation of many copies of DNA, which can be recombined or fragmented (Hadi et al., 1996). Particle bombardment is usually carried out with helium pressure factor guns, which allow for highly accurate control of target tissue penetration and the delivery of the maximum number of particles carrying the DNA to a precise cell membrane. This feature is critical since various explants may need different settings for effective particle diffusion (Christou et al., 1990). In 1992 Finer et al. created the Particle Inflow Gun (PIG), a particle accelerator that has since been effectively employed in the alteration of soybean somatic embryos (Hernandez-Garcia et al., 2009; Homrich et al., 2008a).

Several studies have been published concerning soybean alteration by particle bombardment victimization using the meristems as the target tissue (Vianna et al., 2011; McCabe et al., 1988; Christou et al., 1989; Christou & McCabe, 1992; Aragão et al., 2000). Notably, in soybeans, the original upper plant tissue is made up of many cell membranes (L1, L2, and L3), which are engaged in the formation of the entire shoot; L1 is responsible for the stratum and L2 and L3 control the assembly of various interior tissues (Christou et al., 1990; Christou & McCabe, 1992). This explains why plant growth from shoot meristem revolution is frequently chimerical (Sato et al., 1993) and why the transformation of the inner cell layer is required for transgenic progeny acquisition. By combining multiple shoot induction from embryonic axes of mature seeds, and resistance to the herbicide imazapyr as a selectable marker, and biolistics techniques, Rech et al. (2008) observed a considerable improvement in the revitalization of viable modified soybean plants. The first explanation of somatic embryogenesis in soybeans was given by Christianson et al. in 1983; a method that allows cells to reproduce and rebuild tissues in vitro. Because of their cellular nature, the mutation of primordial somatic embryos led directly to mythical monster plants (Parrott et al., 1989; Finer & Nagasawa, 1988). According to various researches on soybean alteration by particle bombardment exploitation of embryogenic tissue, the use of 2,4-D and permanent culture are capable of generating genetic and epigenetic variations in may plant species (Wu et al., 2008; Hernandez-Garcia et al., 2009; Xing et al., 2010).

4.4.2 **Agrobacterium tumefaciens** ***technique***

Agrobacterium is a Gram-negative phytopathogenic bacterium which infects certain soil flora (De Cleene & De Ley, 1976). Crown gall disease, bushy root disease, and cane gall disease are all caused by phytopathogens (Gelvin, 2010b). Interkingdom horizontal gene switch is responsible for a number of diseases: As soon as virulent Agrobacterium fragments infect plant tissue, bacteria transfer one or more genetic materials (transferred DNA or T-DNA) from the Ti (Tumor inducing) or Ri (Root inducing) plasmids into the host plant cells (Gelvin, 2003). This gene switch system is currently being reviewed (Pitzschke & Hirt, 2010; Gelvin, 2010a, b). Over the most recent years, disarmed Agrobacterium traces have provided a means for producing genetically altered flora. To construct engineered vectors from Ti or Ri plasmids, foreign DNA is used to change genes in the T-DNA region (Gelvin, 2010b). The agrobacterium-mediated gene switch system has certain advantages over particle bombardment including the ability to transfer relatively large fragments of DNA, a lower number of transgene copies integrated into the plant genome, exceptional transgene rearrangement ,a lower frequency of genomic DNA interpersonal, and reduced abnormal transgene expression (Kohli et al., 2003; Gelvin, 2003). In addition, this technique is inexpensive and enables effortless alteration methods (Brasileiro & Lacorte, 2000).

4.5 Impact on the modern agriculture trade

Soybean crops are generally less tolerant to the various herbicides used for weed control. In such a situation, the soybean plant can be damaged by the presence of a herbicide, and yet if the herbicide rate is kept low or no herbicide is used, it could, conversely, be killed by weeds. This situation leads to major losses in soybean production every year. A herbicide-tolerant soybean crop called Roundup Ready was first planted in the USA in 1996. Since then, it has proven its incredible potential to control weeds multiple times. A statistical study of soybeans was carried out from 1968 to 1979, which concluded that there was a 62 percent improvement in yield as the result of the greater herbicide tolerance of the GM soybean as well as a further 13 percent increase in yield due to variety improvement (Carpenter & Gianessi, 2000). As a consequence, higher yield, lower input costs, less crop damage, and improved environmental profiles are some of the benefits obtained as an outcome of improved variety of genetically modified soybean crops.

A study conducted by the Economic Research Service in 1997 recognized the impact of GM crops on the correlation between crop yield and the implementation of herbicide-tolerant and pesticide-tolerant crops (USDA, 1999). They found a significant increase in crop yield after the adoption of herbicide-tolerant cotton and herbicide-tolerant soybean, though, GM cotton yield was relatively higher in comparison to that of GM soybean. The vegetative growth of soybean, its productivity, and seed quality are not only affected by herbicides, pesticides, or insecticides; other factors involved include drought, heat stress, or hindrance to nutrient uptake. In fact, in 2017, the Indian Council of Agricultural Research reported (Bhatia, 2017) that soybean crop productivity had declined due to insufficient rainfall in various districts of Madhya Pradesh, Maharashtra, and Rajasthan. Of the crop damage reported, 10–90 percent was caused by one common disease – Asian Soybean Rust (ASR). However, greater tolerance to all these environmental stresses is achievable with the help of GM technology.

With the additional traits of GM crops, farmers are getting the benefit of obtaining better yields without using additional land. Graham Brookes, co-author of the PG Economics report, stated that GM technology is a boon for the agriculture industry and for farmers in terms of providing better productivity and environmentally friendly farming practices (Brookes & Barfoot, 2020). Yet, despite these benefits, it is assumed that such technology will have a long-term impact on agriculture as a result of the development of resistant weeds and insects.

4.6 Benefits to society

According to US estimates (FAO, 2017), the current world population as of December 2016 was 7.9 billion and it is expected to achieve the level of 9.7 billion by 2050. Currently, a large proportion of the global population is affected by hunger either because of insufficient food availability or limited access to nutritionally adequate food. However, current agriculture practices alone cannot address the concerns of food security or improve the global hunger index.

Its high protein content and good quality amino acids has led to the soybean being increasingly consumed by the vegetarian population in the form of soy-based meat substitutes. Likewise, various factors are responsible for the greater demand for this crop such as rising biofuel adoption in lieu of conventional fuels in Asia-Pacific (APAC) and Africa, leading to increased plantings of energy crops (wheat, sugarcane, corn, and soybean) for production. The Global Soybean Market Report (2020) anticipates the global soybean consumption rate to grow at a rate of 3.4 percent from 2020 to 2025. And yet, in this era of modernization, increased urbanization has led to the destruction of arable land, with an expected reduction from 0.242 ha/person in 2016 to only 0.18 ha/person in 2050 (Raman, 2017).

Nonetheless, GM technology can be successful in mitigating the risk of the global mounting hunger index due to limited agriculture practices, increasing productivity and achieving proportionally higher profits for farmers. For example, in the United States, GM soybeans account for more than 90 percent of all soybeans farmed. Herbicide-resistant genes are usually put into the genetic structure of these beans so that farmers can use herbicides to kill invasive weeds without hurting the soybean plant. These adaptable beans can subsequently be utilized as animal feed, processed into soybean oil, or added to meals like tofu. Farmers generally observe improved yields and fewer losses from bean damage as a result of soybean biotechnology, which means more supplies and more opportunity to feed people (Food Insight, 2021).

As well as improved yield as the result of enhanced tolerance to stress, soybean has even been genetically modified with superior nutritional traits and with certain therapeutic properties. Nutritional improvements have been carried out to cater for the public's current health aspirations, with soybean modified to enhance the quality of protein and essential amino acids. Soybean is an important source of edible oil, and therefore the focus had been to improve the oil content and its quality, and efforts have been made to improve its oxidative stability to improve shelf life. All these additions are of great value to the consumer as well as for industrial applications.

Hence, crop biotechnology has been widely considered to be a strategy that benefits farmers as well as consumers. It is a boon for farmers as they obtain increased yield and experience lower losses from damage, which results in a greater supply and more opportunities to feed people. Although the technology is still relatively new, it has been a part of the food system for the last 20 years with no evidence of safety issues. The products of this technique have also been analyzed multiple times for safety and deemed safe for consumption (both human and livestock).

4.7 Trading on a worldwide scale

Soybean is the foremost consumed edible oil after palm oil. As per the ICAR report (Bhatia, 2017), the area under soybean cultivation is approximately 121 million hectares globally. Major cultivation of this crop carried out in the United States, Brazil, Argentina, India, and China, accounting for more than 85 percent of the total global soybean area. Though production was estimated at 349 million tonnes in 2016–17, it was projected there would be a marginal decline in production due to abiotic and biotic stress (Bhatia, 2017), which turned out to be the case. Various countries such as Mexico and the Netherlands have a high demand for soybean, which is fulfilled by importing soybean from other countries. To fulfill this demand, biotechnology innovation plays a big role in the agriculture industry.

The United States and Argentina are the world-leading countries for GM soybean crops, followed by Canada and China. According to statistics released by the International Service for the Acquisition of Agri-Biotech Applications, genetically modified crops were planted on a record 457.4 million acres worldwide in 2016, a 3 percent increase from the 444 million acres planted in 2015. This follows a 1 percent decrease in acres in 2015 compared to 2014. This decline was primarily due to falling commodity prices, which prompted farmers to reduce genetically modified corn and soybean acres in favor of alternative crops. In 2016, the global market value of biotech crops was US$15.8 billion (Silva, 2017).

Despite soybean demand being high, GM technology has been controversial and consumer groups have raised safety concerns about GM crops (Sobolevsky et al., 2005). This has led to a demand by consumers for the right to know the source and technology used if GM products are deployed. As a result, national policies are in place to protect consumer interest either for cultivation guidelines or labelling guidelines. In this way, such policies have proven to interfere with trade. The European Union regulation on GM food – EC No. 1829/2003 – aims to ensure human, animal, and environmental health concerning genetically modified food and feed. And yet, even with these safety measures, its member states still impose restrictions on the use and sale of genetically modified products.

According to ReportLinker (2021), the United States is a worldwide innovation hub for the cultivation of GM crops, accounting for roughly 30 percent of the global market share in 2020. The federal laws stipulated are not restrictive, leaving it open for innovation and encouraging the consumer to adopt such technology. Following the lead of United States, Canada has established itself as one of the top five cultivators of crop biotechnology. Conversely, Japan is one of the world's largest importers of GM crops, importing 94 percent of their soybean supply (Turnbull et al., 2021).

However, in some countries, such as Peru, there is a ban imposed on the entry of GM seeds as well as their cultivation. And though Egypt and Burkina Faso initially permitted the production of a few GM crops, they later suspended their cultivation.

The reluctance of international trading groups greatly impacts the marketability of these crops on the basis of their acceptability by different countries. This is attributed to the safety concerns raised by consumers and by farmers who fear losing the ability to sell their commodities to more restrictive countries. As a consequence, the overall process has an impact on the belief in and trading of GM crops. These were the reasons put forward for the drop in export prices from \$3 to \$1 observed from 1996 to 2000 (Voora et al., 2020).

4.8 Health and ethical issues/risks

To make herbicide-resistant soybeans, researchers used the Agrobacterium gene 5-enolpyruvylshikimate-3-phosphate synthase. The GM soybean variety is "basically identical" to regular soybeans, according to safety testing. Glyphosate is the active ingredient the nonselective herbicide Roundup. The sensitivity of crop plants to glyphosate has limited its in-season use as a postemergence herbicide. The extension of the use of Roundup herbicide to allow in-season application in major crops such as soybeans [Glycine max (L.) Merr.] would provide new weed control options for farmers (Padgette et al., 1996).

This herbicide was believed to have the same effect on GTS (glyphosate-resistant soybeans) (Taylor et al., 1999). However, the investigation discovered important alterations in the amounts of genistein, which is vital for human health (Lappè et al., 1998), as well as an increase in trypsin inhibitor content between the GM and control lines (Padgette et al., 1996). Two GM lines of glyphosate-resistant soybeans have been tested for nutritional value and probable toxicity in rats, catfish, broiler chickens, and dairy cows (Hammond et al., 1996). However, here were several flaws in the investigation: (a) There were no significant difference in growth, food intake, or weight of liver and spleen, and histopathology of thymus, spleen, liver, mesenteric lymph node and small intestine; (b) since one of the GTS lines developed considerably better than the other, the feeding value of the two GTS lines was not essentially similar; (c) the purpose of the experiments with broiler chickens was insufficiently persuasive; (d) there were also significant differences in milk production due to soybean feed. Teshima et al. (2000) found no significant differences in nutritional concert, organ weights, histology, or IgE and IgG antibody construction in rats and mice given 30 percent toasted GTS or non-GTS.

Gene changeover has sparked a flurry of divisive and emotive debates, particularly in the German-speaking market in Germany. However, some important requirements for the proper labelling of GM foods in the EU have been met, allowing customers to make choices based on their values (religious, ethnic, medical). As a result, on May 17, 1997, the EU introduced a new system of Novel Food classification (Jany, 1996).

4.9 Current status and future prospects

Crop engineering products and technology are seen as promising solutions for addressing food scarcity and other human needs from the agriculture sector. Transgenic crops with novel features and traits can be created using genetic engineering technology to improve output and reduce nutrient deficits and environmental stress in plants (Chaturvedi et al., 2016). Developing countries planted more GM plants than developed countries in 2013 and 2014.

There are 24 developing countries and 5 industrial countries among the 29 countries that have planted transgenic crops (James, 2014; Cowan, 2015; ISAAA, 2020). Transgenic crops were first planted in Bangladesh, Pakistan, and Vietnam in 2015 (Jefferson & Padmanabhan, 2016) and have now been grown on a significant scale in Argentina, Canada, India, Pakistan, Brazil, China, and the USA, with a total area of more than 190 million hectares (Jefferson & Padmanabhan, 2016; ISAAA, 2020). Soybeans, maize, canola, cotton, rice, and a variety of other transgenic crops have also been planted. Transgenic agricultural products are important export products – industrial raw materials, food and feed – in addition to meeting domestic demands. Agricultural products like soybeans, corn, sugar cane, potatoes, and a range of chicken vaccinations were among the 27 GM organisms (GMOs) that Indonesia had licenced till 2019 and a state-owned agricultural firm (PTPN XI) in Indonesia has grown drought-tolerant sugarcane in limited regions (Bahagiawati & Hadiarto, 2020).

The effectiveness of transgenic crops as a means of increasing agricultural production is undeniable. In actuality, however, the crops being cultivated also have the potential to make biotechnology businesses a lot of money. Transgenic development should ideally be focused on producing crops that are resistant to excessive environmental circumstances, and then made available to all farmers (Uzogara, 2000). Nonetheless, transgenic crops are frequently intended for large-scale agricultural companies and are challenging to use in small-scale agriculture. This is a consequence of biotechnological inventors preferring to create more economically profitable products above meeting the needs of today's agriculture.

4.10 Conclusions

A review of the benefits and issues highlighted by GM seeds may lead to the conclusion that neither full-scale adoption nor full-scale rejection is a viable choice. By improving output and lowering reliance on synthetic pesticides and herbicides, GM foods have the potential to solve many of the world's hunger and malnutrition problems, as well help to protect and maintain the environment. Many sectors, such as safety testing, legislation, policies, and food labelling, will face challenges in the future. Yet, many individuals believe that genetic engineering is the way of the future and that we cannot afford to ignore a technology with so many potential benefits.

References

Agarwal, P. K., Agarwal, P., Reddy, M. K., & Sopory, S. K. (2006). Role of DREB transcription factors in abiotic and biotic stress tolerance in plants. *Plant Cell Reports*, 25(12), 1263–1274.

Aragão, F. J. L., Sarokin, L., Vianna, G. R., & Rech, E. L. (2000). Selection of transgenic meristematic cells utilizing a herbicidal molecule results in the recovery of fertile transgenic soybean [Glycine max (L.) Merril] plants at a high frequency. *Theoretical and Applied Genetics*, 101(1), 1–6.

Bahagiawati, B., & Hadiarto, T. (2020). Perkembangan pemanfaatan, regulasi, dan metode deteksi produk rekayasa genetika pertanian di indonesia/Development of utilization, regulation, and detection methods of agricultural genetically modified products in Indonesia. *Jurnal Penelitian dan Pengembangan Pertanian*, 39(1), 61–71.

Bhatia V. S. (2017). *Soybean Monitor / Market Watch*. Indian Institute of Soybean Research, Madhya Pradesh.

Brasileiro, A. C. M., & Lacorte, C. (2000). Agrobacterium: Um sistema natural de transferência de genes para plantas. *Biotecnologia Ciência & Desenvolvimento, ano* III, (15).

Brookes, G., & Barfoot, P. (2020). Environmental impacts of genetically modified (GM) crop use 1996–2018: Impacts on pesticide use and carbon emissions. *GM Crops & Food*, 11(4), 215–241.

Buhr, T., Sato, S., Ebrahim, F., Xing, A., Zhou, Y., Mathiesen, M., … Clemente, T. (2002). Ribozyme termination of RNA transcripts down-regulate seed fatty acid genes in transgenic soybean. *The Plant Journal*, 30(2), 155–163.

Carpenter, J., & Gianessi, L. (2000). *Agricultural Biotechnology: Benefits of Transgenic Soybeans*. Report, National Center for Food and Agricultural Policy (NCFAP), April.

Chaturvedi, S., Srinivas, K. R., & Kumar, A. (2016). Agriculture technology choices and the responsible research and innovation (RRI) framework: Emerging experiences from China and India. *Asian Biotechnology and Development Review*, 18(1), 93–111.

CERA (2010). GM Crop Database. Center for Environmental Risk Assessment (CERA), ILSI Research Foundation, Washington, DC. http://ceragmc.org/index.php?action=gm_crop_database

Chee, P. P., & Hu, C.-Y. (2000). Transgenic soybean (Glycine max). In *Biotechnology in Agriculture and Forestry 46: Transgenic Crops I*, edited by Y. P. S. Bajaj (Berlin: Springer), 268–282.

Chen, M., Xu, Z., Xia, L., Li, L., Cheng, X., Dong, J., … Ma, Y. (2009). Cold-induced modulation and functional analyses of the DRE-binding transcription factor gene, GmDREB3, in soybean (*Glycine max* L.). *Journal of Experimental Botany*, 60(1), 121–135.

Chiera, J. M., Bouchard, R. A., Dorsey, S. L., Park, E., Buenrostro-Nava, M. T., Ling, P. P., & Finer, J. J. (2007). Isolation of two highly active soybean (*Glycine max* (L.) Merr.) promoters and their characterization using a new automated image collection and analysis system. *Plant Cell Reports*, 26(9), 1501–1509.

Cho, M. J., Widholm, J. M., & Vodkin, L. O. (1995). Cassettes for seed-specific expression tested in transformed embryogenic cultures of soybean. *Plant Molecular Biology Reporter*, 13(3), 255–269.

Christianson, M. L., Warnick, D. A., & Carlson, P. S. (1983). A morphogenetically competent soybean suspension culture. *Science*, 222(4624), 632–634.

Christou, P., & McCabe, D. E. (1992). Prediction of germ-line transformation events in chimeric R_O transgenic soybean plantlets using tissue-specific expression patterns. *The Plant Journal*, 2(3), 283–290.

Christou, P., McCabe, D. E., & Swain, W. F. (1988). Stable transformation of soybean callus by DNA-coated gold particles. *Plant Physiology*, 87(3), 671–674.

Christou, P., McCabe, D. E., Martinell, B. J., & Swain, W. F. (1990). Soybean genetic engineering – commercial production of transgenic plants. *Trends in Biotechnology*, 8, 145–151.

Christou, P., Swain, W. F., Yang, N. S., & McCabe, D. E. (1989). Inheritance and expression of foreign genes in transgenic soybean plants. *Proceedings of the National Academy of Sciences*, 86(19), 7500–7504.

Clemente, T. E., & Cahoon, E. B. (2009). Soybean oil: Genetic approaches for modification of functionality and total content. *Plant Physiology*, 151(3), 1030–1040.

Cowan, T. (2015, July). *Agricultural Biotechnology: Background, Regulation, and Policy Issues*, Report 7–5700, 20 July. Congressional Research Service, Washington, DC. https://www.everycrsreport.com/files/20150720_RL32809_e4c1dec6c4c51de14b8315e0ccc8b2bb9df05b85.pdf

De Cleene, M., & De Ley, J. (1976). The host range of crown gall. *The Botanical Review*, 42(4), 389–466.

De Ronde, J. A., Cress, W. A., Krüger, G. H. J., Strasser, R. J., & Van Staden, J. (2004a). Photosynthetic response of transgenic soybean plants, containing an Arabidopsis P5CR gene, during heat and drought stress. *Journal of Plant Physiology*, 161 (11), 1211–1224.

De Ronde, J. A., Laurie, R. N., Caetano, T., Greyling, M. M., & Kerepesi, I. (2004b). Comparative study between transgenic and non-transgenic soybean lines proved transgenic lines to be more drought tolerant. *Euphytica*, 138(2), 123–132.

FAO (2017). *The Future of Food and Agriculture*. Food and Agriculture Organization of the United Nations. http://www.fao.org/publications/fofa/en/

Fernandez-Cornejo, J., Wechsler, S., Livingston, M., & Mitchell, L. (2014). *Genetically Engineered Crops in the United States*. USDA-ERS Economic Research Report No. 162.

Finer, J. J., & Nagasawa, A. (1988). Development of an embryogenic suspension culture of soybean (*Glycine max* Merrill.). *Plant Cell, Tissue and Organ Culture*, 15(2), 125–136.

Finer, J. J., Vain, P., Jones, M. W., & McMullen, M. D. (1992). Development of the particle inflow gun for DNA delivery to plant cells. *Plant Cell Reports*, 11(7), 323–328.

Food Insight (2021, February). Beans and biotech: Why soybeans are the stars of genetically modified foods. *Food Insight*, 10 February. https://foodinsight.org/soybeans-are-the-stars-of-genetically-modified-foods/

Gelvin, S. B. (2003). Agrobacterium and plant transformation: The biology behind the "gene-jockeying" tool. *Microbiology and Molecular Biology Reviews*, 67, 16–37.

Gelvin, S. B. (2010a). Finding a way to the nucleus. *Current Opinion in Microbiology*, 13, 53–58.

Gelvin, S. B. (2010b). Plant proteins involved in Agrobacterium-mediated genetic transformation. *Annual Review of Phytopathology*, 48, 45–68.

GlobalInfoResearch (2020). *Global Soybean Market Report by Manufacturers, Regions, Type and Application, Forecast to 2024*. See https://southeast.newschannelnebraska.com/story/45845469/soybean-market-share-size-global-growth-opportunities-driving-factors-by-manufacturers-regions-type-and-application-revenue-market-forecast-to-2024

Govindarajulu, M., Kim, S. Y., Libault, M., Berg, R. H., Tanaka, K., Stacey, G., & Taylor, C. G. (2009). GS52 ecto-apyrase plays a critical role during soybean nodulation. *Plant Physiology*, 149(2), 994–1004.

Graef, G., LaVallee, B. J., Tenopir, P., Tat, M., Schweiger, B., Kinney, A. J., … Clemente, T. E. (2009). A high-oleic-acid and low-palmitic-acid soybean: Agronomic performance and evaluation as a feedstock for biodiesel. *Plant Biotechnology Journal*, 7(5), 411–421.

Hadi, M. Z., McMullen, M. D., & Finer, J. J. (1996). Transformation of 12 different plasmids into soybean via particle bombardment. *Plant Cell Reports*, 15(7), 500–505.

Hammond, B. G., Vicini, J. L., Hartnell, G. F., Naylor, M. W., Knight, C. D., Robinson, E. H., ... Padgette, S. R. (1996). The feeding value of soybeans fed to rats, chickens, catfish and dairy cattle is not altered by genetic incorporation of glyphosate tolerance. *The Journal of Nutrition*, 126(3), 717–727.

Hernandez-Garcia, C. M., Martinelli, A. P., Bouchard, R. A., & Finer, J. J. (2009). A soybean (*Glycine max*) polyubiquitin promoter gives strong constitutive expression in transgenic soybean. *Plant Cell Reports*, 28(5), 837–849.

Hinchee, M. A., Connor-Ward, D. V., Newell, C. A., McDonnell, R. E., Sato, S. J., Gasser, C. S., ... Horsch, R. B. (1988). Production of transgenic soybean plants using Agrobacterium-mediated DNA transfer. *Bio/technology*, 6(8), 915–922.

Homrich, M. S., Passaglia, L. M. P., Pereira, J. F., Bertagnolli, P. F., Pasquali, G., Zaidi, M. A., ... Bodanese-Zanettini, M. H. (2008). Resistance to Anticarsia gemmatalis Hübner (Lepidoptera, Noctuidae) in transgenic soybean (*Glycine max* (L.) Merrill Fabales, Fabaceae) cultivar IAS5 expressing a modified Cry1Ac endotoxin. *Genetics and Molecular Biology*, 31, 522–531.

Homrich, M. S., Passaglia, L. M. P., Pereira, J. F., Bertagnolli, P. F., Salvadori, J. R., Nicolau, M., ... Bodanese-Zanettini, M. H. (2008). Agronomic performance, chromosomal stability and resistance to velvetbean caterpillar of transgenic soybean expressing cry1Ac gene. *Pesquisa Agropecuária Brasileira*, 43(7), 801–807.

Hong, H. P., Zhang, H., Olhoft, P., Hill, S., Wiley, H., Toren, E., ... Cheng, M. (2007). Organogenic callus as the target for plant regeneration and transformation via Agrobacterium in soybean (*Glycine max* (L.) Merr.). *In Vitro Cellular & Developmental Biology – Plant*, 43(6), 558–568.

Hongyu, Z., Ziniu, Y., & Wangxi, D. (2000). Composition and ecological distribution of Cry proteins and their genotypes of *Bacillus thuringiensis* isolates from warehouses in China. *Journal of Invertebrate Pathology*, 76(3), 191–197.

Hur, Y. S., Shin, K. H., Kim, S., Nam, K. H., Lee, M. S., Chun, J. Y., & CheonIII, C. (2009). Overexpression of GmAKR1: A stress-induced aldo/keto reductase from soybean, retards nodule development. *Molecules and Cells*, 27(2), 217–223.

International Service for the Acquisition of Agribiotech Applications (2020). *ISAAA in 2019: Accomplishment Report*. https:// www.isaaa.org/resources/publications/annualreport/2019/default.asp

ISAAA (2018). Biotech country facts & trends: Brazil. https://www.isaaa.org/resources/publications/biotech_country_facts_and_trends/download/Facts%20and%20Trends%20-%20Brazil-2018.pdf

James, C. (2011). *Global Status of Commercialized Biotech/GM Crops*. ISAAA Brief No. 43 (Ithaca, NY: ISAAA).

James, C. (2014). *Global Status of Commercialized Biotech/GM Crops*. ISAAA Brief No. 49 (Ithaca, NY: ISAAA).

Jany, K. D. (1996). Gentechnik im Ernährungsbereich. *Abc. Ernähr.-Med.*, 21, 114–117.

Jany, K. D. (1999). Gentechnik und Lebensmittel. 6. Aachener Diätetik Fortbildung. 30–34.

Jefferson, D. J., & Padmanabhan, M. S. (2016). Recent evolutions in intellectual property frameworks for agricultural biotechnology: A worldwide survey. *Asian Biotechnology and Development Review*, 18, 47–67.

Kocsy, G., Laurie, R., Szalai, G., Szilágyi, V., Simon-Sarkadi, L., Galiba, G., & De Ronde, J. A. (2005). Genetic manipulation of proline levels affects antioxidants in soybean subjected to simultaneous drought and heat stresses. *Physiologia Plantarum*, 124(2), 227–235.

Kohli, A., Twyman, R. M., Abranches, R, Weget, E., Stoger, E., & Christou, P. (2003) Transgene integration organization and interaction in plants. *Plant Molecular Biology*, 52(1), 247–258.

Kruft, D. (2001). Impacts of genetically-modified crops and seeds on farmers. The Agricultural Law Resource and Reference Center, Pennsylvania State University, Carlisle, PA, 1–11.

Lappè, M. A., Bailey, E. B., Childress, C., & Setchell, K. D. (1998). Alterations in clinically important phytoestrogens in genetically modified, herbicide-tolerant soybeans. *Journal of Medicinal Food*, 1(4), 241–245.

Lee, M. Y., Shin, K. H., Kim, Y. K., Suh, J. Y., Gu, Y. Y., Kim, M. R., ... Cheon, C. I. (2005). Induction of thioredoxin is required for nodule development to reduce reactive oxygen species levels in soybean roots. *Plant Physiology*, 139(4), 1881–1889.

Levy-Booth, D. J., Gulden, R. H., Campbell, R. G., Powell, J. R., Klironomos, J. N., Pauls, K. P., ... Dunfield, K. E. (2009). Roundup Ready® soybean gene concentrations in field soil aggregate size classes. *FEMS Microbiology Letters*, 291(2), 175–179.

Li, X., & Dhaubhadel, S. (2011). Soybean 14-3-3 gene family: Identification and molecular characterization. *Planta*, 233(3), 569–582.

Lundry, D. R., Ridley, W. P., Meyer, J. J., Riordan, S. G., Nemeth, M. A., Trujillo, W. A., ... Sorbet, R. (2008). Composition of grain, forage, and processed fractions from second-generation glyphosate-tolerant soybean, MON 89788, is equivalent to that of conventional soybean (*Glycine max* L.). *Journal of Agricultural and Food Chemistry*, 56(12), 4611–4622.

Manavalan, L. P., Guttikonda, S. K., Phan Tran, L. S., & Nguyen, H. T. (2009). Physiological and molecular approaches to improve drought resistance in soybean. *Plant and Cell Physiology*, 50(7), 1260–1276.

Mathieu, M., Winters, E. K., Kong, F., Wan, J., Wang, S., Eckert, H., ... Clemente, T. (2009). Establishment of a soybean (*Glycine max* Merr. L) transposon-based mutagenesis repository. *Planta*, 229(2), 279–289.

Mazarei, M., Elling, A. A., Maier, T. R., Puthoff, D. P., & Baum, T. J. (2007). GmEREBP1 is a transcription factor activating defense genes in soybean and Arabidopsis. *Molecular Plant-Microbe Interactions*, 20(2), 107–119.

McCabe, D. E., Swain, W. F., Martinell, B. J., & Christou, P. (1988). Stable transformation of soybean (*Glycine max*) by particle acceleration. *Bio/technology*, 6(8), 923–926.

McGloughlin, M. N. (2008). Nutritionally improved agricultural crops. *Plant Physiology*, 147(3), 939–953.

Nunes, A., Vianna, G. R., Cuneo, F., Amaya-Farfan, J., de Capdeville, G., Rech, E. L., & Aragão, F. J. (2006). RNAi-mediated silencing of the myo-inositol-1-phosphate synthase gene (GmMIPS1) in transgenic soybean inhibited seed development and reduced phytate content. *Planta*, 224(1), 125–132.

Padgette, S. R., Kolacz, K. H., Delannay, X., Re, D. B., LaVallee, B. J., Tinius, C. N., ... Kishore, G. M. (1995). Development, identification, and characterization of a glyphosate-tolerant soybean line. *Crop Science*, 35(5), 1451–1461.

Padgette, S. R., Taylor, N. B., Nida, D. L., Bailey, M. R., MacDonald, J., Holden, L. R., & Fuchs, R. L. (1996). The composition of glyphosate-tolerant soybean seeds is equivalent to that of conventional soybeans. *The Journal of Nutrition*, 126(3), 702–716.

Pandey, A. K., Yang, C., Zhang, C., Graham, M. A., Horstman, H. D., Lee, Y., ... Whitham, S. A. (2011). Functional analysis of the Asian soybean rust resistance pathway mediated by Rpp2. *Molecular Plant–Microbe Interactions*, 24(2), 194–206.

Parrott, W. A., Hoffman, L. M., Hildebrand, D. F., Williams, E. G., & Collins, G. B. (1989). Recovery of primary transformants of soybean. *Plant Cell Reports*, 7(8), 615–617.

Peferoen, M. (1997). Progress and prospects for field use of Bt genes in crops. *Trends in Biotechnology*, 15(5), 173–177.

Pitzschke, A., & Hirt, H. (2010). New insights into an old story: Agrobacterium-induced tumour formation in plants by plant transformation. *The EMBO Journal*, 29(6), 1021–1032.

Preiszner, J., Vantoai, T., Huynh, L., Bolla, R., & Yen, H. (2001). Structure and activity of a soybean Adh promoter in transgenic hairy roots. *Plant Cell Reports*, 20 (8), 763–769.

Qinggele, C., Mingchun, L., Dongsheng, W., Yi, C., & Laijun, X. (2007). Isolation and sequencing analysis on the seed-specific promoter from soybean. *Frontiers of Agriculture in China*, 1, 17–23.

Raman, R. (2017). The impact of Genetically Modified (GM) crops in modern agriculture: A review. *GM Crops & Food*, 8(4), 195–208.

Rech, E. L., Vianna, G. R., & Aragao, F. J. (2008). High-efficiency transformation by biolistics of soybean, common bean and cotton transgenic plants. *Nature Protocols*, 3(3), 410–418.

ReportLinker (2021). Global agricultural biotechnology industry. https://www.reportlinker.com/p04838495/Global-Agricultural-Biotechnology-Biotechnology.html

Sanford, J. C. (1990). Biolistic plant transformation. *Plant Physiology*, 79(1), 206–209.

Sanford, J. C. (1988). The biolistic process. *Trends Biotechnology*, 6(1), 299–302.

Sato, S., Newell, C., Kolacz, K., Tredo, L., Finer, J., & Hinchee, M. (1993). Stable transformation via particle bombardment in two different soybean regeneration systems. *Plant Cell Reports*, 12(7), 408–413.

Silva, G. (2017, December). Global genetically modified crop acres increase amid concerns. Michigan State University Extension, East Lansing, MI. https://www.canr.msu.edu/news/global_genetically_modified_crop_acres_increase_amid_concerns#:~:text=Genetically%20modified%20crops%20were%20planted,Acquisition%20of%20Agri%2DBiotech%20Applications

Sobolevsky, A., Moschini, G., & Lapan, H. (2005). Genetically modified crops and product differentiation: Trade and welfare effects in the soybean complex. *American Journal of Agricultural Economics*, 87(3), 621–644.

Stewart Jr, C. N., Adang, M. J., All, J. N., Boerma, H. R., Cardineau, G., Tucker, D., & Parrott, W. A. (1996). Genetic transformation, recovery, and characterization of fertile soybean transgenic for a synthetic *Bacillus thuringiensis* cryIAc gene. *Plant Physiology*, 112(1), 121–129.

Subramanian, S., Hu, X., Lu, G., Odelland, J. T., & Yu, O. (2004). The promoters of two isoflavone synthase genes respond differentially to nodulation and defense signals in transgenic soybean roots. *Plant Molecular Biology*, 54(5), 623–639.

Taylor, N. B., Fuchs, R. L., MacDonald, J., Shariff, A. R., & Padgette, S. R. (1999). Compositional analysis of glyphosate-tolerant soybeans treated with glyphosate. *Journal of Agricultural and Food Chemistry*, 47(10), 4469–4473.

Teshima, R., Akiyama, H., Okunuki, H., Sakushima, J.-i., Goda, Y., Onodera, H., … Toyoda, M. (2000). Effect of GM and non-GM soybeans on the immune system of BN rats and B10A mice. *Journal of the Food Hygienic Society of Japan*, 41(3), 188–193.

Turnbull, C., Lillemo, M., & Hvoslef-Eide, T. A. (2021). Global regulation of genetically modified crops amid the gene edited crop boom – a review. *Frontiers in Plant Science*, 12, 258.

USDA (1999). *Genetically Engineered Crops for Pest Management*. Economic Research Service, June 25. https://www.ers.usda.gov/webdocs/publications/41121/15699_aer786_1_.pdf?v=0

USDA (2020, July). Recent trends in GE adoption. USDA, Economic Research Service. https://www.ers.usda.gov/data-products/adoption-of-genetically-engineered-crops-in-the-us/recent-trends-in-ge-adoption/

Uzogara, S. G. (2000). The impact of genetic modification of human foods in the 21st century: A review. *Biotechnology Advances*, 18(3), 179–206.

Vianna, G. R., Aragão, F. J. L., & Rech, E. L. (2011). A minimal DNA cassette as a vector for genetic transformation of soybean (*Glycine max*). *Genetics and Molecular Research*, 10(1), 382–390.

Voora, V., Larrea, C., & Bermudez, S. (2020). *Global Market Report: Soybeans*. International Institute for Sustainable Development, Winnipeg.

Walker, D. R., All, J. N., McPherson, R. M., Boerma, H. R., & Parrott, W. A. (2000). Field evaluation of soybean engineered with a synthetic cry1Ac transgene for resistance to corn earworm, soybean looper, velvetbean caterpillar (Lepidoptera: Noctuidae), and lesser cornstalk borer (Lepidoptera: Pyralidae). *Journal of Economic Entomology*, 93(3), 613–622.

Wang, L. Z. (1997). Soybean: A world-wide crop. In *Proceedings of the World Soybean Research Conference V*. (Chiang Mai, Thailand: Kasetsart University Press), 517–520.

Wu, C., Chiera, J. M., Ling, P. P., & Finer, J. J. (2008). Isoxaflutole treatment leads to reversible tissue bleaching and allows for more effective detection of GFP in transgenic soybean tissues. *In Vitro Cellular & Developmental Biology – Plant*, 44(6), 540–547.

Xing, A., Moon, B. P., Mills, K. M., Falco, S. C., & Li, Z. (2010). Revealing frequent alternative polyadenylation and widespread low-level transcription read-through of novel plant transcription terminators. *Plant Biotechnology Journal*, 8(7), 772–782.

Yorinori, J. T., Paiva, W. M., Frederick, R. D., Costamilan, L. M., Bertagnolli, P. F., Hartman, G. E., … NunesJr., J. (2005). Epidemics of soybean rust (*Phakopsora pachyrhizi*) in Brazil and Paraguay from 2001 to 2003. *Plant Disease*, 89(6), 675–677.

Zhang, G., Chen, M., Li, L., Xu, Z., Chen, X., Guo, J., & Ma, Y. (2009). Overexpression of the soybean GmERF3 gene, an AP2/ERF type transcription factor for increased tolerances to salt, drought, and diseases in transgenic tobacco. *Journal of Experimental Botany*, 60(13), 3781–3796.

Zhou, G. A., Chang, R. Z., & Qiu, L. J. (2010). Overexpression of soybean ubiquitin-conjugating enzyme gene GmUBC2 confers enhanced drought and salt tolerance through modulating abiotic stress-responsive gene expression in Arabidopsis. *Plant Molecular Biology*, 72(4), 357–367.

5 Golden Rice

Prerna Gupta, Richa Sharma and Manish Bakshi

5.1 Introduction

The growing prevalence of vitamin A deficiency has stimulated interest in enriched foods. 'Golden Rice' is the name given to genetically engineered rice that is enhanced with β-carotene (pro-vitamin A) – an important nutrient for vision, growth, reproduction, cellular differentiation and proliferation, and regulation of the immune system. β-carotene possesses antioxidative constituents and is mainly added to foods and feedstuffs in the form of supplements (Zhu et al., 2018) to improve human nutrition and health globally, especially that of young children and pregnant and lactating woman. Moreover, animals cannot synthesize carotene themselves, and thus need to acquire it either directly or indirectly from external plant-derived sources to fulfil their general requirements for performing various physiological functions (Farre et al., 2011). It is not only vitamin A deficiency that can be solved by using plant-based sources; most macro and micro nutrient deficiencies can be overcome in this way (White & Brown, 2010).

With the advancement in biological knowledge concerning plants and their subspecies, the ability to successfully engineer plants for nutrition is often considered an appropriate use of GM organisms (Moghissi et al., 2015). One such example is 'Golden Rice' – named after the yellow color of the rice grain obtained after milling and polishing (Al-Babili & Beyer, 2005). The term 'C4 Golden Rice' was first coined by German scientists Ingo Potrykus and Peter Beyer in 1990 (Stokstad, 2019) to prevent vitamin A deficiency where rice is considered a staple food and forms the major part of the diet. Therefore, Ingo Potrykus first tried to introduce a gene into rice to curb the problem of malnutrition among growing children and to solve the problem of food security.

For example, in Bangladesh rice is a major dietary staple and according to 2019 data nearly 21 percent of children and 23.7 percent of pregnant women were found to be deficient in vitamin A. Even in Southeast Asian countries, the average per capita intake of rice is as much as 130 kg/year. In 2005 the UN estimated that worldwide there were 140 million pre-school children suffering from a deficiency in macro nutrients especially vitamin A; of these

DOI: 10.4324/9781003278566-5

more than 35 million children were from different Indian states (UN SCN, 2005). This rose to 190 million children and 19 million pregnant women in 2009; of which around 1 million children were suffering from eye-related issues like losses to their eyesight (WHO, 2009).

Ingo Potrykus, joined by Peter Beyer, focussed their study on the biochemical pathway related to gene modification (Ye et al., 2000). It has been observed that the photosynthetic leaves of the plants help in promoting the synthesis of β-carotene in the endosperm part of the plant. (Swamy et al., 2019). Yet it was not until July 2021 that rice enriched with provitamin A – 'Golden Rice' – was cultivated in the Philippines and was considered to be the world's first commercialized GM bio-fortified crop (Steur et al., 2022). This was because, a similar problem with vitamin A deficiency had been seen in the Philippines as well and, after many trials, Golden Rice was developed by the Department of Agriculture – Philippine Rice Research Institute (DA-PhilRice) in partnership with the International Rice Research Institute (IRRI) to contain additional levels of beta-carotene, which the body converts into vitamin A. Bangladesh and the Philippines are, therefore, the principal markets targeted for the cultivation and sale of Golden Rice or C4 rice by the International Rice Research Institute (IRRI).

Since its introduction, Golden Rice has become a major source of vitamin A supplementation not only to prevent childhood blindness but also to prevent deaths from infectious diseases such as measles, to counter anemia (iron metabolism), raise host resistance to infection, aid cell maintenance, reduce the risk of maternal mortality, maintain membrane integrity, support red blood cell (RBC) production, bolster the immune system, and play a key role reproduction and cell differentiation (Swamy et al., 2021; Sellappan et al., 2009).

Yet, because there are lot of questions concerning the acceptance of Golden Rice as the result of it being genetically modified, a sustainable production system needs to be created. A basic approach would be to include modified rice in the diet along with other vitamin A enriched products like cassava or yellow maize or in the form of supplements, fortification, and/or optimal breast feeding practices.

5.2 Why are genetic modifications required?

To meet the demands of the future global population, an inevitable increase in rice (crop) productivity is required. Rice is a predominant, staple crop in many developing countries providing 20 percent dietary protein, 3 percent dietary fat, and other micronutrients. Various agricultural practices like post-harvest processing, cultivation, and breeding have been implemented to enhance the nutrient content of rice (Das et al., 2020). Yet refinement of micronutrient-dense crops is not always a practical solution as not all the traits are desirable. Moreover, infertility matters and the core genetic makeup of rice makes it unsuitable for traditional breeding processes (Lee et al., 2021). Thus, in such cases, genetic modification remained the only viable option to modify these plant species. Even so, this was still not easy because of the major issue of bacterial blight affecting large areas of Asia and sub-Saharan Africa, leading to rice spoilage. Solutions were needed, and thus to fulfil the demands of the society and to maintain sustainable agriculture the IRRI (2019) came up with various technological advancements such as farm mechanization, crop rotation, pest management etc. Various efforts have been made to address and fulfil the micronutrient deficiency by different means such as: supplementation, fortification, and dietary diversification. In fact, it was during the 1960s and 1970s, that semi-dwarf and photoperiod insensitive rice varieties were able to feed the growing population in different regions of the country as well as reducing food shortages (Lu & Snow, 2005).

In rice, micronutrients are present in the endosperm at minimal concentrations. According to the findings of Huang et al. (2008) GM rice represents the major symbolic step in breaking the world's plant biotechnology logjam. Rice has a 3-carbon pathway, i.e. C3 photosynthesis, which is found to be less effective than the 4-carbon pathway (C4 photosynthesis) of crops such as corn or sugar cane. This has led scientists to focus on gene expression in the metabolic pathway of C4 photosynthesis, especially in relation to rice plants. This resulted in an increased uptake of CO_2 (carbon dioxide) by plants which is then accumulated under specialized cells within the leaves, thereby making the whole process of photosynthesis more efficient. This mechanism of enriching plant cells with CO_2 is also known as GOC bypass (Anonymous, 2019).

5.3 Genes involved in agronomic trait development

Practically, the genetic improvement of rice with increased crop productivity is currently needed to feed a growing global population and to minimize the use of cultivable soil. With the increasing population growth, agricultural commodities are generally divided into three phases. In the first phase, crop utilization of efficiently grown plants in a selected area is made to feed the growing population. In the second phase, the various limitations that prevent the growth or lessen the suitability of plants in a particular area are analysed. This phase involves better utilization of cross breed plants. The third phase evolves and introduces new varieties by the use of GM organisms (Moghissi et al., 2015). This led to the production of GM crops.

The chief genetic modification carried out in Golden Rice concerns photosynthesis, where carbon dioxide is mainly converted into oxygen and energy-rich sugar compounds. Here two methods can be used. The first involves photorespiration – a light-dependent process in which CO_2 is released and O_2 is taken up. This can increase the photosynthesis process by up to 55 percent, but the operational cost is quite high as the intense light produced requires tools to control it. The second method is via genetic modification. This has resulted in an increase in the photorespiration process and a molecule called glycolate has been produced by using three rice enzymes: glycolate oxidase, oxalate oxidase, and catalase (FNRI, 2016).

Basically two genes, one from *zea mays* – phytoene synthase (zmpsy1) – and the other from soil bacterium (*pantoea ananatis, syn. erwinia uredovora*) – carotene desaturase (crtI) – were used. In an experiment by Swamy et al. (2021), the GM GR2E Kaybonnet rice variety was crossbred with three different rice varieties – IR64, PSBRc82, and BR29 – which were popular in Asia, the Philippines, and Bangladesh, respectively. These gave different recombinant traits for β-carotene expression, as can be seen in Figure 5.1.

To study the process of enhancement during carotenoid synthesis, Gayen et al. (2016) carried out the proteomic and metabolic analysis of modified rice. They observed that it was the carbohydrate metabolic pathway where alteration in the transgenic gene was observed involving basic carbohydrate metabolic enzymes such as pullulanase, UDP-glucose pyrophosphorylase, and glucose-1-phosphate adenylyltransferase. In fact, the pyruvate biosynthetic pathway was also enhanced by the enzyme pyruvate phosphate dikinase (PPDK). This enzyme is also the precursor of the carotenoid biosynthetic pathway. The main function of this enzyme is to catalyse the ATP to produce AMP:

ATP + pyruvate + phosphate → AMP + phosphoenolpyruvate + diphosphate

Elevated levels of glyceric acid, fructo-furanose, and galactose along with a reduction in galactonic acid and gentiobiose were also observed in the GM rice. Carotenoids are principally synthesized from the precursor of the methylerythritol 4-phosphate (MEP) pathway. MEP is responsible for isoprenoid precursors isopentenyl diphosphate (IDP) and dimethylallyl diphosphate (DMADP) in most plant cells, bacterial species, and eukaryotic parasites (Banerjee & Sharkey, 2014). Overall carotenoid formation is enhanced by the expression of carotene desaturase and phytoene synthase that express the enzyme PPDK (Farre et al., 2013). In addition to this, the phosphoglycerate kinase enzyme that plays an important role in the glycolysis pathway was also found to increase by 2.9-fold as compared with conventional rice (Decourcelle et al., 2015; Kim et al., 2012). An increase in the mineral composition of the modified rice was also found (Lu et al., 2008). Other research includes that by Singh et al. (2017), who tried to improve iron, zinc, and β-carotene in a single locus of rice endosperm as a GM trait for better expression of *Arabidopsis* nicotianamine synthase1 and a similar study carried out by Zhu et al. in 2019.

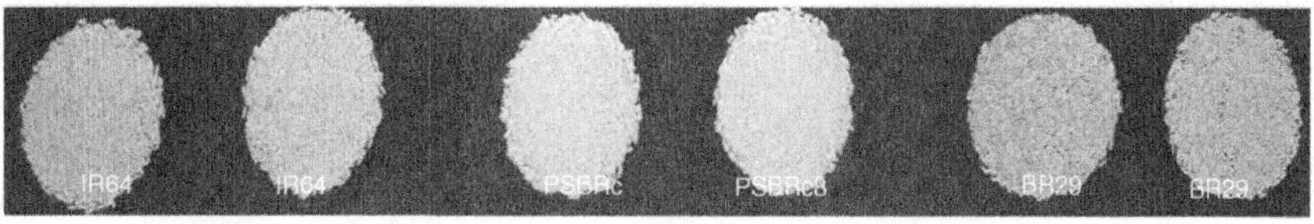

Figure 5.1 Different recombinant traits for β-carotene.
Source: Authors.

5.4 Mechanism of genetic modification leading to desired trait

Conventional rice plants contain β-carotene in their leaves and stems but not in their kernels. Bio-fortification is a sustainable and economical approach for increasing micro or phytonutrients in crops via conventional breeding and/or genetic engineering (Bouis & Saltzman, 2017). Golden Rice contains β-carotene in the part of the plant that people eat (Regis, 2019).

The photosynthetic tissues of all plants can naturally produce and store β-carotene. The biosynthetic pathway of carotenoids in the immature part of the rice endosperm directs the synthesis of geranylgeranyl diphosphate. Because this compound is not only responsible for carotenogenesis but can also be used as a substrate for producing the uncolored carotenoid phytoene, it has the capacity to express the heterologous enzyme phytoene synthase (psy) in the endosperm of a rice plant (Swamy et al., 2019). As an example, grains of astaxanthin rice were enriched with astaxanthin (a red pigment that belongs to the carotenoid group) in the endosperm and with higher antioxidant activity. This pigment is responsible for the reddish-pink color and is extracted from certain algae. The results showed that the introduction to a minimal four pairs of transgenes enables *de novo* biosynthesis of astaxanthin in the rice endosperm. Such work provides a successful example of synthetic biology in plants and crop modification, showing that modified rice products produced in this way could be consumed for health-promoting purposes and processed to produce dietary supplements.

Rice has been chosen as a focus of such research as it is the major stable crop for many countries. Moreover, the endosperm of the rice is suitable for the transformation and activation of the biosynthetic pathway for the completion of provitamin-A (β-carotene) by genetic engineering (Beyer et al., 2002); the success of the genetic modification can be easily recognized by their strong color which indicates greater β-carotene concentration (Welch & Graham, 2004). The modified β-carotene in Golden Rice has a conversion efficiency (of β-carotene to vitamin A) of 3.8:1 for normal adults and 2.3:1 for children, which can be equally compared to cassava with a conversion efficiency of 4.4:1 and yellow maize with a β-carotene conversion efficiency of 3.2:1 (Frano et al., 2013; Tang et al., 2012).

In other research, identification and isolation of the psy (phytoene synthase) gene from daffodil flowers (*Narcissus pseudonarcissus*) and crtI (carotene or phytoene desaturase) from the bacterial species *Erwinia uredovora* have been mixed together to form a Ti plasmid. Rice (the japonica variety) was modified with three genes – two genes from the daffodil plant (psy and lcy) and a third from a bacterium species (crtI) – that are important in enabling the rice grain to form and store beta-carotene. The GM plasmid is removed from the agrobacterium and the T-DNA is cut by a restriction enzyme (Ti plasmid). Foreign DNA is also reduced by the same enzyme. The insertion of genes allows the rice plant to promote specific metabolic ways in its cells that can help to generate the precursors of Vitamin A. The Ti plasmid moves into the plant cells and inserts DNA into the plant chromosomes. The plasmid is then reinserted into the bacterium. The bacterium is afterwards used to insert the T-DNA along with the foreign gene into the chromosome of a plant cell (Figure 5.2). These plant cells are then allowed to grow in a culture and now carry a foreign gene of the new trait. Currently, with the advancements made in the field of recombinant technology, overexpression of provitamin A has been made possible in many crops other than rice such as potato, canola, tomato, and maize (Zhu et al., 2008; Datta et al., 2014; Fraser et al., 2009; Paine et al., 2005; Parkhi et al., 2005; Romer et al., 2002).

Molecular analyses of the new DNA within GR2E have assured the presence of a single copy of the inserted DNA at a particular site that has been stably inherited over variable generations in the form of a particular genetic locus as suggested by Mendelian rules of inheritance. The ZmPSY1 and CRTI proteins were expressed in a limited manner in the rice endosperm, with higher concentrations than normal rice in fully grown grain of approximately 0.245 ppm and 0.03 ppm, respectively (Bai et al., 2016). In rice tissues, the PMI protein was expressed, measured, and accumulated to a higher value of 1.89 and 0.796 ppm in fully grown grain and straw, respectively, as compared to normal rice. A tiered "weight-of-evidence" methodology was followed for assuring the safety of the ZmPSY1, CRTI, and PMI proteins present in GR2E rice. The ZmPSY1 and CRTI proteins have not shown significant amino acid sequencing resemblance with known allergens or were there any basic sequence structural alerts for any potential toxicity based on similarity researches against a database of known and putative protein toxins. Both ZmPSY1 and CRTI were quickly and completely digested in the presence of simulated gastric fluid containing pepsin, and the enzymatic activity of both the proteins was halted at temperatures well below those used during cooking (Bai et al., 2014).

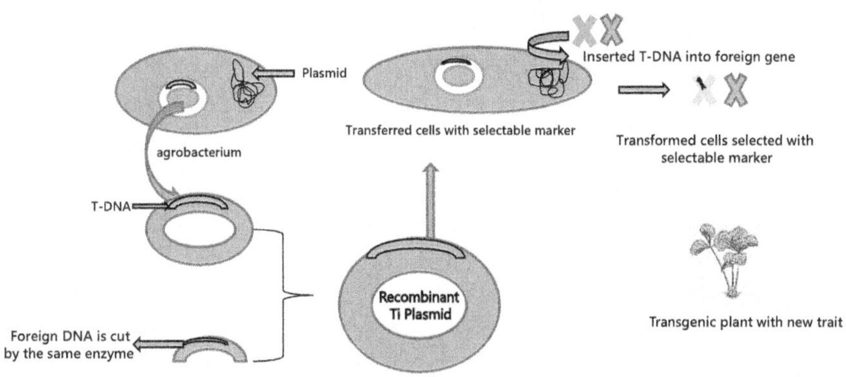

Figure 5.2 Production of Golden Rice and gene identification and isolation. Source: Authors.

5.5 Impact on modern agriculture – trade of GM crops

Today there are more than 25 countries that are growing GM crops, including 16 developing countries (Table 5.1). The major rice promoting countries are China, India, Indonesia, Bangladesh, Vietnam, Thailand, Myanmar, Japan, the Philippines, Korea, and Pakistan. In all these countries, information concerning geographic, ecological, and biological distribution was also taken into consideration for the production of the transgenic trait and for better delivery of the modified trait.

GR2E was developed by using recombinant DNA technology to complement existing efforts for increased vitamin A enhancement in food, especially for the poor. The first experiment on GR2E in the field was performed at IRRI in 2015 for the production of individual homogenous plant varieties (Swamy et al., 2019).

A trial on two varieties of rice was conducted for up to a period of two months wherein significant variation in the carotenoid content was found as

Table 5.1 Increased production rate of genetically modified rice

Year	*Area (million hectares)*	*References*
1996	1.7	James (2008)
2009	134	IRRI (2019)
2011–12	160	Khush et al. (2012)
2016	185.1	Nonero (2018)

Source: Authors.

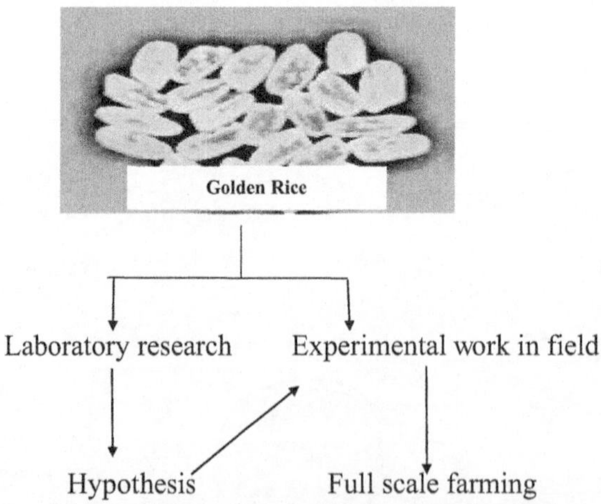

Figure 5.3 Process stages for Golden Rice production.
Source: Authors.

the result of seasonal, locational and genotypic effects. Two varieties of rice – one for beta carotene enhancement and another for a reduction in glutelin content for sake brewing – were modified (Ye et al., 2000). A special long-grain rice variety called Kaybonnet was engineered through genetic engineering with a temperate japonica genetic background (Paine et al., 2005). Kaybonnet possesses good milling features and is blast resistant and contains 11.4 μg/g total carotenoid (Swamy et al., 2019). In the USA more than 200 transgenic rice varieties have been used for experimentation purposes but the only type that has been accepted worldwide is the herbicide-resistant variety known commonly as Liberty Link rice (Lu & Snow, 2005; Jia et al., 2004). Fundamentally, in all these trials, low nutrient rice varieties were studied and their particular deficiency was identified by phenotype or genotype. Traits were then chosen – such as the enhancement of vitamins, amino acids, or minerals – to address these deficiencies (Malik & Chaudhary, 2002).

5.6 Benefits to society

Initially the genetic modification in crops was carried out to improve the farmers' profits by lowering costs, increasing yields, and reducing land use. Since then, plants have started to be engineered with the consumer in mind; i.e. Golden Rice – rice that has been genetically modified to have greater levels of vitamin A (Anderson et al., 2005). GM rice is often a type of Asian transgenic rice that is grown on a large scale for humanitarian purposes. Free distribution of Golden Rice to subsistence farmers, with an additional benefit of being able to keep and replant seed, has been made as well as a free licence given for the commercialization of Golden Rice in developing countries (Roger, 2000). Generally, overall, genetic modification has led to an increased plant growth rate with increased agronomic enhancement. It has also helped to prevent crop spoilage before harvesting and has, overall, improved crop condition during storage. Additionally, it has helped crops develop drought, insect, and pest resistance as well as improving their nutritional quality (Wieczorek & Wright, 2012).

5.7 Health and ethical issues/risks

GM rice poses no risk to humankind and/or the environment. It also doesn't biodegrade. Stein et al. (2008) calculated the adverse effect of vitamin A deficiency in India and found an increased rate of mortality, increased cases of night blindness, corneal scarring, and measles among children. They also observed increased cases of blindness among pregnant and lactating women. Other problems, related to respiratory, urinary, and gastrointestinal infections, have also been observed (Sahoo, 2010). Therefore, in remote rural areas, Golden Rice can act as a stable source for vitamin A delivery.

Because donor genes from external sources are inserted into Golden Rice (GR2E), many questions and statements regarding its acceptability have been raised. In order to prove the validity of GM rice varieties they were compared with the normal rice on the basis of molecular characterization, nutritional stability, stability of the genes inserted, and potential allergenicity and toxicity. During 2013, the International Rice Research Institute (IRRI) planted an experimental plot of Golden Rice that was uprooted by the protestors (Slezak, 2013; Harmon, 2013). But such protests did not stop scientists working on rice modification. And in 2018, the US FDA, in joint collaboration with Australia, Canada, and New Zealand, recommended the use of Golden Rice (GR2E) for human consumption based on the reports and data provided by the IRRI – the third international assessment for GM varieties (Lynas, 2018).

Many parameters are taken into consideration during the safety assessment of GM foods. These include: identity, source, and composition; the effects of processing/cooking; the transformation process; the recombinant DNA (e.g. stability of insertion, potential for gene transfer); the protein expression product of the novel DNA; effects on function; their potential toxicity and/or allergenicity; possible secondary effects from gene expression or the disruption of the host DNA or metabolic pathways; and the potential intake and dietary impact of the introduction of the GM food (FAO/WHO, 2000). The main focus is on the preparation of GM foods and what methodologies can be used to find unintended effects and ways of analyzing their biological relevance and impact on food security (Schilter et al., 1996). As a single method cannot be relied upon for finding such unintended effects i.e. proper digestion of GM foods, changes in the metabolic function, weight gain in the body (NRC, 2004).

According to the IFT (2000), a basic plan of substantial equivalence is required to perform a safety assessment of GM foods by which they can be compared with their traditional counterparts on the basis of genes, phenotypic behavior, and physico-chemical composition (Kok & Kuiper, 2003). To confirm the intended effect, compositional responses were compared between GR2E rice and a control, unmodified, rice grain and straw taken from PSB Rc82 rice. GR2E and the near-isogenic control PSB Rc82 rice were grown at four separate sites in the Philippines during 2015 and 2016. The samples were then analysed for proximate composition, fiber, and minerals. All the parameters were found to be similar in both the modified forms and the normal rice. This is a clear indication that no biologically meaningful change in composition or nutritive value of the grain or straw had occurred as a consequence of the genetic modification. Altogether, the studies performed on GR2E rice have not found it to have any potential health risks. Grune et al. (2010) showed that the β-carotene found in modified rice possesses no harmful toxicity and is equally as good as normal rice. Sorbitol levels were also found to be higher in transgenic traits of rice (Kim et al., 2010; Decourcelle et al., 2015). According to Goodman and Wise (2006) no toxicity symptoms or potential health risks associated with Golden Rice were found.

In fact, in 2016 Congress passed a standard law wherein directions were given to the US FDA to set mandatory standards for commercializing GM foods via the National Bioengineered Food Disclosure. The Disclosure defined GM food products as those wherein the genetic material had been modified under proper laboratory conditions and that GM foods cannot be grown by conventional sowing of seeds under soil. This norm was eventually implemented in 2020 for larger food manufacturing units and in 2021 for smaller manufacturing units (US FDA, 2020). The term modern biotechnology may alternatively be called recombinant DNA (rdNA) technology, genetic engineering and or bio-fortification (FDA, 2019).

5.8 Current status and future prospects

In the present COVID-19 pandemic where food safety and security are major challenges, Golden Rice can be of benefit to many vulnerable communities worldwide. In fact, the World Bank has approved the bio-fortification of micronutrients in most of the world's staple crops (Wesseler & Purnhagen, 2020). In 2016, more than 100 Nobel Laureates recommended the use of Golden Rice as it can produce around 23 times more carotene as compared to normal rice and is generally recognized as safe (Kettenburg et al., 2018). Recently, on July 21, 2021, the DA-BPI (Director of the Philippines Department of Agriculture's Bureau of Plant Industries) signed an agreement on the Commercial Propagation Permit for Golden Rice in the Philippines (Anonymous, 2022). Many supporters like the NSF (National Science Foundation) and many charitable organizations have supported experiments to be conducted for the growth and production of C4 rice. The FDA also assessed the safety of GM crops by focussing on three basic parameters. These were:

1 The allergies related to the organisms used for modifying genes and the related modified crops were analysed (Moghissi & Swetnam, 2010).
2 Toxicities related to genetically modified crops were evaluated.
3 Comparative analysis with the original crop was made from nutritional point of view (McHughen & Smyth, 2008).

5.8.1 Applications of GM rice

Genetic modification is not a wonder science but a solution to many problems. Genetic engineering can contribute to the continuous production and breeding of new rice varieties. GM food products are prepared to provide sustainable growth, proper nutrition, and health, and to combat poverty. These modified varieties are easily assessable to the farmers – the ultimate producers. It also provides opportunity of a cost-free solution for farmers as there is less or no use of costly fertilizers and chemicals and therefore additional restrictions (lack of access to credit and funds) to the subsistence farmers is neutralized. Thus doors and new horizons of opportunity for poor and disadvantaged farmers can be maintained. Also the negative side effects of the green revolution can be avoided. Rice can be grown easily without any additional inputs and the added gene does not change the quality of original rice (Qaim, 2010; James, 2008). The most commercial applications of genetic modification include drought resistance, pesticide resistance, insect resistance, herbicide tolerance, reduction in biotic and abiotic stress, and improved soil condition etc.

5.9 Conclusions

To improve the nitrogen efficiency of the soil and to increase the sustainability of agricultural crops, genetic engineering is the most suitable approach. The most desirable product from this approach is β-carotene which carries the capacity to improve the health of people in regions where rice is a staple crop, especially for the poor. As GM crops can be grown on often infertile or barren land without the use of special techniques etc, this can lead to improved soil conditions and ultimately help to increase farm productivity. Also, as well as it is helping to solve the problem of global food security, overall, not only the farmers but the environment, in general, benefit from the reduction in land use. On average, it is estimated that a rise of 60–70 percent yield is needed to feed future generations by 2050 and genetic modification is currently considered the best possible approach to boost food security and, at the same time, lessen the environmental impact of food production. Thus, in order to fulfil the aim of global food production, all stakeholders, decision makers, and government regulatory bodies have to work all together to make this technology a success.

References

Al-Babili, S., & Beyer, P. (2005). Golden Rice – five years on the road – five years to go? *Trends in Plant Science*, 10(12). doi:10.1016/j.tplants.2005.10.006.

Anderson, K., Jackson, L. E., & Nielson, C. P. 2005. Genetically modified rice adoption: Implications for welfare and poverty allevation. *Journal of Economic Integration*, 20(4), 771–788.

Anonymous (2019). Genetically engineered rice plants that produce high yields developed. *The Economic Times*. https://economictimes.indiatimes.com/news/economy/agriculture/genetically-engineered-rice-plants-that-produce-high-yields-developed/articleshow/67526777.cms

Anonymous (2022). Golden rice humanitarian board. 2005–2022. Maintained by Jorge Mayer. https://www.goldenrice.org/Content1-Who/who5_contact.php

Banerjee, A., & Sharkey, T. D. (2014). Methylerythritol 4-phosphate (MEP) pathway metabolic regulation. *Natural Product Reports*, 31(8), 1043–1055.

Bai, C., Capell, T., Berman, J., Medina, V., Sandmann, G., Christou, P., & Zhu, C. (2016). Bottlenecks in carotenoid biosynthesis and accumulation in rice endosperm are influenced by the precursor–product balance. *Plant Biotechnology Journal*, 14, 195–205.

Bai, C., Rivera, S. M., Medina, V., Alves, R., Vilaprinyo, E., Sorribas, A., … Zhu, C. (2014). An in vitro system for the rapid functional characterization of genes involved in carotenoid biosynthesis and accumulation. *Plant Journal*, 77, 464–475.

Beyer, P., Babili, S. A., Ye, X., Lucca, P., Schaub, P., Welsch, R., & Portykus, I. (2002). Golden Rice: Introducing the β-carotene biosynthesis pathway into rice endosperm by genetic engineering to defeat vitamin a deficiency. *The Journal of Nutrition*, 132(3), 506–510.

Bouis, H. E., & Saltzman A. (2017). Improving nutrition through biofortification: A review of evidence from Harvest Plus, 2003 through 2016. *Global Food Security*, 12, 49–58.

Datta, K., Sahoo, G., Krishnan, S., Ganguly, M., & Datta, S. K. (2014). Genetic stability developed for β-carotene synthesis in BR29 rice line using dihaploid homozygosity. *PLoS One*, 9, e100212. doi:10.1371/journal.pone.0100212.

Das, P., Adak, S., & Majumder, A. L. (2020). Genetic manipulation for improved nutritional quality in rice. *Frontiers in Genetics*, 11. doi:10.3389/fgene.2020.00776.

Decourcelle, M., Perez-Fons, L., Baulande, S., Steiger, S., Couvelard, L., Hem, S., … Sandmann, G. (2015). Combined transcript, proteome, and metabolite analysis of transgenic maize seeds engineered for enhanced carotenoid synthesis reveals pleotropic effects in core metabolism. *Journal of Experimental Botany*, 66, 3141–3150.

FAO/WHO (2000). *Safety Aspects of Genetically Modified Foods of Plant Origin*. Report of a Joint FAO/WHO Expert Consultation on Foods Derived from Biotechnology.

Farre, G., Bai, C., Twyman, R. M., Capell, T., Christou, P., & Zhu, C. (2011). Nutritious crops producing multiple carotenoids – a metabolic balancing act. *Trends in Plant Science*, 16, 532–540.

Farre, G., Maiam Rivera, S., Alves, R., Vilaprinyo, E., Sorribas, A., Canela, R., … Christou, P. (2013). Targeted transcriptomic and metabolic profiling reveals temporal bottlenecks in the maize carotenoid pathway that may be addressed by multigene engineering. *Plant Journal*, 75, 441–455.

FDA (2019). *Guidance for Industry: Voluntary Labeling Indicating Whether Foods Have or Have Not Been Derived from Genetically Engineered Plants*. Guidance Document. https://www.fda.gov/regulatory-information/search-fda-guidance-documents/guidance-industry-voluntary-labeling-indicating-whether-foods-have-or-have-not-been-derived

FNRI (2016). Vitamin A deficiency (VAD) among Filipino preschool children, pregnant, and lactating women: 1993–2013. 42nd FNRI Seminar Series, First 1000 Days.

Frano, L. M. R., Woodhouse, L. R., Burnett, D. J., & Burri, B. J. (2013). Biofortified cassava increases β-carotene and vitamin A concentration in the TAG-rich plasma layer of American women. *British Journal of Nutrition*, 110(2), 310–320.

Fraser, P. D., Enfissi, E. M., & Bramley, P. M. (2009). Genetic engineering of carotenoid formation in tomato fruit and the potential application of systems and synthetic biology approaches. *Archive of Biochemistry and Biophysics*, 483, 196–204.

Gayen, D., Ghosh, S., Paul. S., Sarkar, S. N., Datta, S. K., & Datta, K. (2016). Metabolic regulation of carotenoid-enriched golden rice line. *Frontiers in Plant Science*. doi:10.3389/fpls.2016.01622.

Goodman, R. E., & Wise, J. (2006). *Bioinformatic Analysis of Proteins in Golden Rice 2 to Assess Potential Allergenic Cross-Reactivity*. Preliminary Report, Food Allergy Research and Resource Program, University of Nebraska.

Grune, T., Lietz, T.G., Palou, A., Ross, C., Stahl, W., Tang, G., … Biesalski, H. K. (2010). Carotene is an important vitamin A source for humans. *Journal of Nutrition*, 140, 2268–2285.

Harmon, A. (2013). Golden Rice: Lifesaver. *The New York Times*. https://www.nytimes.com/2013/08/25/sunday-review/golden-rice-lifesaver.html

Huang, J., Hu, R., Rozelle, S., & Pray, C. (2008). Genetically modified rice, yields and pesticides: Assessing farm-level productivity effects in China. *Economic Development and Cultural Change*, 56(2). doi:10.1086/522898.

IFT (Institute of Food Technologists) (2000). IFT expert report on biotechnology and foods. Human food safety of rDNA biotechnology-derived foods. *Food Technology*, 54, 53–61.

IRRI (2019). *Race for Impact*. Annual Report, International Rice Research Institute. http://books.irri.org/AR2019_content.pdf

James, C. (2008). *Global Status of Commercialized Biotech/GM Crops: 2008*. ISAAA Brief No. 39 (Ithaca, NY: ISAAA).

Jia, H., Jayaraman, K. S., & Louet, S. (2004). China ramps up efforts to commercialize GM rice. *Nature Biotechnology*, 22, 642.

Khush, G. S., Lee, S., Cho, J.-I., & Jeon, J.-S. (2012). Biofortification of crops for reducing malnutrition. *Plant Biotechnology Reports*, 6, 195–202.

Kim, J., Kong, M. K., Lee, S. Y., & Lee, P. C. (2010). Carbon sources-dependent carotenoid production in metabolically engineered *Escherichia coli*. *World Journal of Microbiology and Biotechnology*, 26, 2231–2239.

Kim, Y. J., Yeu, S. Y., Park, B. S., Koh, H.-J., Song, J. T., & Seo, H. S. (2012). Protein disulfide isomerase-like protein 1–1 controls endosperm development through regulation of the amount and composition of seed proteins in rice. *PLoS One*, 7, e44493. doi:10.1371/journal.pone.0044493.

Kettenburg, A. J., Hanspach, J., Abson, D. J., & Fischer, J. (2018). From disagreements to dialogue: Unpacking the Golden Rice debate. *Sustainability Science*, 13(5), 1469–1482.

Kok, E. J., & Kuiper, H. A. (2003). Comparative safety assessment for biotech crops. *Trends in Biotechnology*, 21, 439–443.

Lee, S., Park, J., & Yim, Y. (2021). Genetic modification of rice for efficient nitrogen utilization. *Plant Biotechnology Reports*, 15, 573–583.

Lu, B. R., & Snow, A. A. (2005). Gene flow from genetically modified rice and its environmental consequences. *BioScience*, 55(8), 669–678.

Lu, K., Li, L., Zheng, X., Zhang, Z., Mou, T., & Hu, Z. (2008). Quantitative trait loci controlling Cu, Ca, Zn, Mn and Fe content in rice grains. *Journal of Genetics*, 87, 305–310.

Lynas, M. (2018, May). US FDA approves GMO Golden Rice as safe to eat. Genetic Literacy Project. geneticliteracyproject.org

Malik S., & Chaudhary P. (2002). Non-conventional tools in the improvement of aromatic rices. In *Specialty Rices of the World: Breeding, Production and Marketing*, edited by R. C. Chaudhary, D. V. Tran, & R. Duffy (Rome: FAO), 207–222.

McHughen, A., & Smyth, S. (2008). US regulatory system for genetically modified organism (GMO), rDNA or transgenic crop cultivars. *Journal of Plant Biotechnology*, 6, 2–12.

Moghissi, A. A., Pei, S., & Liu, Y. (2015). Golden rice: Scientific, regulatory and public information processes of a genetically modified organism. *Critical Reviews in Biotechnology*. Early Online: 1–7.

Moghissi, A. A., & Swetnam, M. S. (2010). The role of peer review in policy decision. *Technological Innovation*, 12, 203–211.

Nonero, D. (2018, February). GMO crops have been increasing yield for 20 yrs, with more progress ahead. Alliance for Science, February 23. https://allianceforscience.cornell.edu/blog/2018/02/gmo-crops-increasing-yield-20-years-progress-ahead/

National Research Council (US) (2004). *Safety of Genetically Engineered Foods. Approaches to Assessing Unintended Health Effects.* Committee on Identifying and Assessing Unintended Effects of Genetically Engineered Foods on Human Health, Washington, DC (Washington, DC: National Academies Press (US)).

Paine, J. A., Shipton, C. A., Chaggar, S., Howells, R. M., Kennedy, M. J., & Vernon, G. (2005). Improving the nutritional value of golden rice through increased provitamin A content. *Natural Biotechnology*, 23, 482–487.

Parkhi, V., Rai, M., Tan, J., Oliva, N., Rehana, S., Bandyopadhyay, A., … Dhatta, S. K. (2005). Molecular characterization of marker free transgenic lines of indica rice that accumulate carotenoids in seed endosperm. *Molecular Genetics and Genomics*, 274, 325–336.

Qaim, M. (2010). Benefits of genetically modified crops for the poor: Household income, nutrition, and health. *New Biotechnology*, 27(5), 552–557.

Regis, E. (2019). *Golden Rice: The Imperiled Birth of a GMO Superfood* (Baltimore, MD: Johns Hopkins University Press).

Roger, D. (2000). Royalty-free licenses for genetically modified rice made available to developing countries. *Bulletin of the World Health Organization*, 78(10), 1281.

Romer, S., Lubeck, J., Kauder, F., Steiger, S., Adomat, C., & Sandmann, G. (2002). Genetic engineering of a zeaxanthin-rich potato by antisense inactivation and co-suppression of carotenoid epoxidation. *Metabolic Engineering*, 4, 263–272.

Sahoo, J. P. (2010). The story of Golden Rice in India. Technical Report. *AGROBIOS Newsletter*, XIX(5), 57–58.

Schilter, B., Holzhauser, D., Cavin, C., & Huggett, A. C. (1996). An integrated in vivo/in vitro strategy to improve food safety evaluation. *Trends Food Science and Technology*, 7, 327–332.

Sellappan, K., Datta, K., Parkhi, V., & Datta, S. K. (2009). Rice caryopsis structure in relation to distribution of micronutrients (iron, zinc, β-carotene) of rice cultivars including transgenic indica rice. *Plant Science*, 177, 557–562.

Slezak, M. (2013, August). Militant Filipino farmers destroy Golden Rice GM crop. *New Scientist Newsletter*, August 9. https://www.newscientist.com/article/dn24021-militant-filipino-farmers-destroy-golden-rice-gm-crop/

Stein, A. J., Sachdev, H. S., & Qaim, M. (2008). Genetic engineering for the poor: Golden Rice and public health in India. *World Development*, 36(1), 144–158.

Steur, H. D., Stein, A. J., & Demont, M. (2022). From Golden Rice to golden diets: How to turn its recent approval into practice. *Global Food Security*, 32, 100596.

Stokstad, E. (2019). After 20 years, golden rice nears approval. *Plant Science*, 366 (6468). doi:10.1126/science.366.6468.934.

Swamy, B. P. M., MarundanJr, S., Samia, M., Ordonio, R. L., Rebong, D. B., Miranda, R., … MacKenzie, D. J. (2021). Development and characterization of GR2E Golden rice introgression lines. *Nature: Scientific Reports*, 11, no. 2496.

Swamy, B. P. M., Samia, M., Boncodin, R., Marundan, S., Rebong, D. B., Ordonio, R. L., & MacKenzie, D. J. (2019). Compositional analysis of genetically engineered GR2E "Golden Rice" in comparison to that of conventional rice. *Journal of Agricultural and Food Chemistry*, 67, 7986–7994.

Singh S. P., Gruissem W., & Bhullar N. K. (2017). Single genetic locus improvement of iron, zinc and β-carotene content in rice grains. *Scientific Reports*, 7, no. 6883.

National Academies (2010). Genetically engineered crops benefit many farmers, but the technology needs proper management to remain effective. News Release, April 13. https://www.nationalacademies.org/news/2010/04/genetically-engineered-crops-benefit-many-farmers–but-the-technology-needs-proper-management-to-remain-effective

Tang, G., Hu, Y., Yin, S., Wang, Y., Dallal, G. E., Grusak, M. A., & Russell, R. M. (2012). β-carotene in Golden Rice is as good as β-carotene in oil at providing vitamin A to children. *American Journal of Clinical Nutrition*, 96, 658–664.

UN SCN Fifth Report (2005). *World Nutrition Situation for Improved Development Outcomes* (Geneva: United Nations System, Standing Committee on Nutrition).

US FDA (2020). *New Plant Variety Regulatory Information in Bioengineering/GMOs*. https://www.fda.gov/food/food-new-plant-varieties/new-plant-variety-regulatory-information

Welch, R. M., & Graham, R. D. (2004). Breeding for micronutrients in staple food crops from a human nutrition perspective. *Journal of Experimental Botany*, 55, 353–364.

Wesseler, J., & Purnhagen, K. (2020). Is the Covid-19 pandemic a game changer in GMO regulation? [La pandémie Covid-19 va-elle changer la donne dans la réglementation des OGM ?]. *EuroChoices*, 19(3), 49–52.

White P. J., & Brown P. H. (2010). Plant nutrition for sustainable development and global health. *Annals of Botany*, 105, 1073–1080.

WHO (2009). *Global Prevalence of Vitamin A Deficiency in Populations at Risk 1995–2005*. WHO Global Database on Vitamin A Deficiency (Geneva: WHO).

Wieczorek, A. M., & Wright, M. G. (2012). History of agricultural biotechnology: How crop development has evolved. *Natural Education Knowledge*, 3, 9–10.

Ye, X., Al-Babili, S., Klöti, A., Zhang, J., Lucca, P., Beyer, P., & Potrykus, I. (2000). Engineering the provitamin A (β-carotene) biosynthetic pathway into (carotenoid-free) rice endosperm. *Science*, 287, 303–305.

Zhu, C., Naqvi, S., Breitenbach, J., Sandmann, G., Christou, P., & Capell, T. (2008). Combinatorial genetic transformation generates a library of metabolic phenotypes for the carotenoid pathway in maize. *Proceedings of the National Academic of Sciences USA*, 105, 18232–18237.

Zhu, Q., Zeng, D., Yu, S., Cui, C., Li, J., Li, H., ... Liu, Y.-G. (2018). From Golden Rice to a STARice: Bioengineering astaxanthin biosynthesis in rice endosperm. *Molecular Plant*, 11, 1440–1448.

6 Genetically modified maize

Bharati Sharma, Kanchan Sharma, Priya Sharma, Isha Gupta, Jasmeet Kour and Kawaljeet Kour

6.1 Introduction

Maize (*Zea mays* L.) or corn is one of the most widely cultivated grains and has been used as a staple food for a major section of the population, specifically those belonging to lower income strata (Ala-Kokko et al., 2021), as well as being used as feed for animals (Zou et al., 2018). However, several pests of the order 'Lepidoptera', *viz.* maize borer, spodoptera, armyworm, cotton bollworm, and various common maize pests, pose a serious threat to the quality and productivity of the crop. Additionally, the growth of weeds along with maize plants can lead to the increased chances of transmission of diseases and pests/insects that in turn cause reduced growth. These issues can often only be controlled by the use of chemical pesticides and herbicides but they have certain drawbacks such as environment pollution, an increased cost of production, and herbicide/pesticide residue (Hu et al., 2020). Thus, the development of transgenic crops using genetic modification (GM) techniques offers an alternative solution for increasing crop yield by improving tolerance to pesticides and herbicides and providing more flexibility in farm management (Zou et al., 2018).

GM crops have been recognized for their remarkable potential in improving global food security and reducing ecological and environmental impact in addition to causing a spike in species diversity. In order to introduce genetic modification in a crop, the desired genes (having certain traits) are transferred to the host plant cell which is then regenerated by the method of plant tissue culture (Zhang et al., 2016). Generally, two methods are employed for the transfer of DNA into a plant cell: the first includes coating the surface of small metal particles with the relevant DNA fragment and then bombarding these metal particles into the plant cells; a second, widely used method is to use a bacterium or virus, of which the most frequent used in GM plants is *Agrobacterium tumefaciens* (Fleischer et al., 2021).

6.2 Why are genetic modifications required?

Several challenges are being faced by the world that have paved way for resorting to the technology of modifying the genetic makeup of plants. The

DOI: 10.4324/9781003278566-6

biggest hurdle is the expansion of population and malnutrition combined with a decrease in cultivable land accompanied by increased urbanization, climate change, and land desertification. The growing global demand for food with improved nutritional status could possibly be achieved by means of optimizing crop genetics along with improvement in agricultural systems and increased crop yield (Ray et al., 2013). In addition, genetic modifications can make the crop resistant to insects, pests, bacteria, fungus and viral infections as well as give abiotic stress (cold, drought, salinity) resistance. GM crops have been associated with a reduced cost of food/drug production and a reduced need for pesticides, insecticides, and herbicides (James, 2002). Genetic modification could also aid in providing essential micronutrients by the process of enrichment and could prove to be another relevant method for achieving global food security (Thomas et al., 2003). Another reason behind the need for genetic modification is that it reduces the production timeline for new strains unlike the conventional breeding process which takes much longer. The adoption of GM insect-resistant and herbicide-tolerant crops led to reduced pesticide spraying which in turn lowered the environmental impact related to usage of herbicide and insecticide on these crops. Moreover, the technology has also facilitated significant cuts in fuel use and tillage changes, leading to a reduction in the release of greenhouse gas emissions from GM cropping areas.

6.3 Genes involved in agronomic trait development

Plants are genetically engineered to express Bt crystalline (Cry) protein and are helpful in controlling pest invasion because the protein is harmful to some insects (Hu et al., 2020). Insects belonging to the order Lepidoptera, Diptera, and Coleoptera, that generally include the European corn borer (*Ostrinia nubilalis*), corn earworm (*Helicoverpa zea*), cotton bollworm (*Helicoverpa armigera*), fall armyworm (*Spodoptera frugiperda*), are found to be susceptible to Cry proteins (Adang et al., 2014).

The insect-resistant gene (Cry1Ab) and herbicide-tolerant gene (epsps) were generated using an agrobacterium-mediated transformation technique in order to protect crops from Lepidopterous pests/insects and also to give corn glyphosate tolerance, respectively. The Cry1Ab gene (from *Bacillus thuringiensis*) makes pores on the epithelial cells of the intestine of lepidoptera, causing cells to rupture and eventually causing the death of the insect (Betz et al., 2000). Glyphosate is the most widely used and potentially effective herbicide and works by inhibiting the activity of 5-enolpyruvylshikimate-3-phosphate synthetase (epsps) in the chloroplast (Cao et al., 2012). To avoid the damage caused by glyphosate on maize crops, the epsps gene from *Agrobacterium tumefaciens*, which is not susceptible to glyphosate, was genetically transferred to generate crops with glyphosate-resistant characteristics.

6.4 Mechanism of genetic modification leading to the desired trait

Maize is genetically modified in order to enhance the quality and quantity of production while overcoming many biotic and abiotic stressors such as pest infestation, the effect of weeds, and the transmission of various diseases that lead to adverse expansion in maize plants. However, the different chemicals like pesticides, herbicides, etc. used to combat the various problems encountered throughout the growth and production of maize can affect the environment by causing environmental pollution, leaving residues of herbicides and pesticides and, ultimately, increasing production costs. Many strong, vital weed plants compete with the maize for different production resources: e.g. light, water, minerals, and other nutrients. They also increase the chances of transmission of diseases and pests and thus inhibit plant growth. The most predominant insects and pests affecting the growth and production of maize plants include spodoptera, frugiptera, armyworm, cotton bollworm, etc.

There are many transgenic maize crops with improved traits available across the globe at present (De Schrijver et al., 2015). As mentioned above, GM maize having insect-resistant gene (Cry1Ab) and herbicide-tolerant gene (epsps) was formed from Agrobacterium to combat Lepidopterous pests. Another genetically engineered maize was developed by incorporating genes obtained from *Agrobacterium tumefaciens* using line NP2222 immature embryos with plasmid vector pSYN17629. Additionally, a helper plasmid (non-oncogenic) pSB1 was also used in the agrobacterium-mediated transformation process. Three expression cassettes present between the right and left ends of the T-DNA of the pSYN17629 plasmid, namely ecry3.1Ab, mcry3A, and pat08, were used as a selectable marker during the transformation process (Negrotto et al., 2000).The other genetic elements present are the promoter region obtained from cestrum yellow leaf curling virus and the chimeric ecry3.1Ab gene which consists of two fusion elements as modified cry3A (mcry3A) and a synthetic plant codon optimized cry1Ab obtained from *Bacillus thuringiensis* (kurstaki strain HD-1) and a terminator sequence from the nopaline synthase (nos) gene from *A. tumefaciens*. Moreover, the pat expression cassette is incorporated with the promoter region (cauliflower mosaic virus), the codon pat gene (Streptomyces viridochromogenes, Tu494), and a terminator gene i.e., nos (*Agrobacterium tumefaciens*). The elements necessary for the maintenance of bacteria plasmid are present in the vector backbone. These vectors are also known as plant transformation constructs (Naegeli et al., 2019).

There are many factors that affect the outcome of these genetic modifications in maize: the varieties of medium utilized at different phases of propagation, the kind of genotype introduced, and the explant material used to initiate the process of transformation. Various methods are used for the transformation of genes in maize plants: protoplast transformation, particle bombardment, the use of silicon carbide whiskers, conversions using Agrobacterium, in planta transformation, etc. (Yadava et al., 2017). The extensively utilized alleles for genetic modification of maize specifically for commercial purpose include Hi-II (High type II callus production). The induction of callus is a crucial aspect in the genetic engineering of maize. Type I callus produced by most of the maize genotypes is loosely spaced and more revivable than Type II callus which is friable, transformable, and embryogenic (Green, 1982; Armstrong et al., 1991; Hansen & Wright, 1999). The most preferred targets for transgenic maize are embryogenic callus initiated in cell suspension cultures and immature embryos (Armstrong, 1999; Hansen & Wright, 1999; Torney et al., 2007).

6.4.1 *Protoplast transformation*

Protoplast transformation is done by the electroporation method. Pectinase and/or cellulase is used to remove the cell wall from the protoplast and is then purified. This purified protoplast is mixed with the vectors and the carrier DNA to introduce the desired DNA into the host cell. This mixture is then exposed to a short and continuous electric pulse of high voltage, temporarily forming pores in the plasma membrane. From these pores, the vector carrying the desired DNA enters the host cell. The major drawback of this method is that the regenerated plant may have fertility problems. It is also a time consuming and labour-intensive method.

6.4.2 *Particle bombardment*

In this method, the target DNA is incorporated with the help of plasmid DNA coated tungsten or gold particles. These particles are accelerated to very high speeds and are used to target into the desired maize tissue. The target can be cell suspension culture, type I & II callus, immature embryos, callus from seedlings, or shoot meristem cultures etc. Most commonly, the target tissues for the transformation of cry1Ab gene by this method are immature embryos (Koziel et al., 1993). Research has shown that more fertile transgenic plants from embryonic callus are produced by the particle bombardment method than by protoplast transformation methods (Koziel et al., 1993). Here, steps like processing of T-DNA and integration are not involved, thus making the vector backbone superfluous and simplifying the whole process.

6.4.3 *Silicon carbide whiskers*

As the name indicates, these are needle-like structures (20 μm long) and they transfer the desired DNA into the plasma membrane of the target cell by penetrating the cell wall (Southgate et al., 1998). Black Mexican Sweet, a transgenic non-regenerable maize, was developed using silicon carbide whiskers (Kaeppler et al., 1992). This method was also used to obtain transgenic fertile maize plants from Type II callus and cell suspension cultures (Frame et al., 1994; Petolino et al., 2000).

6.4.4 *Agrobacterium-mediated transformation*

A. *tumefaciens*, a soil pathogen, is capable of transforming plants and has been used extensively for more than a decade to transform dicots. This method has several advantages over direct gene transfer methods, including integration of significantly larger DNA segments into the host plant with minimal rearrangement, resulting in high quality. A modified version of this method is known as Sonication-Assisted Agrobacterium-Mediated Transformation (SAAT) and has greater efficiency of DNA transfer in different plants. This method involves exposing the targeted species to agitated sound waves at specific periods in the presence of A. *tumefaciens* (Trick & Finer, 1997).

6.4.5 *In planta transformation*

This method involves vacuum infiltration of the whole plant and floral dip (Bechtold et al., 1993; Clough & Bent, 1998). Using plumular meristems of sprouting seedlings as explants and subsequently converting them using the A. *tumefaciens* technique means that translation can be achieved independently of tissue culture. However, this method is not as widely used for commercial purposes as the previous methods.

6.5 Impact on modern agriculture (trade in GM crops)

A study has shown that most farmers were cultivating Bt maize mainly due to its higher yield. Those who had not adopted Bt maize stated that the reason was its high cost in spite of their willingness to adopt it in their farming practices. A very few farmers had no knowledge about the Bt seed or their properties and thus were still using conventional maize seeds. Research also revealed that farmers cultivating Bt maize with properties like herbicide resistance, insecticide resistance, and weed resistance, etc., reduced their labour hours throughout the season; a result due to the time saved concerning the spraying of various chemicals and removal of weeds. Most of the farmers preferred green GM maize varieties, particularly HT & stacked ones, rather than their conventional counterparts. And of these, the majority preferred stacked varieties (both Bt and HT traits) over Bt maize (Gouse et al., 2016).

Transgenic maize varieties have replaced many of their conventional counterparts. One of the most widely and successful applications of genetic engineering is the use of genes from the soil bacterium (*Bacillus thuringiensis*) for incorporation into the target plant to give desired properties. It provides protection of the crop against insects and pests which may harm the crop yield. These plants, particularly those having cry proteins, are able to combat insects and pest infestation due to the presence of certain toxic chemicals (Roh et al., 2007). The increased yield, along with the reduced costs due to decreased use of chemicals like pesticides, results in profit to the farmers that compensates for the higher seed costs (Betz et al., 2000). For instance, different chemicals like glyphosate, terbuthylazine, and atrazine, etc. are used as herbicides while cultivating conventional maize varieties; these are rarely used for GM maize.

6.6 Benefits to society

Bt crops help to improve the household income of farmers cultivating these crops as they have a higher yield. Research has shown that in India farmers' profit has increased by 50 percent due to the use of Bt cotton and that there was a reduction in food insecurity. It also helped in reducing pest damage, leading to the protection of yield that ultimately has increased agriculture output both in quantity, quality, and economic terms (Carzoli et al., 2018). Other beneficial effects – its labour-saving qualities, the reduced use of chemicals such as pesticides, etc., and environmental benefits – have also been reported in various studies (Ahmed et al., 2020). Additionally, it narrows the yield gaps because of accelerated yield growth, enhanced weed control, timely growth, better insect resistance, and tolerance to abiotic stress like drought (Edmeades et al., 2010; Parisi et al., 2016; Svitashev et al., 2016). And, most importantly, after testing for potential effects on human health and the environment, the GMO panel concluded that transgenic maize MZIR098 is as safe as its conventional counterparts (Naegeli et al., 2020).

6.7 Trade in the global market

GM crops have been commercialized at a rapid rate throughout the world from 2017–2021. In Canada and United States, GM crops such as maize, soybean, canola, and cotton were cultivated on over 44 million hectares in 2000, which represents 75 percent of the total cultivated global area under plantation of GM crops (Clive, 2000). Corn, soybean, canola, and cotton are the top four GM crops and have been cultivated on a very large scale. Among these, GM maize is the second largest crop in world representing 15 percent of all transgenic crops planted globally.

The European Union does not export any genetically engineered crops. GE corn is produced in the EU for the production of animal feed and biogas production. In 2021, there was a decline in Bt corn in the EU by 8.5 percent to 101,000 hectares. Spain is the leading (96 percent) in total area coverage whereas Portugal holds 4 percent only (De Belder, 2021). In 2019, there was a plantation of about 33.17 million hectares of GM corn by the United States. From 2004 to 2017, the average US corn export has been estimated at US$6.2 million. From 2000–2017, there has been an upsurge in the production of GM maize from 20 percent to 90 percent respectively (Addey, 2021). The United States Department of Agriculture Economic Research Service (USDA ERS) indicated that GM corn and soybean covered about 90 percent of their respective productions in 2017 (USDA, 2020).

Of maize importers, Japan is the topmost maize importing country. Other countries such as Mexico, Egypt, Spain, The Republic of Korea, and Taiwan also import large amounts of GM maize (Wu & Guclu, 2013).

6.8 Health and ethical issues/risks

Genetic engineering is an important tool of biotechnology that alters the genetic makeup of various living organisms such as animals, plants, and micro organisms (Ralley et al., 2016). The production of these GM crops involves the introduction of foreign genes into the desired organism to gain various qualities such as increasing the productivity and qualities of the product, increasing the content of various desired nutrients, and increasing the resistance of the plant against insects, pests, and drought (Hu & Xiong, 2014; Barrows et al., 2014). In 2018, 92 percent of the maize which we consumed was grown from GM maize seeds. GM maize has a gene that produces a protein that is toxic to many insects and pests and protects the maize from damage. The gene presenting these properties is taken from the soil bacterium i.e. *Bacillus thuringiensis*. It is therefore mandatory to evaluate the risk issues of this genetically engineered maize on human health and environment. A snapshot of evaluative studies are listed below.

a By using GM maize, the number of sprays needed of pesticides and insecticides is reduced; between 1999 and 2008 farmers reduced their use of these chemicals by 775.4 million kg (Brookes & Barfoot, 2020).
b The introduction of Bt in GM crop has shown no negative effects (WHO, 1999; Koch et al., 2015).
c The Monsanto Company revealed that this GM maize is active against pests and insects and also confirmed no negative effects on the environment and/or humans (Mendelsohn et al., 2003).
d Research studies at Cornell University (1999) found that pollen of this GM maize containing Cry protein had a deleterious effect on the larvae of *Cithoronia regalis* (Monarch butterfly) (Losey et al., 1999). However, in 2001, various laboratory studies revealed that the impact of this negative effect was not adequate to cause harm to the insects (Sears et al., 2001).

6.9 Current status and future prospects

The use of GM crops for commercial purposes began in the mid-1990s. Since then, the technology has swiftly spread throughout the world, both in developing and developed nations. GM maize cultivars have been effectively introduced to the market over the last three decades, bringing attributes such as herbicide tolerance and insect resistance to farms. The traditional way of genetic modification that involved the use of a single gene expressing single trait has now been replaced by a method that involves the development of a gene with multiple useful traits. Farmers have had a lot of success with this so-called stacking variety. From 1996 to 2018, the economic benefit at the global level from GM crops totaled US$225 billion, with developing nations accounting for 52 percent of these benefits (Yassitepe et al., 2021).

In 2019, an area of 190.4 million hectares was used for the cultivation of transgenic plants in 29 countries for use as food and feed. This area represents a 112-times increase from the area of 1.7 million hectares cultivated in 1996. Soyabean is the most widely accepted GM crop, followed by maize, cotton, and canola. In 2019, GM maize crops were cultivated on 31 percent (60.9 million hectares) of the 193.4 million hectares of worldwide maize cultivation areas across 14 nations. Maize is now the first commercial GM insect-resistant crop with 148 GM maize events possessing herbicide and insect resistance certified for commercial usage around the world (Yassitepe et al., 2021). Although genetic modification has been proven to have the potential of developing new crops with better traits, GM crops remain a hot topic with regards to their safety. Consumers acceptability for GM crops is likely continue at a low pace. The public are not sure whether or not GM crops are as safe to use as non-GM crops. Hence, another important limiting factor in the way of the development and commercialization of GM crops is the high cost of risk assessment to ascertain their safety.

6.10 Conclusions

GM crops are those crops whose genome or genetic material is modified using genetic engineering technologies to improve the inherent trait of the crop or to introduce a new trait into that crop. Among GM crops, soybean, maize, canola, and cotton are the main crops that are produced worldwide at a commercial level. Globally, maize is the most important and richest crop of genomic sources for conducting biotechnological research. To date, many herbicide- and insect-resistant GM maize cultivars have been developed. In 32 crops (including maize), nearly 525 different genes with useful traits have been approved for cultivation in different parts of the world. In 2019, GM maize varieties accounted for over 30 percent of the world's maize cultivated area (ISAAA database, 2021). Yet, the useful traits of modified crops are no longer restricted to herbicide and insect resistance but also include abiotic stress tolerance, disease resistance, high yield, and nutritional quality improvement (ISAAA database, 2021). The adverse weather conditions and the explosion in human population pose a great problem of food security worldwide. To overcome these conditions, food safety and food security need to be increased by promoting agricultural yield. Developing countries should therefore be encouraged not only to adopt GM technologies for improving crop yield but also to act as a source of income for food-insecure farmers. Currently, Monsanto is the biggest and financial secured player of genetically modified crops. The company has been credited with the development of a range of genetically engineered crops producing higher yield – including the controversial Roundup Ready crops, which impart pest resistance. Monsanto apart from generating huge revenue from selling high yield seed, also generate the associated products that need to be sprayed on them in order to attain maximum output.

References

Adang, M. J., Crickmore, N., & Jurat-Fuentes, J. L. (2014). Diversity of *Bacillus thuringiensis* crystal toxins and mechanism of action. In *Advances in Insect Physiology* volume 47, edited by T. S. Dhadialla & S. S. Gill (Cambridge, MA: Academic Press), 39–87.

Addey, K. A. (2021). The cost of partners' genetically modified organisms regulatory index on US corn and soybean exports. *Food and Energy Security*, 10(1), e265.

Ahmed, Z., Asghar, M. M., Malik, M. N., & Nawaz, K. (2020). Moving towards a sustainable environment: The dynamic linkage between natural resources, human capital, urbanization, economic growth, and ecological footprint in China. *Resources Policy*, 67, 101677.

Ala-Kokko, K., Nalley, L. L., Shew, A. M., Tack, J. B., Chaminuka, P., Matlock, M. D., & D'Haese, M. (2021). Economic and ecosystem impacts of GM maize in South Africa. *Global Food Security*, 29, 100544.

Armstrong, C., Green, C. E., & Phillips, R. L. (1991). Development and availability of germplasm with high type II culture response. *Maize Genetics Cooperation Newsletter*, 65, 92–93.

Armstrong, C. L. (1999). The first decade of maize transformation: A review and future perspective. *Maydica*, 44(1), 101–109.

Barrows, G., Sexton, S., & Zilberman, D. (2014). Agricultural biotechnology: The promise and prospects of genetically modified crops. *Journal of Economic Perspectives*, 28(1), 99–120.

Bechtold, N., Ellis, J., & Pelletier, G. (1993). In planta Agrobacterium mediated gene transfer by infiltration of adult *Arabidopsis thaliana* plants. *Comptes Rendus de L'Académie des Sciences Serie III-Sciences De La Vie/Life Sciences*, 316, 1194–1199.

Betz, F. S., Hammond, B. G., & Fuchs, R. L. (2000). Safety and advantages of Bacillus thuringiensis-protected plants to control insect pests. *Regulatory Toxicology and Pharmacology*, 32(2), 156–173.

Brookes, G., & Barfoot, P. (2020). Environmental impacts of genetically modified (GM) crop use 1996–2018: Impacts on pesticide use and carbon emissions. *GM Crops & Food*, 11(4), 215–241.

Cao, G., Liu, Y., Zhang, S., Yang, X., Chen, R., Zhang, Y., ... Wang, G. (2012). A novel 5-enolpyruvylshikimate-3-phosphate synthase shows high glyphosate tolerance in *Escherichia coli* and tobacco plants. *PLoS One*, 7(6), e38718.

Carzoli, A. K., Aboobucker, S. I., Sandall, L. L., Lübberstedt, T. T., & Suza, W. P. (2018). Risks and opportunities of GM crops: Bt maize example. *Global Food Security*, 19, 84–91.

Clive, J. (2000). The current global status of transgenic crops. *Növényvédelem*, 36(5), 257–264.

Clough, S. J., & Bent, A. F. (1998). Floral dip: A simplified method for Agrobacterium-mediated transformation of *Arabidopsis thaliana*. *The Plant Journal*, 16(6), 735–743.

De Belder, T. (2021). *Biotechnology and Other New Production Technologies Annual*. Biotechnology and Other New Production Technologies, USDA, Foreign Agricultural Service.

De Schrijver, A., De Clercq, P., de Maagd, R. A., & van Frankenhuyzen, K. (2015). Relevance of Bt toxin interaction studies for environmental risk assessment of genetically modified crops. *Plant Biotechnology Journal*, 13(9), 1221–1223.

EFSA Panel on Genetically Modified Organisms (EFSA GMO Panel), Naegeli, H., Bresson, J. L., Dalmay, T., Dewhurst, I. C., Epstein, M. M., … Koukoulanaki, M. (2019). Assessment of genetically modified maize Bt11 × MIR162 × MIR604 × 1507 × 5307 × GA21 and subcombinations, for food and feed uses, under Regulation (EC) No 1829/2003 (application EFSA-GMO-DE-2011-103). *EFSA Journal*, 17(4), e05635.

EFSA Panel on Genetically Modified Organisms (EFSA GMO Panel), Naegeli, H., Bresson, J. L., Dalmay, T., Dewhurst, I. C., Epstein, M. M., … Raffaello, T. (2020). Assessment of genetically modified maize MZIR 098 for food and feed uses, under Regulation (EC) No 1829/2003 (application EFSA-GMO-DE-2017-142). *EFSA Journal*, 18(6), e06171.

Edmeades, G. O., Fisher, R. A., & Byerlee, D. (2010, January). Can we feed the world in 2050? In *Proceedings of the New Zealand Grassland Association*, XXXIV–XLII.

Fleischer, S. J., Hutchison, W. D., & Naranjo, S. E. (2021). Sustainable management of insect-resistant crops. In *Plant Biotechnology*, edited by A. Ricroch, S. Chopra, & M. Kuntz (Cham: Springer), 111–125.

Frame, B. R., Drayton, P. R., Bagnall, S. V., Lewnau, C. J., Bullock, W. P., Wilson, H. M., … Wang, K. (1994). Production of fertile transgenic maize plants by silicon carbide whisker-mediated transformation. *The Plant Journal*, 6(6), 941–948.

Gouse, M., Sengupta, D., Zambrano, P., & Zepeda, J. F. (2016). Genetically modified maize: Less drudgery for her, more maize for him? Evidence from smallholder maize farmers in South Africa. *World Development*, 83, 27–38.

Green, C. E. (1982). Somatic embryogenesis and plant regeneration from the friable callus of Zea mays. In *Plant Tissue Culture 1982: Proceedings, 5th International Congress of Plant Tissue and Cell Culture held at Tokyo and Lake Yamanake, Japan, July 11–16, 1982*, edited by Akio Fujiwara (Tokyo: Japanese Association for Plant Tissue Culture), 107–108.

Hansen, G., & Wright, M. (1999). Recent advances in the transformation of plants. *Trends in Plant Science*, 4, 226–231.

Hu, H., & Xiong, L. (2014). Genetic engineering and breeding of drought-resistant crops. *Annual Review of Plant Biology*, 65, 715–741.

Hu, Y., Guo, M., Zhuo, Q., Han, C., Shi, L., Mao, H., … Yang, X. (2020). Three-generation reproductive toxicity of genetically modified maize with Cry1Ab and epsps genes in rats. *Journal of Agricultural and Food Chemistry*, 68(39), 10912–10919.

ISAAA (2021). GM approval database. Available at: https://www.isaaa.org/gmapprovaldatabase/default.asp (accessed August 10, 2021).

James, C. (2002). *Global Review of Commercialized Transgenic Crops: 2001 Feature: Bt Cotton*. ISAAA Brief No. 26. (Ithaca, NY: ISAAA).

Kaeppler, H. F., Somers, D. A., Rines, H. W., & Cockburn, A. F. (1992). Silicon carbide fiber-mediated stable transformation of plant cells. *Theoretical and Applied Genetics*, 84(5), 560–566.

Koch, M. S., Ward, J. M., Levine, S. L., Baum, J. A., Vicini, J. L., & Hammond, B. G. (2015). The food and environmental safety of Bt crops. *Frontiers in Plant Science*, 6, 283.

Koziel, M. G., Beland, G. L., Bowman, C., Carozzi, N. B., Crenshaw, R., Crossland, L., ... Evola, S. V. (1993). Field performance of elite transgenic maize plants expressing an insecticidal protein derived from *Bacillus thuringiensis*. *Bio/technology*, 11(2), 194–200.

Losey, J. E., Rayor, L. S., & Carter, M. E. (1999). Transgenic pollen harms Monarch larvae. *Nature*, 399(6733), 214–214.

Mendelsohn, M., Kough, J., Vaituzis, Z., & Matthews, K. (2003). Are Bt crops safe? *Nature Biotechnology*, 21(9), 1003–1009.

Negrotto, D., Jolley, M., Beer, S. R. W. A., Wenck, A. R., & Hansen, G. (2000). The use of phosphomannose-isomerase as a selectable marker to recover transgenic maize plants (*Zea mays* L.) via Agrobacterium transformation. *Plant Cell Reports*, 19 (8), 798–803.

Parisi, C., Tillie, P., & Rodríguez-Cerezo, E. (2016). The global pipeline of GM crops out to 2020. *Nature Biotechnology*, 34(1), 31–36.

Petolino, J. F., Hopkins, N. L., Kosegi, B. D., & Skokut, M. (2000). Whisker-mediated transformation of embryogenic callus of maize. *Plant Cell Reports*, 19(8), 781–786.

Ralley, L., Schuch, W., Fraser, P. D., & Bramley, P. M. (2016). Genetic modification of tomato with the tobacco lycopene β-cyclase gene produces high β-carotene and lycopene fruit. *Zeitschrift für Naturforschung* C, 71(9–10), 295–301.

Ray, D. K., Mueller, N. D., West, P. C., & Foley, J. A. (2013). Yield trends are insufficient to double global crop production by 2050. *PLoS One*, 8(6), e66428.

Roh, J. Y., Choi, J. Y., Li, M. S., Jin, B. R., & Je, Y. H. (2007). *Bacillus thuringiensis* as a specific, safe, and effective tool for insect pest control. *Journal of Microbiology and Biotechnology*, 17(4), 547–559.

Sears, M. K., Hellmich, R. L., Stanley-Horn, D. E., Oberhauser, K. S., Pleasants, J. M., Mattila, H. R., ... Dively, G. P. (2001). Impact of Bt corn pollen on Monarch butterfly populations: A risk assessment. *Proceedings of the National Academy of Sciences*, 98(21), 11937–11942.

Southgate, E. M., Davey, M. R., Power, J. B., & Westcott, R. J. (1998). A comparison of methods for direct gene transfer into maize (*Zea mays* L.). *In Vitro Cellular & Developmental Biology – Plant*, 34(3), 218–224.

Svitashev, S., Schwartz, C., Lenderts, B., Young, J. K., & Mark Cigan, A. (2016). Genome editing in maize directed by CRISPR–Cas9 ribonucleoprotein complexes. *Nature Communications*, 7(1), 1–7.

Thomas, S., Burke, D., Gale, M., Lipton, M., Weale, A. et al. (2003). The use of genetically modified crops in developing countries. A follow-up Discussion Paper, Nuffield Council on Bioethics, London.

Torney, F., Moeller, L., Scarpa, A., & Wang, K. (2007). Genetic engineering approaches to improve bioethanol production from maize. *Current Opinion in Biotechnology*, 18(3), 193–199.

Trick, H. N., & Finer, J. J. (1997). SAAT: Sonication-assisted Agrobacterium-mediated transformation. *Transgenic Research*, 6(5), 329–336.

USDA (2020, July). Recent trends in GE adoption. USDA, Economic Research Service. https://www.ers.usda.gov/data-products/adoption-of-genetically-engineered-crops-in-the-us/recent-trends-in-ge-adoption/

World Health Organization (1999). *Microbial Pest Control Agent: Bacillus Thuringiensis*. Environmental Health Criteria 217. https://apps.who.int/iris/bitstream/handle/10665/42242/WHO_EHC_217.pdf?sequence=1&isAllowed=y

Wu, F., & Guclu, H. (2013). Global maize trade and food security: Implications from a social network model. *Risk Analysis*, 33(12), 2168–2178.

Yadava, P., Abhishek, A., Singh, R., Singh, I., Kaul, T., Pattanayak, A., & Agrawal, P. K. (2017). Advances in maize transformation technologies and development of transgenic maize. *Frontiers in Plant Science*, 7, 1949.

Yassitepe, J. E. d C. T., da Silva, V. C. H., Hernandes-Lopes, J., Dante, R. A., Gerhardt, I. R., Fernandes, F. R., … Arruda, P. (2021). Maize transformation: From plant material to the release of genetically modified and edited varieties. *Frontiers in Plant Science*, 12, 766702. doi:10.3389/fpls.2021.766702.

Zhang, C., Wohlhueter, R., & Zhang, H. (2016). Genetically modified foods: A critical review of their promise and problems. *Food Science and Human Wellness*, 5(3), 116–123.

Zou, S., Lang, T., Liu, X., Huang, K., & He, X. (2018). Safety evaluation of genetically modified DAS-40278–9 maize in a subchronic rodent feeding study. *Regulatory Toxicology and Pharmacology*, 96, 146–152.

7 Bt brinjal

Amrit Sudershan, Shivali Panjgotra, Priya Sharma and Vishal Sharma

7.1 Introduction

Biotechnology appears to be a buzzword these days and is heard everywhere, from classrooms to cafeterias, and widely found in journals, newspapers, magazines, and numerous media venues. Karl Erkey, a Hungarian engineer, introduced the phrase "biotechnology" for the first time in 1919 and during the last two decades, the term "biotechnology" has gained great prominence and significance. Biotechnology is described as "the application of engineering and biological science concepts to the creation of new products from biological raw materials, such as vaccines or food". In other words, it is the "use of living organisms or their products to improve or enhance health and the human environment" (Verma et al., 2011).

There are a lot of science fiction (sci-fi) movies such as "Rampage," "Rise of the Planet of the Apes," "Resident Evil," "Elysium," and many others that if you haven't seen you should. All of these movies are focused on altering DNA's coded information to create a better organism/life/human that is much more competent/superior than their parents. In this science fiction, Frankenstein, a doctor and well-known research scientist created a human life that turned into a monster, and this monster is ultimately responsible for Frankenstein's death.

For about 10,000 years, humans have been interested in changing crops, animals, and microorganisms to produce desirable traits or characteristics which can be used for betterment of life and/or food enhancement. In modern times the world's scientific community has practically all of the fundamental instruments it needs for their applications, and most of the key ideas have been clarified, paving the way for big scientific breakthroughs. Trait evolution and natural selection have resulted in competent species that are genetically diverse from their ancestors. This has inspired scientists and technologists to design cutting-edge genetic modification (GM) methods for identifying, selecting, and evaluating GM organisms (Key et al., 2018; National Research Council, 2004).

Some strategies for introducing variety into organisms/cells include somatic hybridization, soma-clonal variation, mutant breeding, cell selection,

DOI: 10.4324/9781003278566-7

interspecies crossing, embryo rescue, and genetic engineering. An incredible example of recombinant DNA (rDNA) technology is genetic engineering, which is a type of genetic modification that includes a purposefully targeted gene modification in a plant or animal by means of a bacteria gene sequence to accomplish a specific result. There are several methods for inserting genes into cells, including electroporation, microinjection, transposable elements, microprojectile bombardment, and microbial vectors (*agrobacterium mediated gene transfer*) (National Research Council, 2004). All of these genetic changes are significant because they result in a superior, more evolved, and capable life that humans may utilize for the advancement and development of the human race. We have been successful in modifying the gene sequence of diverse plant species resulting in increased food production, reliability, yields, better flavor, and nutritional content, as well as reduced losses due to biotic and abiotic issues such as fungal and bacterial infections.

Converging the scatterings of the narrative to a specific example, a commonly used vegetable which is popularly knowns as brinjal or eggplant (*Solanum melongena* L.), also known by various names such as aubergine or guinea, is a non-tuberous, agronomically and economically important plant species. *Solanum* is a large and species-rich genus belonging to the *Solanaceae* family consisting of over 1400 species in which brinjal eggplant belongs. It is considered to be the second-most important solanaceous fruit species after tomato (S. *Lycopersicum* L.) and belongs to the Leptostemonum Clade (the "spiny" solanums) (Knapp et al., 2013). Eggplants are considered an Old-World crop that was domesticated in Africa, Asia, and Europe and is the fifth most commercially significant vegetable in the *Solanaceae* family after tomato, potato, pepper, and tobacco. Among the *Solanaceae* family, eggplant is ranked third in harvested area and production after tomato and potato (Oladosu et al., 2021).

7.1.1 Statistics of brinjal eggplants

In many regions of the world, eggplant is an extensively grown vegetable; notably in Africa, the subtropics countries such as India, Bangladesh, Central America, the Middle East, and Southeast Asia (Taher et al., 2017). The southern United States and the Mediterranean are some warm temperate climate countries where eggplant is also grown. According to 2014 estimates, global eggplant production was estimated to be about 50 million tons annually, with a net value of over US$10 billion. Cultivation of Bt brinjal helps farmers in saving up to US$115 per hectare as an input cost due to the minimum use of insecticides as compared to non Bt brinjal. The cost of insecticide use could also be reduced up to 61 percent as compared to conventional varieties (Paul et al., 2022). Brinjal production in various countries is shown in Figure 7.1.The main producer is China with an average of around 35.5 million tons which represents 57 percent of total global production, and India (12.6 million; 27 percent of global total), followed by Egypt (1.2 million tons), Turkey (0.8 million tons), Iran (0.75 million tons),

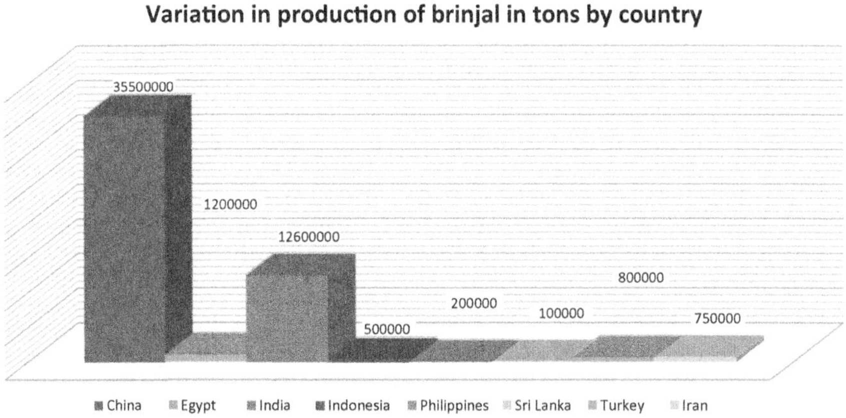

Figure 7.1 Graph showing the variation in brinjal production by different country in millions of tons, 2016.
Source: Authors.

Indonesia (0.5 million tons), the Philippines (0.2 million tons), and Sri Lanka (0.1 million tons). In Mexico, eggplant production was 172,112 tons in 2016 (Taher et al., 2017).

As shown in Figure 7.2, 2019–2020 brinjal production in India was estimated at around 12.777 million tons in an area of around 736,000 hectares Figure 7.2. (Horticulture Statistics Division).

In India's neighboring country, Bangladesh, eggplant is the second most significant crop. Here, in 2018 around 150,000 resource-poor farmers cultivated 50,955 hectares of land and produced 507,000 metric tons of brinjal (Bangladesh Bureau of Statistics (BBS), 2018). The same study also states that for 2018 brinjal represented 4.7 percent of all winter vegetable production and 9.6 percent of all summer vegetable production (Bangladesh Bureau of Statistics (BBS), 2018).

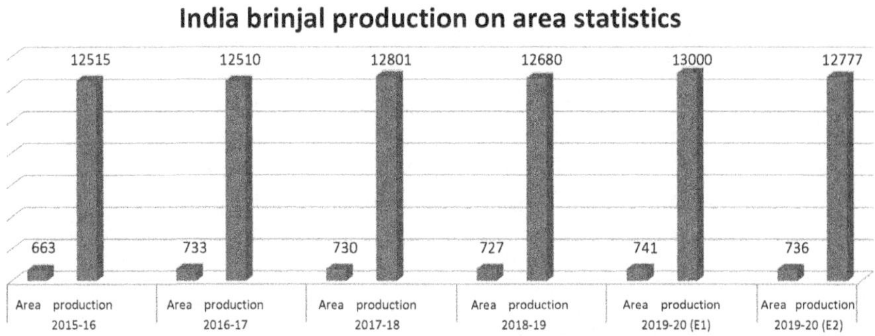

Figure 7.2 Bar graph depicting total brinjal production in India over time.
Source: Authors.

7.1.2 Nutritional value of brinjal eggplants

Brinjal's widespread acceptance and consumption are due to many factors including its ability to be cultivated practically year-round, its affordable cost, and, most importantly, its nutritional benefit. The nutritional value of eggplant per 100 g comprises 0.3 percent minerals, 0.3 percent fat, 1.3 percent fiber, 1.4 percent protein, 4 percent of various vitamins and carbohydrates, and 92.7 percent moisture. According to the US Department of Agriculture it is a good source of potassium, phosphate, calcium, iron, and vitamin B (Nwanna et al., 2019).

Phytochemical analysis of the fruits and leaves of eggplant has revealed that it is a fruit with a high moisture content and low in calories and is a rich source of various essential compounds such as aspartic acid, steroid alkaloids, oxalic acid, glycoalkaloids, flavonoids, lanosterol, tropane, histidine, ascorbic acid, solasodine, nasunin, gramisterol, and tryptophan (Naeem et al., 2020). In addition, phenolic-enriched eggplant extracts with mild oxygen radical scavenging-linked antioxidant activity showed strong alpha-glucosidase inhibitory activity and, in certain cases, moderate to high angiotensin I-converting enzyme (ACE) inhibitory activity. Inhibiting these enzymes provides a strong biological basis for type 2 diabetes therapy by regulating glucose absorption and lowering related hypertension (Kwon et al., 2008).

Furthermore, eggplant has potential therapeutic applications that are only now being identified. It has been discovered that eggplant has potential pharmacological applications, including its powerful cardioprotective molecules, which have been shown to boost left ventricular function, reduce myocardial infarct size, and reduce cardiomyocyte apoptosis. In experiments, the antioxidant vitamins, including vitamin A, vitamin C, and beta carotene, were reduced in grilled eggplants, while some polyphenolic compounds, especially nasunin concentration, were increased, but they were unable to exhibit improved cardioprotective properties when compared to raw fruit (Das et al., 2011).

The fiber in eggplant supports proper digestion, which aids in the removal of waste items from the body (Dhingra et al., 2012). In type 2 diabetes, eggplant may reduce hyperglycemia, hypertension, and oxidative stress, with *Solanum Kumba* (African eggplant) having the maximum activity (Nwanna et al., 2016). Sukprasansap and colleagues found that eggplant extracts inhibited DNA human lymphocytes in response to hydrogen peroxide, with the ML (Solanum aculeatissimum *Jacq*. 'Ma-khuea-lueang') extract exhibiting a larger proportion of suppression than other samples, inferring that eggplants are safe to consume and their compounds could guard against DNA damage (Sukprasansap et al., 2019). Nwanna and colleagues discovered that those who consumed a supplemented brinjal diet had a substantial reduction in lipid profile, lowered leakages of liver and kidney function enzymes, and restored depleted endogenous antioxidant enzymes (Nwanna et al., 2018). Nishimura and colleagues provided the first proof of eggplant's BP (blood pressure) and PS (psychological state) improving effects in people. The functional component responsible for the effects was identified as Ach – an eggplant-derived choline ester (acetylcholine) (Nishimura et al., 2019).

To summarise, eggplant has many potential therapeutic applications such as in type 2 diabetes therapy, a cardioprotective function, lessening of blood pressure/hypertension, improved psychological state, as a guard against DNA damage, and a reduction in hyperglycemia, etc.

7.2 Why is genetic modification required?

Humans are the most superior organisms on this lovely life-supporting medium i.e., Earth, and they always strive to be better and more widely recognized than other species. To achieve supremacy and excellence, humankind is dependent on a multitude of factors, with agriculture being one of the most notable examples. Agriculture has had a role in both the growth and demise of civilizations. Neolithic towns grew rapidly and excessively because of their dependence upon agriculture, yet when natural calamities happened, the food supply and population collapsed. Crop failure resulted in the deaths of up to 60 percent of the population in some cases. Historical examples of such societal collapse have happened in a variety of locations and across a range of time periods (Cumming & Peterson, 2017; Downey et al., 2016). The present environmental catastrophe has sparked interest in social collapse among researchers. Overpopulation and overconsumption are widely acknowledged as the primary dangers to environmental protection and sustainability (Paz et al., 2020).

Many factors can negatively affect agricultural growth and subsequently human life. These include, among others, major crop destruction by pests, larval feeding leading to leaf damage, weed infestation inhibiting growth etc. To address such factors, an integrated control approach was developed in 1959 to acknowledge the numerous ecological and practical benefits of combining biological and chemical pest control strategies. IPM (Integrated Pest Management), a cornerstone of Integrated Production (IP), is described as a crop-protection approach that meets the demands of sustainable growth by allowing farmers to control pests in a cost-effective, environmentally friendly, and regionally suitable manner.

The foundation of an IPM approach is based on the use of indirect or preventive crop protection practices, such as choosing suitable agricultural varieties for the geographical area, understanding the surroundings, soil, nutrient, and water management, knowing natural enemy biology, and using sustainable pest suppression practices, as well as the implementation of methods that foster abundance, the use of optimized farming practices to manage pests, and encouraging the diversity of beneficial species (Dara, 2019). The globe, as well as the majority of regions and countries, is undergoing unparalleled demographic transformation and has experienced a massive increase in human population since 1950: four billion people have been added. Therefore, an integrated strategy to control pests will become increasingly crucial for food security, agricultural sustainability, and environmental protection (Bongaarts, 2009).

A mixture of materials or chemicals usually known as "pesticides" (from 'pest' – any animal or plant harmful to humans or human concerns and *cides* – a Latin word which means "killer") are used to protect plants against pests, weeds, or infections, as well as humans from vector-borne diseases including malaria, dengue fever, and schistosomiasis. Typical examples of pesticides include insecticides, fungicides, herbicides, rodenticides, and various chemicals with plant growth properties, regulators, etc. The primary benefit of pesticides is that they lead to a direct gain in growth. There are also various less immediate or less obvious – secondary – benefits that have been identified, ranging from fitter/healthier people to conserving biodiversity (Aktar et al., 2009). The use of pesticides has been the key source of improved yields and better crop quality. However, in the 1970s concerns were raised about overuse of chemical pesticides having major effects on human health and the environment (Dhawan & Peshin, 2009; Nicolopoulou-Stamati et al., 2016). Exposure to pesticides can be through direct contact with the skin, direct ingestion, or inhalation (Damalas & Koutroubas, 2016) and this may be linked to several detrimental health impacts, including dermatological (Śpiewak, 2001), gastrointestinal (Yuan et al., 2019), neurological, carcinogenic (Bassil et al., 2007), respiratory (such as respiratory symptoms, asthma, and chronic bronchitis) (Mamane et al., 2015), reproductive (Bretveld et al., 2006; Frazier, 2007), and endocrine effects (Mnif et al., 2011). The kind of pesticide, the length and method of exposure, as well as the individual's health state (e.g., nutritional inadequacies and healthy/damaged skin) all have a role in the potential health consequences. Pesticides can also be digested, expelled, stored, or bioaccumulated in the body fat of humans and animals (Nicolopoulou-Stamati et al., 2016).

There are several reasons why we need GE crops including the increasing population which is demanding a higher quality and greater quantity of food. During the last 50 years farming has enhanced the world's capacity to supply its people by increasing output, diversifying foods, and reducing seasonal dependency. Food availability has also increased as income levels have risen and food prices have fallen. This has resulted in significant shifts in food consumption patterns (Kearney, 2010).

Genetically Engineered (GE) plants that can provide insect and disease resistance or herbicide tolerance are valuable additions to varied IPM strategies. GE crops, however, have hurdles in terms of successful introduction and long-term usage, despite the benefits they may provide in terms of simplifying the approach and boosting the effectiveness of insect and weed management (Anderson et al., 2019).

GE crops have been cultivated on a large scale since 1996, with global acreage topping 190 million hectares in 2016. Defense against numerous agricultural detrimental variables, such as lepidopterans, coleopteran pest growth suppression, and herbicide resistance (e.g., glyphosate, glufosinate-ammonium, dicamba), or a combination of these features, are all important characteristics of GE crops. Different examples of GE crops with herbicide tolerance (HT) characteristics include cotton, sugar beet, soybean, maize, canola, and alfalfa. This HT phenotype/character is mostly provided by an insecticidal protein produced from *Bacillus thuringiensis* (Bt), and it is seen in cotton, soybean (lepidopteran pests), and maize (lepidopteran and coleopteran pests). This recipient plant defense characteristic is a cornerstone of IPM and a powerful tool when used in conjunction with other pest management strategies, whether it is developed through traditional breeding or genetic engineering.

The brinjal crop is severely damaged by the brinjal fruit and shoot borer (BFSB) (*Leucinodes orbonalis* Guenee) (Lepidoptera: Crambidae) as shown in Figure 7.3. BFSB is a major problem because *L. orbonalis* larvae eat the leaf and flower of the brinjal plant, causing the fruit to be deformed and bloom to drop; and, due to their great reproductive potential leading to fast generational turnover, these larvae can extensively damage crops in both wet and dry seasons. Fruit is damaged by caterpillars that bore into the petiole and midrib of leaves and delicate shoots, causing withering and stem desiccation. However, the most major economic harm produced by BFSB is to the fruit, which may become unmarketable and unsuitable for human eating due to the holes, feeding tunnels, and larval waste (Prodhan et al., 2018). To manage BFSB, farmers rely on pesticide sprays, which they can apply up to 84 times during the growing season (Prodhan et al., 2018); a practice that is unhealthy and that severely impacts consumers, farmers, and the environment in addition to being expensive for farmers.

Figure 7.3 (a) Hole created in the brinjal fruit by the larvae of the *Leucinodes orbonalis*; (b) Larvae of *Leucinodes orbonalis* feeds on the leaf; (c) Vertical section represents the presence of larvae of the *Leucinodes orbonalis* inside the fruit of brinjal; (d) *Leucinodes orbonalis* larva on the leaf of the brinjal plant.

Source: Authors.

In India, the Maharashtra Hybrid Seed Company (Mahyco) took the initiative to prevent BFSB feeding damage with the development of a GE crop. This GE crop was formed by the introduction of the Cry1Ac gene which is obtained from *Bacillus thuringienesis* (Bt) into the eggplant genome (termed 'event' EE-1) whose expression is under the control of the constitutive 35S CaMV promoter. This initiative was taken to bypass extensive pesticide application. Mahyco, Cornell University, the United States Agency for International Development (USAID), and public sector partners in India, Bangladesh, and the Philippines formed a collaboration in 2003 under the Agricultural Biotechnology Support Project II.

In Bangladesh, the Bangladesh Agricultural Research Institute (BARI) takes the responsibility for integrating EE-1 events into local varieties of brinjal. BARI then requests approval to distribute Bt eggplant from the National Technical Committee on Crop Biotechnology (NTCCB). The proposal for Bt crop release is then sent to the National Technical Committee on Crop Biotechnology (NTCCB) Core Committee, which subsequently refers it to the National Committee on Biosafety based on the NTCCB's recommendation.

On October 30, 2013, the Bangladesh government approved the distribution of four varieties of brinjal, and on 22 January 2014, Bt seedlings were distributed to 20 farmers in four districts, making Bangladesh a global pioneer for allowing the commercial cultivation of a GE vegetable crop (Shelton et al., 2017). To expand the spectrum of GE Bt crops, additional herbicide-tolerant features that confer resistance to other herbicidal active components (e.g., isoxaflutole) and other insect active traits (using RNAi, other non-Bt insecticidal proteins, etc.) are being developed (ISAAA, 2019).

7.3 Mechanism of genetic modification leading to the desired trait

Despite fewer farmers, limited land availability, and other technological barriers, in the twenty-first century – a time of freedom of scientific growth – agriculture feeds and fuels billions of people. Humans have always found ways to improve agricultural skills by building on prior knowledge, and it is these advancements that have led to increased output and availability (Turnbull et al., 2021).

A lot of work has been carried out on various organisms by changing their DNA sequences via different methods and creating phenotypically different and well-developed adaptations. For example, to generate the desired traits needed for food enhancement, humankind has been interested in modifying plants, animals, and microorganisms for almost 10,000 years. These goals continue to drive biologists and food scientists to develop cutting-edge genetic modification tools for discovering, selecting, and evaluating genetically altered species (Key et al., 2008). Some strategies for adding variety into organisms/cells include crossing, interspecies crossover, somatic hybridization, somaclonal variation, and genetic engineering.

Genetic engineering is the application of recombinant DNA (rDNA) technology and is a type of genetic modification that includes a purposefully targeted gene modification in a plant or animal genetic sequence to obtain a specific result. There are several methods for inserting genes into cells, including electroporation, microinjection, transposable elements, microprojectile bombardment, and microbial vectors (*agrobacterium mediated gene transfer*). Trait evolution and natural selection have resulted in common food species that are genetically diverse from their progenitors. There are a lot of advantages of these genetic alterations, including higher food production, greater dependability, increased yields, increased flavor, and better nutritional content, and reduced losses from biotic and abiotic issues such as fungal and bacterial infections.

7.3.1 Gene construction, agrobacterium strain culture

Agrobacterium tumefaciens (NCBI genome ID: 177) is used as a vector to create transgenic plants. It is a most suitable vector for gene transfer by natural methods, is known as "nature's genetic engineer", and has become the major, widely accepted vector in plant genetic engineering (Nester, 2015). *Agrobacterium tumefaciens* possesses a binary vector that contains the DNA of interest which we want to insert into the genome of the organism/plant of interest. The vector is characterized by the presence of a selection marker e.g. neomycin phosphotransferase (nptII) whose expression is under the control of a NOS (nopaline synthase) promoter (Yenofsky et al., 1990). The entire element is then fused with the Cry1Ac gene which is tightly regulated by the cauliflower mosaic virus 35S (CaMV35S) constitutive upstream promoter element (Li et al., 2001).

First, however, we need *Agrobacterium tumefaciens*, and we get it from freshly grown bacteria culture inoculated in a culture medium known as the Luria-Bertani medium which is characterized by the presence of antibiotics namely kanamycin and rifampicin, and grown overnight at 28°C on a rotary shaker.

The resulting *A. tumefaciens* containing medium is then centrifuged at room temperature for the harvesting process and the cells are resuspended in 20 ml MS (Murashige and Skoog's) medium – or the inoculation medium.

7.3.2 Culture condition and media

As mentioned above, the primary culture medium which is used after harvesting for the growth of our vector (i.e. *Agrobacterium tumefaciens*) is known as Murashige and Skoog's medium (MS medium) and consists of a basal salt with added sucrose (glucose source). The factors which govern the bacterial growth are shown in Figure 7.4. Half-strength MS salt was employed for aseptic seed germination, and the medium was autoclaved (121°C for 20 min). In all experiments, the cultures were incubated at 25°C with a 16-hour photoperiod given by cold white fluorescent light, which generated a photosynthetic proton flow.

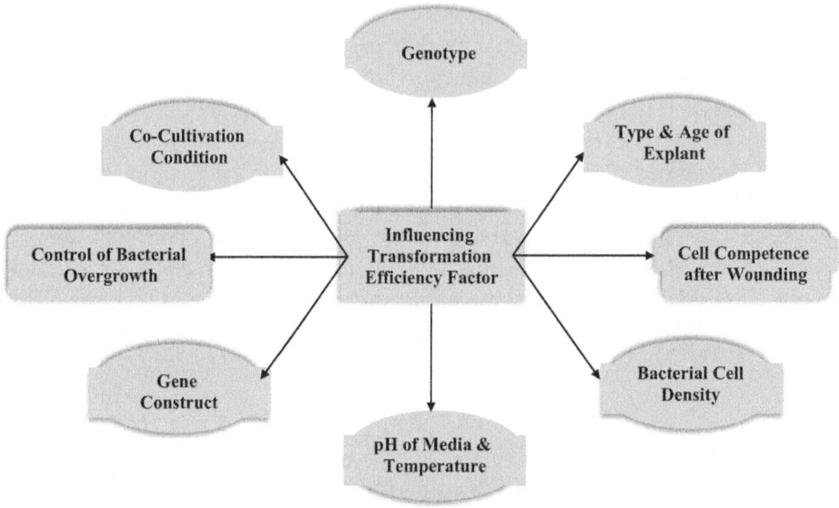

Figure 7.4 Flow diagram depicting factors governing bacterial growth.
Source: Authors.

The selective regeneration medium was made composed of an MS medium supplemented with 6-benzyladenine, kinetin, and kanamycin (SRM). Cefotaxime was added to the MS culture medium to stop the development of the agrobacterium cell.

In a rooting medium (RM) comprised of growth regulators, free MS basal salts, kanamycin, and cefotaxime, roots were induced from in-vitro developed shoots. Rooted shoots were then acclimatized and watered with sterile water in a sterilized soil and sand combination.

7.3.3 Donor material and explant preparation

The complete mechanism followed during the genetic modification in brinjal is shown in Figure 7.5. First, running water is used to clean the fruit and then it is sterilized with ethanol and sliced inside the laminar airflow. After this, brinjal seeds are removed from the mature fruit to form hypocotyl explants. These seeds undergo surface sterilization by placing them in a beaker containing sodium hypo-chlorite (NaOCl) for a few minutes before rinsing them with sterile double distilled water. After the seed sterilization, a half-strength MS basal medium is required for the seed inoculation and then they are incubated in the dark for the initial 3 days and thereafter transferred to the light.

7.3.4 Transformation and plant regeneration

Both the tip and the base of hypocotyls were cut and then placed in a petri dish with pH5.8 pre-culture and co-culture media (PCM). Explants from the

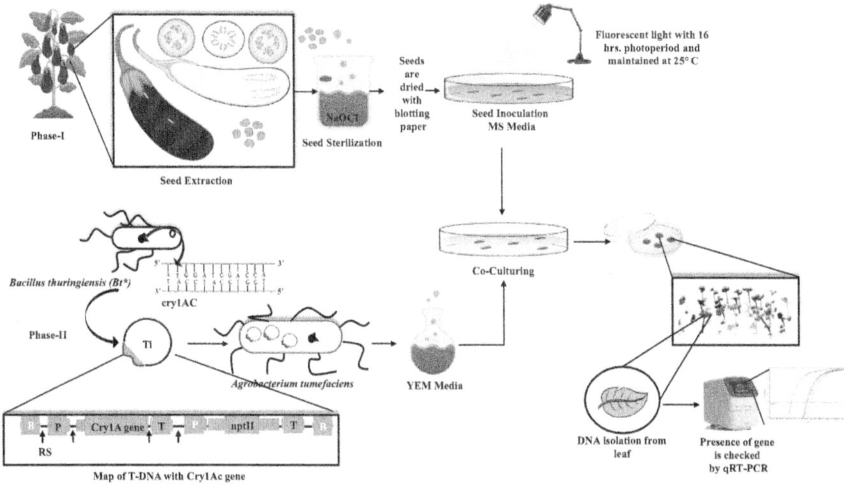

Figure 7.5 Mechanism of genetic modification in brinjal leading to the desired trait. Source: Authors.

PCM plates were inserted in the Agrobacterium suspension for 10 minutes after the pre-culture phase and cultivated at 26°C in the dark. The most essential thing to remember during following the transformation is to provide "control" by using IM (Inoculating Media) without agrobacterium cells.

The next step is to remove the excess bacterial suspension after inoculation, and the agrobacterium inoculated explants were reintroduced to the same PCM plates for co-cultivation. Before being cefotaxime-treated, rewashed, and blotted dry, it is necessary to wash the explant from the co-cultivation medium with moderate shaking in a washing medium i.e., MS salt alone. After washing, the explants were cultured in an SRM medium with an antibiotic solution such as kanamycin and cefotaxime. Every 14 days, hypocotyls with regenerating shoot buds were sub-cultured onto a new SRM-containing culture plate. Regenerated shoots were then removed from putatively altered explants and cultured on RM (Rooting Medium) for 30 days.

7.3.5 PCR analysis

The presence of a transgene in the newly transformed shoot was confirmed by PCR amplification of the Cry and nptII genes with gene-specific primers. To achieve this, first we need to isolate the genomic DNA from the leaf of the transformed plant and also from the wild plant. A sequence-specific primer for the Cry gene and nptII genes is used to amplify the gene, and after specific amplification the amplified product is separated and confirmed using gel electrophoresis (UV transilluminator).

7.3.6 Mechanism of action of Cry protein

Bacillus thuringiensis (Bt) parasporal inclusion bodies contain insecticidal proteins (δ-endotoxins), which are frequently employed as the basis for microbial larvicides. The Cry gene from the *Bacillus thuringiensis* acts as a kind of endotoxin (see Figure 7.6).

The δ-endotoxins cause cytotoxicity by generating lytic lesions in the epithelial membranes of the larval midgut. Protoxins are transformed to toxins by trypsin-like proteases in the midgut, and then the active toxins bind to particular receptors on the apical brush border membrane of columnar cells. The toxin inserts permanently into the cell's plasma membrane after binding to a particular receptor and producing a hole. Toxin-induced holes in the apical membrane's columnar cells allow for rapid ion fluxes, and the pores are K+ selective, allowing cations, anions, or small solutes like sucrose to pass through. The compromise of gut integrity causes the insect to die from hunger or septicemia (Figure 7.6) (Biswas et al., 2018).

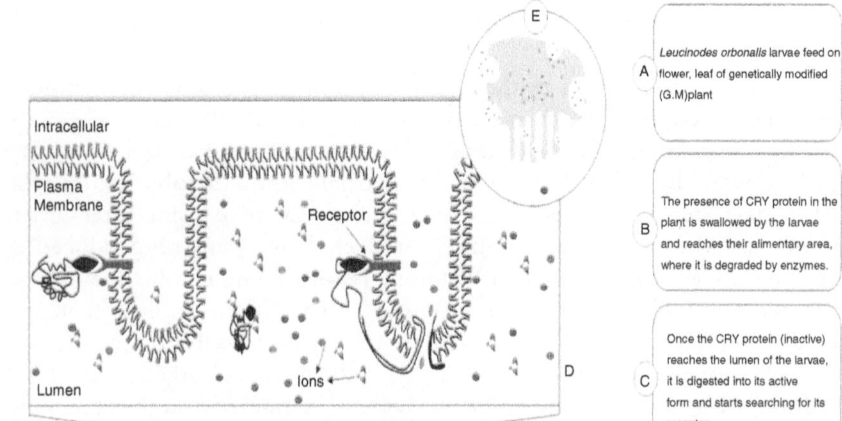

Figure 7.6 Mechanism of the action of Cry protein.
Source: Authors.

7.4 Genes involved in agronomic trait development

The main priority of plant breeders has always been to breed plants with improved agronomic qualities such as yield, uniformity, and resistance. With population expansion, climate change, and soil and irrigation management methods all contributing to a decrease in the quantity and quality of arable land, crop production and disease resistance remain breeding goals. Figure 7.7 illustrates the integration of different members of the Cry family of genes in the production of novel genetically modified crops

7.5 Impact on modern agriculture – trade of GM crops

The Flavr Savr tomato, which had been genetically engineered to halt the ripening process and prevent softening and rotting, was the first GE crop to be released in the USA (1994). Subsequently, since the mid-1990s, there has been a significant surge in the production of GM crops. According to the recent report by the International Service for the Acquisition of Agri-Biotech applications (ISAAA), only 1.7 million hectares (M ha) of GM crops were planted worldwide in 1996; however, by 2015, GM crops had been planted on 179.7 M ha, accounting for more than 10 percent of the globe's agricultural land which further increased to 190.4 M ha in 2019. Around 29 countries were growing these 190 M ha of GM crops in 2019, of which the United States, Brazil, Argentina, Canada, and India accounted for 91 percent of the global GM crop cultivation area. In 2021, about 3 M ha of GM crops were planted in Africa alone (Endale et al., 2022). GM crops grown

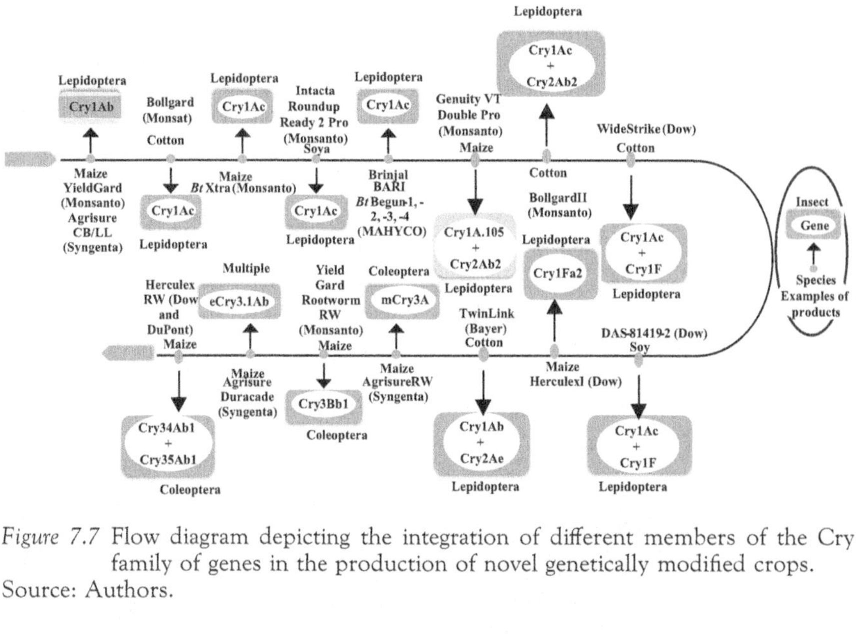

Figure 7.7 Flow diagram depicting the integration of different members of the Cry family of genes in the production of novel genetically modified crops.
Source: Authors.

commercially in the USA include squash/pumpkin, potato, alfalfa, papaya (China also), sugar beet (Canada also), as well as oilseed rape, maize, soya beans (11 countries), and cotton (15 countries).

Soybean (92.1 M ha) was the most widely planted GM crop in 2015, followed by maize (53.6 M ha), cotton (24 M ha), and oilseed rape (canola) (8.5 M ha). This accounts for 83 percent of the world's soybean output and 75 percent of global cotton production. Genetically engineered crops produced 29 percent of the world's maize and nearly a quarter of the world's oilseed rape (James, 2015).

The United States (70.9 M ha), Brazil (44.2 M ha), Argentina (24.5 M ha), India (11.6 M ha), and Canada (11 M ha) are the top users of genetically modified crops. Five European nations – Spain, Portugal, the Czech Republic, Romania, and Slovakia – cultivate GM maize, with Spain the leading country (0.1 M ha). South Africa (2.3 M ha), Burkino Faso (0.4 M ha), and Sudan (0.1 M ha) are among the African countries that cultivate GM crops, with GM cotton being the most common (James, 2015).

Ahmed et al. (2019) evaluated the impacts of the Bt brinjal technology on production systems, marketability, and human health. The study found that farmers experienced significantly lower pesticide use, a reduction in overall production costs, increased yields, and provided higher profits by growing Bt brinjals.

7.6 Benefits to society

GM crops are effective in preventing the spread of some diseases, and they also grow more quickly than conventionally cultivated foods. As a result of their increased output, more food is available to the population. Furthermore, genetically engineered food crops can often be grown in locations where the climate is unfavorable. Genetically modified crops are believed to be richer in micronutrients and contain more vitamins and minerals than traditional crops, as well as have a superior flavor. Another reason consumers like genetically altered crops/foods is because they have a longer shelf life, which means they are less prone to deterioration. A typical crop can only be grown during a given season or under certain climatic circumstances but this is often not the case with GM crops. Thus, even if the seeds for GM crops are relatively expensive, the production costs are reported to be lower than those of conventional crops due to their natural resistance to pests. This reduces the need for GM crops to be subjected to hazardous herbicides and insecticides, resulting in more chemical-free and ecologically friendly meals (Hefferon, 2015; Smyth, 2020; Hirschi, 2020).

7.7 Health and ethical issues/risks

As already mentioned, there are several benefits associated with genetically modified organisms, but as we all know, there are often drawbacks as well as benefits. This controversial subject has many negative aspects which include the effect of GM crops on livestock. Livestock which feed on the genetically modified crops such as Bt maize, GM cottonseed, and certain biotech corn have experienced complications including a negative effect on young sperm, early deliveries, abortions, infertility, and death (Maghari & Ardekani, 2011).

According to recent reports from India, hundreds of villagers and cotton workers developed skin sensitivities after working with GM crops (Dimetry, 2016). According to another study, Bt maize produces an allergenic protein that alters the overall immunological response in the body (Then & Bauer-Panskus, 2017). It is believed, therefore, that consuming these genetically engineered crops could result in the development of antibiotic-resistant infections.

Many people prefer to avoid these GM foods products because their health implications are unknown. Many religious and cultural communities oppose such foods because they believe they are produced unnaturally. Many people also have reservations about transplanting animal DNA into plants and vice versa (Bawa & Anilakumar, 2013). Adequate knowledge regarding GM food safety testing is therefore essential to give the public the option of making an informed decision regarding its consumption. But unfortunately, due to a variety of reasons, such information is scarce. Therefore, all agencies responsible for producing GM crops must thus investigate the safety and long-term impacts of GM crops before their distribution into the food chain.

It is, however, difficult to compare the nutritional contents of GM and traditional crops, for example, because crop composition varies based on the growth and agronomic conditions in different places. Although developments in modification technologies can reduce the time it takes to bring new foods to market, a lengthier evaluation period that raises the possibility of finding adverse features and eliminating harmful new varieties before commercial distribution is necessary. Identifying potentially dangerous products before they become commercially available is both prudent and desirable, and with few exceptions, standard plant breeding processes have proven to be highly effective in this regard.

7.8 Current status and future perspective

The words genomic engineering, genetic manipulation, and gene editing all relate to changes (insertions, deletions, and replacements) introduced to a living organism's genome. Clustered Regularly Interspaced Short Palindromic Repeats and associated protein 9 is the most widely used method of genome editing today (CRISPR-Cas9). CRISPR-Cas9 is a naturally occurring adaptive immune system that protects cells from DNA viral infections. The identification of target genes that produce phenotypes of interest, such as resistance to pesticides/herbicides or negative climate variables, vulnerability to viruses and other diseases, is a vital step in the genome editing process. Guide RNAs are synthetic RNAs that are precisely engineered to direct Cas9 to the target gene to be edited (Knott & Doudna, 2018; El-Mounadi et al., 2020).

7.9 Trade in the global market

The first commercialized GM crops were planted in 1996 and covered approx. 1.7 million hectares of land (ISAAA, 2017). Since then, the use of GM crops has risen exponentially and, in 2017, the global total had risen to almost 190 M Ha, with emerging countries accounting for 53 percent of all GM crop hectares (ISAAA, 2017).

Since the commercialization of GM crops, both the economy and the environment have benefitted. The global food crop yield has increased by > 370 million tonnes over a relatively small acreage area. Also, GM crops have been recorded to reduce environmental and ecological impacts, leading to increases in species diversity (Zang et al. 2016).

Bt crops affect productivity by lowering the quantity of produce lost to pest infestation. Bt maize yielded 0.27 to 0.75 tonnes per hectare more than equivalent non-transgenic lines in the early stages of adoption in the United States in 1998–1999 (Carzoli et al., 2018). In addition, a study of the worldwide economic effects of agricultural biotechnology from 1996 to 2015 revealed a benefit to agricultural income of $167.8 billion, with 49 percent and 51 percent of farmers in developed and developing nations enjoying these economic gains, respectively (Carzoli et al., 2018).Crops can raise farm revenue, reduce poverty, and enhance food security for smallholder farmers, and in India the use of Bt cotton has increased farmer earnings by 50 percent and reduced food insecurity by 15–20 percent (Kathage & Qaim, 2012).

The price of brinjal varies depending on the type of market where it is sold. Bt variations are different from the popular local types, yet Bt brinjal varieties always earn more revenue per hectare than non-Bt brinjal varieties; Bt brinjal variants earning approximately US$3,720 per hectare compared with non-Bt brinjal varieties earning around US$3,056 per hectare (Shelton et al., 2020).

7.10 Conclusion

Biotechnology has got us to the current situation; the question now is, where will it take us? Biotechnology has both positive and negative aspects. The phrase 'biotechnology' was scientifically formed from two simple scientific terms, namely 'Biology' and 'Technology.' When we try to understand these two statements, we find that it means that technology makes our lives easier and more comfortable through the use of biological resources. Yet, biotechnological principles, in addition to their beneficial uses, have the potential for devastation; the most prominent example of this is 'bioterrorism'. WE therefore must determine how to use this technology to benefit humankind rather than harm it.

References

Aktar, M. W., Sengupta, D., & Chowdhury, A. (2009). Impact of pesticides use in agriculture: Their benefits and hazards. *Interdisciplinary Toxicology*, 2(1), 1. doi:10.2478/v10102-009-0001-7.

Ahmed, A. U., Hoddinott, J. F., Islam, K. M. S., Khan, A. S. M. M. R., Abedin, N., & Hossain, N. Z. (2019). *Impact of Bt Brinjal (Eggplant) Technology in Bangladesh.* Project Report prepared for the US Agency for International Development (USAID). (Dhaka: International Food Policy Research Institute). Available online at: http://www.ifpri.org/

Anderson, J., Ellsworth, P. C., Faria, J. C., Head, G. P., Owen, M. D., Pilcher, C. D., … Meissle, M. (2019). Genetically engineered crops: Importance of diversified integrated pest management for agricultural sustainability. *Frontiers in Bioengineering and Biotechnology*, 7, 24. doi:10.3389/fbioe.2019.00024.

Bangladesh Bureau of Statistics (BBS) (2018). *Bangladesh Agricultural Statistics Yearbook*. Government of Bangladesh. Available online at: https://databd.co/resources/agricultural-statistics-yearbook

Bassil, K. L., Vakil, C., Sanborn, M., Cole, D. C., Kaur, J. S., & Kerr, K. J. (2007). Cancer health effects of pesticides: Systematic review. *Canadian Family Physician*, 53(10), 1704–1711.

Bawa, A. S., & Anilakumar, K. R. (2013). Genetically modified foods: Safety, risks and public concerns – a review. *Journal of Food Science and Technology*, 50(6), 1035–1046.

Biswas, R., Banerjee, A., Halder, U., & Bandopadhyay, R. (2018). Transgenic research in vegetable crops with special reference to brinjal. In *Genetic Engineering of Horticultural Crops*, edited by G. R. Rout & K. V. Peter (London: Academic Press), 155–167.

Bongaarts, J. (2009). Human population growth and the demographic transition. *Philosophical Transactions of the Royal Society of London. Series B, Biological Sciences*, 364 (1532), 2985–2990.

Bretveld, R. W., Thomas, C. M., Scheepers, P. T., Zielhuis, G. A., & Roeleveld, N. (2006). Pesticide exposure: The hormonal function of the female reproductive system disrupted? *Reproductive Biology and Endocrinology*, 4(1), 1–14.

Carzoli, A. K., Aboobucker, S. I., Sandall, L. L., Lübberstedt, T. T., & Suza, W. P. (2018). Risks and opportunities of GM crops: Bt maize example. *Global Food Security*, 19, 84–91.

Cumming, G. S., & Peterson, G. D. (2017). Unifying research on social-ecological resilience and collapse. *Trends in Ecology & Evolution*, 32(9), 695–713.

Damalas, C. A., & Koutroubas, S. D. (2016). Farmers' exposure to pesticides: Toxicity types and ways of prevention. *Toxics*, 4(1), 1. doi:10.3390/toxics4010001.

Dara, S. K. (2019). The new integrated pest management paradigm for the modern age. *Journal of Integrated Pest Management*, 10(1), 12. doi:10.1093/jipm/pmz010.

Das, S., Raychaudhuri, U., Falchi, M., Bertelli, A., Braga, P. C., & Das, D. K. (2011). Cardioprotective properties of raw and cooked eggplant (*Solanum melongena* L). *Food & Function*, 2(7), 395–399.

Dhawan, A. K., & Peshin, R. (2009). Integrated pest management: Concept, opportunities and challenges. In *Integrated Pest Management: Innovation-Development Process*, edited by R. Peshin & A. K. Dhawan (Dordrecht: Springer), 51–81.

Dhingra, D., Michael, M., Rajput, H., & Patil, R. T. (2012). Dietary fibre in foods: A review. *Journal of Food Science and Technology*, 49(3), 255–266.

Dimetry, N. Z. (2016). Are genetically engineered crops safe or dangerous? *American Academic Scientific Research Journal for Engineering, Technology, and Sciences*, 20(1), 157–178.

Downey, S. S., Haas, W. R., Jr, & Shennan, S. J. (2016). European Neolithic societies showed early warning signals of population collapse. *Proceedings of the National Academy of Sciences of the United States of America*, 113(35), 9751–9756.

El-Mounadi, K., Morales-Floriano, M. L., & Garcia-Ruiz, H. (2020). Principles, applications, and biosafety of plant genome editing using CRISPR-Cas9. *Frontiers in Plant Science*, 11, 56. doi:10.3389/fpls.2020.00056.

Endale, G. K., Karim, M., Joseph, G., & Muffy, K. (2022). Commercialization of genetically modified crops in Africa: Opportunities and challenges. *African Journal of Biotechnology*, 21(5), 188–197.

Frazier, L. M. (2007). Reproductive disorders associated with pesticide exposure. *Journal of Agromedicine*, 12(1), 27–37.

Department of Agriculture & Farmers Welfare (2018). Horticultural statistics at a glance. Ministry of Agriculture & Farmers Welfare, Government of India. https://agricoop.nic.in/en/statistics/horticulture

Hefferon, K. L. (2015). Nutritionally enhanced food crops: Progress and perspectives. *International Journal of Molecular Sciences*, 16(2), 3895–3914.

Hirschi, K. D. (2020). Genetically modified plants: Nutritious, sustainable, yet underrated. *The Journal of Nutrition*, 150(10), 2628–2634.

ISAAA Briefs (2017). *Global Status of Commercialized Biotech/GM Crops in 2017: Biotech Crop Adoption Surges as Economic Benefits Accumulate in 22 Years*. ISAAA Brief No. 53, 25–26.

ISAAA Briefs (2019). *Global Status of Commercialized Biotech/GM Crops in 2019*. ISAAA Brief No. 55.

James, C. (2015). *Global Status of Commercialized Biotech/GM Crops: 2015*. ISAAA Brief No. 51 (Ithaca, NY: ISAAA).

Kathage, J., & Qaim, M. (2012). Economic impacts and impact dynamics of Bt (*Bacillus thuringiensis*) cotton in India. *Proceedings of the National Academy of Sciences*, 109(29), 11652–11656.

Kearney, J. (2010). Food consumption trends and drivers. *Philosophical Transactions of the Royal Society B: Biological Sciences*, 365(1554), 2793–2807.

Key, S., Ma, J. K., & Drake, P. M. (2008). Genetically modified plants and human health. *Journal of the Royal Society of Medicine*, 101(6), 290–298.

Knapp, S., Vorontsova, M. S., & Prohens, J. (2013). Wild relatives of the eggplant (*Solanum melongena* L.: Solanaceae): New understanding of species names in a complex group. *PLoS One*, 8(2), e57039. doi:10.1371/journal.pone.0057039.

Knott, G. J., & Doudna, J. A. (2018). CRISPR-Cas guides the future of genetic engineering. *Science*, 361(6405), 866–869.

Kwon, Y. I., Apostolidis, E., & Shetty, K. (2008). In vitro studies of eggplant (*Solanum melongena*) phenolics as inhibitors of key enzymes relevant for type 2 diabetes and hypertension. *Bioresource Technology*, 99(8), 2981–2988.

Li, Z., Jayasankar, S., & Gray, D. J. (2001). Expression of a bifunctional green fluorescent protein (GFP) fusion marker under the control of three constitutive promoters and enhanced derivatives in transgenic grape (*Vitis vinifera*). *Plant Science*, 160(5), 877–887.

Maghari, B. M., & Ardekani, A. M. (2011). Genetically modified foods and social concerns. *Avicenna Journal of Medical Biotechnology*, 3(3), 109.

Mamane, A., Baldi, I., Tessier, J. F., Raherison, C., & Bouvier, G. (2015). Occupational exposure to pesticides and respiratory health. *European Respiratory Review*, 24 (136), 306–319.

Mnif, W., Hassine, A. I. H., Bouaziz, A., Bartegi, A., Thomas, O., & Roig, B. (2011). Effect of endocrine disruptor pesticides: A review. *International Journal of Environmental Research and Public Health*, 8(6), 2265–2303.

Naeem, M. Y., & Ugur, S. (2019). Nutritional content and health benefits of eggplant. *Turkish Journal of Agriculture-Food Science and Technology*, 7(sp3), 31–36.

Naeem, M., Basit, A., Ahmad, I., Mohamed, H. I., & Wasila, H. (2020). Effect of salicylic acid and salinity stress on the performance of tomato plants. *Gesunde Pflanzen*, 72(4), 393–402.

National Research Council (2004). *Safety of Genetically Engineered Foods: Approaches to Assessing Unintended Health Effects* (Washington, DC: National Academies Press).

Nester, E. W. (2015). Agrobacterium: Nature's genetic engineer. *Frontiers in Plant Science*, 5, 730. doi:10.3389/fpls.2014.00730.

Nicolopoulou-Stamati, P., Maipas, S., Kotampasi, C., Stamatis, P., & Hens, L. (2016). Chemical pesticides and human health: The urgent need for a new concept in agriculture. *Frontiers in Public Health*, 4, 148. doi:10.3389/fpubh.2016.00148.

Nishimura, M., Suzuki, M., Takahashi, R., Yamaguchi, S., Tsubaki, K., Fujita, T., ... Nakamura, K. (2019). Daily ingestion of eggplant powder improves blood pressure and psychological state in stressed individuals: A randomized placebo-controlled study. *Nutrients*, 11(11), 2797. doi:10.3390/nu11112797.

Nwanna, E. E., Ibukun, E. O., & Oboh, G. (2016). Effect of some tropical eggplant fruits (*Solanum* spp) supplemented diet on diabetic neuropathy in experimental male Wistar rats in-vivo. *Functional Foods in Health and Disease*, 6(10), 661–676.

Nwanna, E. E., Ibukun, E. O., & Oboh, G. (2019). Nutritional content of selected species of tropical eggplant fruit (*Solanum* spp) diet attenuates hepatic inflammation in high-fat fed male Wistar rats induced with streptozotocin. *Food Science & Nutrition*, 7(1), 109–119.

Oladosu, Y., Rafii, M. Y., Arolu, F., Chukwu, S. C., Salisu, M. A., Olaniyan, B. A., ... Muftaudeen, T. K. (2021). Genetic diversity and utilization of cultivated eggplant germplasm in varietal improvement. *Plants*, 10(8), 1714. doi:10.3390/plants10081714.

Paul, S. K., Mahmud, N. U., & Islam, T. (2022). Impacts of Bt brinjal on economic benefit of farmers and environmental sustainability in Bangladesh. In *Bacilli in Agrobiotechnology*, edited by M. T. Islam, M. Rahman, & P. Pandey (Cham: Springer), 539–560.

Paz, D. B., Henderson, K., & Loreau, M. (2020). Agricultural land use and the sustainability of social-ecological systems. *Ecological Modelling*, 437, 109312. doi:10.1016/j.ecolmodel.2020.109312.

Prodhan, M. Z. H., Hasan, M. T., Chowdhury, M. M. I., Alam, M. S., Rahman, M. L., Azad, A. K., ... Shelton, A. M. (2018). Bt eggplant (*Solanum melongena* L.) in Bangladesh: Fruit production and control of eggplant fruit and shoot borer (Leucinodes orbonalis Guenee), effects on non-target arthropods and economic returns. *PLoS One*, 13(11), e0205713. doi:10.1371/journal.pone.0205713.

Rodriguez-Jimenez, J. R., Amaya-Guerra, C. A., Baez-Gonzalez, J. G., Aguilera-Gonzalez, C., Urias-Orona, V., & Nino-Medina, G. (2018). Physicochemical, functional, and nutraceutical properties of eggplant flours obtained by different drying methods. *Molecules*, 23(12), 3210. doi:10.3390/molecules23123210.

Royal Society (n.d.). https://royalsociety.org/topics-policy/projects/gm-plants/what-gm-crops-are-currently-being-grown-and-where/

Shelton, A. M., Hokanson, K. E., Hautea, D. M., Hossain, M. J., Hossain, M. A., Paranjape, … Sarwer, S. H. (2017). Bt eggplant: A genetically engineered 'minor' crop comes of age in Bangladesh and the Philippines. ISB News Report, August 2017. https://vtechworks.lib.vt.edu/bitstream/handle/10919/78874/SHHHHPHMCS.pdf?sequence=1&isAllowed=y

Shelton, A. M., Sarwer, S. H., Hossain, M. J., Brookes, G., & Paranjape, V. (2020). Impact of Bt brinjal cultivation in the market value chain in five districts of Bangladesh. *Frontiers in Bioengineering and Biotechnology*, 8, 498. doi:10.3389/fbioe.2020.00498.

Smyth, S. J. (2020). The human health benefits from GM crops. *Plant Biotechnology Journal*, 18(4), 887. doi:10.1111/pbi.13261.

Śpiewak, R. (2001). Pesticides as a cause of occupational skin diseases in farmers. *Annals of Agricultural and Environmental Medicine*, 8(1), 1–5.

Sukprasansap, M., Sridonpai, P., & Phiboonchaiyanan, P. P. (2019). Eggplant fruits protect against DNA damage and mutations. *Mutation Research/Fundamental and Molecular Mechanisms of Mutagenesis*, 813, 39–45.

Taher, D., Solberg, S. Ø., Prohens, J., Chou, Y. Y., Rakha, M., & Wu, T. H. (2017). World vegetable center eggplant collection: Origin, composition, seed dissemination and utilization in breeding. *Frontiers in Plant Science*, 1484. doi:10.3389/fpls.2017.01484.

Then, C., & Bauer-Panskus, A. (2017). Possible health impacts of Bt toxins and residues from spraying with complementary herbicides in genetically engineered soybeans and risk assessment as performed by the European Food Safety Authority EFSA. *Environmental Sciences Europe*, 29(1), 1–11.

Turnbull, C., Lillemo, M., & Hvoslef-Eide, T. A. (2021). Global regulation of genetically modified crops amid the gene edited crop boom – a review. *Frontiers in Plant Science*, 12, 258. doi:10.3389/fpls.2021.630396.

Verma, A. S., Agrahari, S., Rastogi, S., & Singh, A. (2011). Biotechnology in the realm of history. *Journal of Pharmacy and Bioallied Sciences*, 3(3), 321. doi:10.4103/0975-7406.84430.

Yenofsky, R. L., Fine, M., & Pellow, J. W. (1990). A mutant neomycin phosphotransferase II gene reduces the resistance of transformants to antibiotic selection pressure. *Proceedings of the National Academy of Sciences*, 87(9), 3435–3439.

Yuan, X., Pan, Z., Jin, C., Ni, Y., Fu, Z., & Jin, Y. (2019). Gut microbiota: An underestimated and unintended recipient for pesticide-induced toxicity. *Chemosphere*, 227, 425–434.

Zhang, C., Wohlhueter, R., & Zhang, H. (2016). Genetically modified foods: A critical review of their promise and problems. *Food Science and Human Wellness*, 5(3), 116–123.

8 Genetically engineered potato

Bharti Mittu, Mahaldeep Kaur, Abida Bhat, Jasmeet Kour and Kawaljeet Kour

8.1 Introduction

Potato (*Solanum tuberosum* L.), after rice, wheat, and corn, is the fourth most common crop with respect to yield and the eighth in relation to global cultivation area (FAOSTAT, 2020). Potato farming is mostly carried out for its tuber, which is a major source of energy and meal staple with a high yield per unit area all over the world. As a result, potato has great potential for tackling food shortages. Potato is an annual herbaceous plant belonging to the *Solananceae* family with white to purple flowers and yellow stamens. Potato tubers are underground stems that are high in starch and proteins, and which are morphologically modified stolons carrying scale leaves and auxiliary buds. The carbohydrate content of potato tubers ranges from 1.0 to 7.0 g/kg (Jama-Rodzenska et al., 2021). The content of reducing sugars (glucose and fructose) is higher in young tubers but decreases dramatically towards the end of the growing season. The starch component (the storage form of carbohydrate) in potato dry matter is composed of amylopectin and amylose. In the food sector, starch has traditionally been used to improve the functional qualities of numerous meals. The functional and physiochemical features of the starch differ from each other according to its biological origin and, in potato starch, the amylose-to-amylopectin ratio differs between cultivars. Compared with other food crops, in developing nations it provides many essential biomolecules (carbohydrates, minerals, vitamins, and proteins) per unit area of land. These factors have prompted scientists to focus their efforts on improving the nutritional value of potatoes while preserving yield. Figure 8.1 describes, in diagrammatic form, information related to improvement in the nutrient content in GM potato and the genomic mechanism of its development.

8.2 Why is genetic modification required?

Genetic modification (GM) of animals and plants in order to improve their quality has become an important method in modern times; subsequently leading to better standards of living and succeeding in meeting the

DOI: 10.4324/9781003278566-8

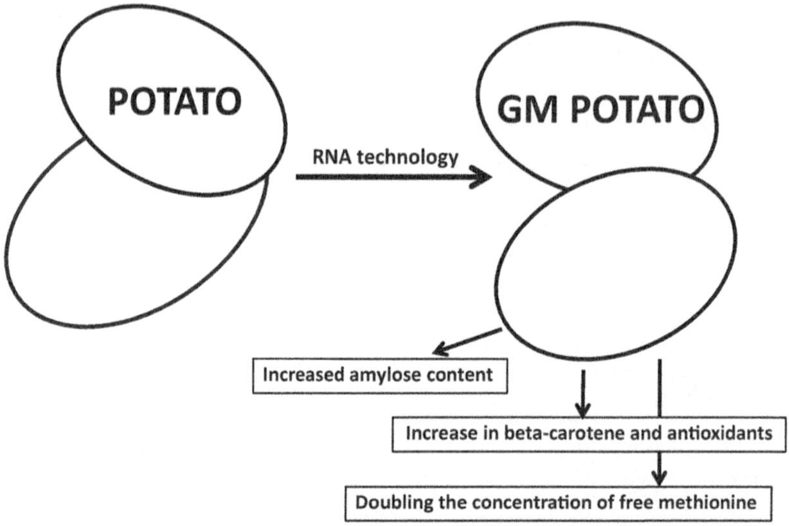

Figure 8.1 Role of molecular biology in enhancing benefits of GM potato.
Source: Authors.

cultivation needs of the population. Plants are the most commonly used higher organisms in genetic modification and potatoes are considered an excellent model plant for GM research. GM plants have proven to have the potential to aid global food security, give increased crop yield, provide increased nutritional value, and be employed in the development of clinically significant pharmaceuticals. Potato plants are now available in a variety of GM species, with increased resistance against biological agents and external environmental factors and with improved nutritional value. In the future, GM plants have the potential to synthesize recombinant vaccines such as plant-derived polymers used in surgery. Modified potatoes that produce biomaterials with potential industrial applications are also a significant category (Karalis et al., 2020; Kumar et al., 2020; Kramkowska et al., 2013; Pribylova et al., 2006). Effective utilization of genetically modified organisms (GMOs) is, however, dependent on minimizing the risks associated with genetic modification by the controlled engineering of genes with a negligible effect on human health.

8.3 Genes involved in agronomic trait development

Conventional/inbreed breeding methods of generating pathogen-resistant cross-hybrid lines with wild species of potato (*S. bulbocastanum*) is very limited and time-consuming (Zhang et al., 2021). Cultivation, quality, and yield of potato farming is drastically affected by phytopathogens: viral (potato leafroll virus, potato virus X, and potato virus Y), bacterial (*Dickeyasolani*, *Streptomyces scabies*, *clavibacter* spp, *pectobacterium* spp), and fungal (*phytophthora infestans*,

Rhizoctonia solani, spongospora subterranean). New breeding technologies (NBTs) including CRISPR/Cas9 (Clustered Regularly Interspaced Short Palindromic Repeats), TALENs (Transcription Activator-Like Effector Nucleases), and Zinc-Finger Nucleases (ZFNs) provide a genome editing mechanism for trait, quality, and yield improvement in potato production (Hameed et al., 2018). Commercial GM potatoes can therefore acquire agronomic traits such as disease-resistance and abiotic stress-tolerance (like thermotolerance) by means of key engineered genes: R-genes (RB/Rpi-blbl, Rpi-vnt1.1), heat-tolerant genes (StSP5G, StGI, StCO, StUGT), and have superior genotypes with combined conventional and NBTs (Li et al., 2018; Singh et al., 2020).

A complete understanding of a gene's biochemical role and function is important for generating the desired agronomic traits in GM potatoes. Experimental studies employing TALENs and CRISPR/Cas9 gene editing tools using Streptococcus SpCas9, tracrRNA (trans-activating crRNA), ceRNA (CRISPR RNA), sgRNAs (single guide RNAs), and GVR (gemini virus replicon) have been shown to generate stable transgenic potato species with DSB (double strand break) mechanisms (Das Dangol et al., 2019). Since the publication of the potato genome sequence by the PGSC in 2011, new technical approaches are being utilized, such as Illumina sequencing and ONT (Oxford Nanopore Technologies), for generating accurate potato genome sequences (DM1–3 516 R44, DMv4.03, DM v4.04, and DM v6.1) and thus further improving the understanding of potato genomics (Kyriakidou et al., 2020; Pham et al., 2020). New potato genomics have helped us in understanding plant N (nitrogen) metabolism by identifying the key genes involved in this process such as NR genes (NRT2.1 (nitrate transporter), AMT1.1 (ammonium transporter)); knowledge that can then be used for generating N-efficient GM potatoes (Tiwari et al., 2018; Getahun et al., 2020).

Quantitative Trait Loci (QTL) is a general method for analyzing the gene control of agronomic traits in vegetative plants including potato. Potato tuber starch content (TSC) is one of the important parameters of its production, and the genes responsible for tuber quality are compelling targets for NBTs (Manrique-Carpintero et al., 2018; Ahmed et al., 2018). Bioactive compounds in potato such as flavonoids (anthocyanins) with anti-oxidant properties have been shown to have a beneficial effect on health and aid against disease; the genome analysis of potato tubers shows seven QTLs (on chromosomes I, II, X, XI, and XIX) associated with it, among which there is a strong signal and stable QTL associated with chromosomes I and X (Parra-Galindo et al., 2019). NBTs generating GM potato with improved tuber quality and stress resistance have also focused on other agronomic traits such as tuber shape, eye depth, and tuber weight. Morphological analysis of QTLs is present for tuber shape (on chromosomes X and II), eye depth (on chromosomes I, III, IV, and V), and tuber weight (on chromosomes I, IV, V, and VI) (Hara-Skrzypiec et al., 2018).

Pathogen resistance is one of the pivotal targets of genetic engineering in GM plants. Colorado potato beetle (CPB), one of the major pests affecting potato cultivation, has encouraged the development of transgenic potato lines (Agria and Lady) incorporated with dsRNA EcR transgenes utilizing a new method (RNAi mechanistic pathway) and giving improved cultivation (Hussain et al., 2019). By using introgressive hybridization to identify genes and their allelic diversity, WR (wild relative) species of potato serve as donors of desirable traits to cultivated potato varieties, which can be then mined using NBTs (Gaiero et al., 2018). Development of diploid inbred potato lines are a promising approach for genetic gain. An experimental method utilizing the Sli (S-locus inhibitor) gene by inbreed hybridization and CRISPR/Cas9 methods has successfully resulted in the generation of self-compatible potato inbreed genotypes (Eggers et al., 2021). Another study has noted that under normal and stress conditions enhanced tuber yield is seen in transgenic potato hybrid plants, showing expression of the Arabidopsis ABF4 gene (35S:ABF4) (Muñiz García et al., 2018). So transgenic potato varieties with induced/controlled expression of targeted genes using different editing methods yield better plant varieties with drought- and pathogenic-resistant properties such as cry gene (resistance to Colorado Potato Beetle), StGA2ox1 gene (tolerance to abiotic stress), CP4-EPSP synthase gene (resistance to glyphosate), etc which are listed in Table 8.1

8.4 Mechanism of genetic modification leading to the desired trait

Plant breeding methods such as interspecies crossing, somatic hybridization, mutation breeding, and genetic engineering have long been utilized by breeders to generate disease-resistant plant lines (van Esse et al., 2020). Genetic engineering was first introduced in plants by the use of *Agrobacterium tumefaciens* mechanism which consists of employing modified bacteria (Ti plasmid) to insert the gene of interest by means of two methods – trans-genesis and cis-genesis (del Mar Martínez-Prada et al., 2021). To date, various genes have been identified that add resistance capability to the plants. Data from various experimental studies support that when resistant genes (R-genes) are added via genetic modification (cis-genesis) or RNA silencing of S-genes (susceptibility genes), F1-progeny potato plants are produced that are resistant to *P. infestans* (potato blight fungus) (Goutam et al., 2018). Transferring target oligonucleotide sequences into the plant genome when propagating the F1-progeny incorporates new agronomic traits to it when expressed. Endonuclease-based genome editing methods involving different synthetic/engineered enzymes for inserting double strand breaks (DSBs) at the target locus include meganucleases, TALENs (Transcription Activator-Like Effector Nucleases), CRISPR/Cas associated nucleases, and Zinc-Finger Nucleases (Kadam et al., 2018). The TALEN mechanism integrates the gene of interest (such as the Ubi7 gene) by inducing DSBs at 5'UTR in the potato genome which result in expression of the promoterless marker gene ALS (acetolactate synthase) in transgenic potato plants resistant to herbicide (Forsyth et al., 2016). The genome editing in plants involving the CRISPR/Cas9 method consists of guide RNAs to carry the Cas9 protein to target gene locus points which are then

Table 8.1 Genes involved in the development of transgenic potato

Transgenic gene	*Genome editing method*	*Agronomic trait*	*References*
StPPO2	CRISPR/Cas9	Reduce the enzymatic browning of potato tubers	González et al. (2020)
StSOD1	Agrobacterium mediated transformation	Tolerance to low temperature	Che et al. (2020)
Human Lactoferrin Gene (hlf)	Agrobacterium mediated transformation	Resistance to bacterial and fungal pathogens	Buziashvilli et al. (2020)
SSR2	TALEN and CRISPR/Cas9	Regulate the levels of toxic metabolites in potato	Yasumoto et al. (2019)
StCYS1	Agrobacterium mediated transformation	Increased salt resistance	Liu et al. (2020)
StGA2ox1	Agrobacterium mediated transformation	Tolerance to dehydration stress	Shi et al. (2019)
PDX-II	Agrobacterium mediated transformation	Vitamin B6 accumulation	Bagri et al. (2018a)
AtCDPK1	Agrobacterium mediated transformation	Multiple abiotic stress tolerance	Nie et al. (2018)
St16DOX	CRISPR/Cas9	Free hairy roots potato	Nakayasu et al. (2018)
OsmWS	Agrobacterium mediated transformation	Resistance to early blight disease	Kaur et al. (2020)
StProDH1	Agrobacterium mediated transformation	Tolerance to drought stress	Li et al. (2020)
AtHKT1	Agrobacterium mediated transformation	Reduced Na+ accumulation, K+/Na+ homeostasis and increased plant productivity	Wang et al. (2019)
StERF94	Agrobacterium mediated transformation	Resistance to fungal diseases	Charfeddine et al. (2019)
StRFP2	Agrobacterium mediated transformation	Tolerance to drought stress	Qi et al. (2020)

Source: Authors.

transferred by *Agrobacterium tumefaciens* in F1 transgenic plants (El-Mounadi et al., 2020). Where the CRISPR/Cas9 method is used to edit the coilin gene, transgenic potatoes are produced that are resistant to virus Y and tolerant towards salt and osmotic stress (Makhotenko et al., 2019). Modern genetic techniques have an added advantage over traditional breeding for improving nutritional value, productivity, and resistance traits of plants in that changes can be made quickly and efficiently (Figure 8.2).

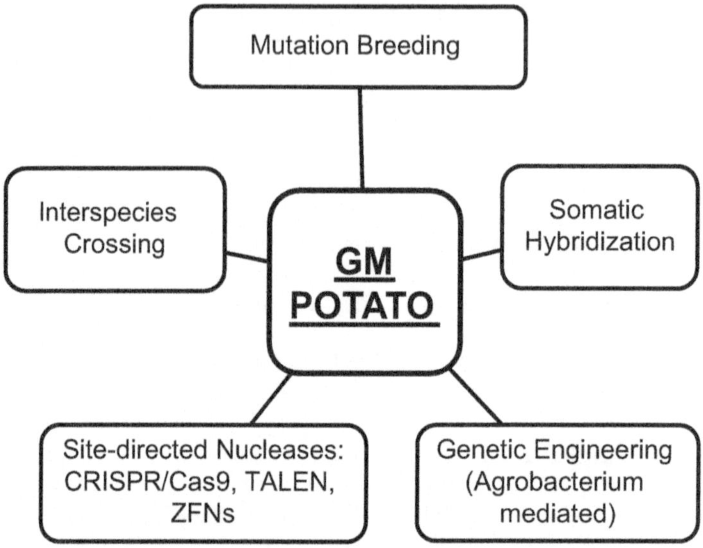

Figure 8.2 Mechanism of genetic modification leading to the desired trait.
Source: Authors.

8.5 Impact on modern agriculture – trade of GM crops

As GM plants have better resistance to phytopathogens and greater tolerance to abiotic/biotic stress, farmers are able to overcome the problem of plant loss. GM crops have been developed with the main aim of benefitting consumers. Their commercialization has also benefitted the environment, ecology, and economy by increasing the yield of the global food crop and enhancing species diversity. The CRISPR system has been successfully used to improve the traits of several economically important crops, including wheat, maize, cotton, cassava, rice, soybean, and potato (Puchta, 2017). Viral/bacterial/fungal disease resistance, herbicide tolerance, drought tolerance, and increased shelf life are among the traits that have been introduced, resulting in improved quality and production.

Southwest Asia was the first region where plant genetic modification was first documented more than 10,000 years ago; where humans performed techniques such as artificial selection and selective breeding. Since then, the current GM crop revolution has resulted in advances in agriculture science and technology. With proven case studies in Indian cotton and Australian canola, GM crops have the potential to mitigate current and future problems in commercial agriculture. However, some controversial studies, as well as current issues with resistance against insects and possible health risks, have harmed their perception among both policymakers and the public, leading to bans in some countries. Despite this, the GM seed market is growing at a 9.83–10 percent compound annual growth rate and has promising prospects (Hameed et al., 2018).

8.6 Benefits of GM potato to society

As mentioned above, potato is a perennial herbaceous plant with white to purple flowers and yellow stamens that belongs to the Solanaceae family. Potato tubers are high in protein and storage starch. Starch is the primary carbohydrate component of dry matter of potato. Potato tubers have carbohydrates ranging from 1.05 to 7.0 g/kg. The concentrations of reducing sugars are higher in young tubers and decrease dramatically near the end of the growing season. In the food sector, starch has traditionally been utilised to improve the functional qualities of numerous meals. The structural features of potato starch, amylose-to-amylopectin ratio differ between cultivars. In addition to it, it can provide more carbohydrates, minerals/vitamins and a protein per unit amount of land, and it is an important staple food in many developing nations All these factors have motivated scientists to focus their efforts on improving the nutritional value of potatoes while preserving yield. Several efforts have been made in the last two decades to improve the nutritional characteristics of potatoes. Information regarding the improvement of the nutrient content in potato is given in the sections below (Hameed et al., 2018; Islam et al., 2020).

8.6.1 Genetically engineered potato with enhanced protein content

Protein deficiency is more prevalent in areas where protein-deficient diets are taken as a staple food by many people (Chakraborty et al., 2010). Unfortunately, farmed potatoes have a lower protein content (0.8–4.21 percent) when compared to other food crops, and are deficient in tyrosine, lysine, and other necessary amino acids (Burlingame et al., 2009). To overcome these limitations, research scientists created a potato with increased protein content by expressing a tuber-specific gene called *Amaranthus hypochondriacus 1* (AmA1) in the tuber (Chakraborty et al., 2000, 2010). The AmA1 gene produces albumin – a non-allergic protein, which has the necessary amino acids, and that is also safe for animal as well as human consumption. The transgenic potato not only contains higher protein content, but it is also associated with a faster rate of photosynthesis, which results in a higher overall yield/biomass of plants. The amino acid (methionine), which is a necessary amino acid that is involved in several biological pathways, has recently been found to be considerably higher in the transgenic potato cultivar Desiree (Kumar & Jander, 2017). Increased expression of an exogenous gene *Arabidopsis thaliana* cystathionine g-synthase (AtCGS) in combination with suppression of a host gene *S. tuberosum* methionine g-lyase (StMGL) using RNAi technology resulted in increasing the content of free methionine inside transgenic tubers compared to control tubers (Islam et al., 2020; Raman, 2017).

8.6.2 Genetically engineered potato with enhanced carotenoid and vitamin content

Different experimental approaches have been followed to enhance the vitamin content of potatoes. One such example – expression of an exogenous gene (*A. thaliana* L-galactose phosphorylase (GDP)) – resulted in a three-fold increase in ascorbate (vitamin C) content (Bulley et al., 2012). As carotenoids found in a variety of fruits and vegetables are associated with various activities such as anti-oxidative, phytonutritive, lipophilic chemicals (precursors to vitamins) (DellaPenna & Pogson, 2006), and they give nutritional benefits by enhancing vitamin uptake. The introduction of the Cauliflower Orange (Or) gene in cold stored tubers resulted in an enhanced (pro-vitamin A) carotenoid content (Li et al., 2012). In comparison to rest of the carotenoids (zeaxanthin, lutein, violaxanthin, and neoxanthin), the concentration of beta-carotene in potatoes is very low (Ezekiel et al., 2013). In the transgenic potato the beta-carotene hydroxylase (bch) gene was silenced using an RNAi method, which resulted in a large increase in beta-carotene and other antioxidants (Hameed et al., 2018).

8.6.3 Genetically engineered potato with enhanced calcium content

Traditionally cultivated potato is a low source of calcium despite being nutritive in other ways (Weaver et al., 1999). Park et al. (2005) used a transgenic technique in potato cv. Russet Norkotah to overcome this defect by expressing an external gene, *Arabidopsis* H+/Ca2+ transporter (sCAX1). In comparison to controls, the calcium concentration in tubers of regenerated plants expressing the sCAX1 gene increased significantly (up to three-fold). Field trials and morphological data showed that the elevated calcium (Ca) trait was successfully integrated into three crop generations with no change in tuber yield and other growth and morphological characteristics. Potatoes with higher calcium levels could be given as a dietary supplement, particularly in regions where potatoes are used as a staple food (Lombardo & Grando, 2020; Park et al., 2005).

8.6.4 Genetically engineered potato with enhanced phenolic contents

Caffeoyl quinic acids (CQAs) make up 80 percent of the phenolic chemicals found in potatoes (Brown, 2005). A recent study showed an increased amount of CQAs in GM potato tubers. Total flavonoid and CQA content were significantly increased more than three-fold because of tuber-specific constitutive expression of an exogenous gene, flavonol-specific transcription activator AtMYB12 (derived from *A. thaliana*). Importantly, the researchers used a selectable marker-free strategy to help with regulatory approvals further down the line (Daniell, 2002). Additionally, enhanced phenolic content confers antibacterial qualities to plants, which aids in the reduction of fungal diseases (Li et al., 2016; Hameed et al., 2018; Akyol et al., 2016).

8.6.5 Genetically engineered potato with enhanced starch contents

Potato tubers, being a rich source of dietary starch, provide sufficient calories for food-deficient countries. Amylose and amylopectin are two main structural components of starch which are produced through the combined actions of numerous enzymes. Engineering different phases of the starch production pathway has been used to boost the digestible amylose level in potatoes (Hofvander et al., 2004). Silencing of two host genes of tuber-specific RNA, which are involved in the starch branching process (SBEI, SBEII), further resulted in the production of potatoes with an enhanced amylose content (Andersson et al., 2006). When a recessive allele (gene marker: IAm) from wild potato (*S. sandemanii*) was inserted into a cultivated potato (*S. tuberosa*), it resulted in dramatically enhanced amylose content (25–60 percent) in non-genetically modified potatoes (Patil et al., 2016; Upadhyaya et al., 2018).

8.6.6 Confers resistance against insects, microorganisms, and abiotic factors

In comparison with their progenitors, a lot of modified cultivars of agricultural plants are less resistant naturally. However, widespread use of chemical pesticides has resulted in the emergence of a new type of pest resistance in the population. The cultivation of pest-resistant crops that contribute to a reduction in the use of chemical insecticides may be advantageous.

The potato beetle (*Leptinotarsa decemlineata*) is one of the most damaging pests to potato plants and it is commonly resistant to chemical and synthetic insecticides. Modified potatoes with the Cry3A gene from *Bacillus thuringiensis* are capable of controlling the beetle. The toxic protein formed in the plants leaves as a gene product passes to the intestines of the beetle after ingestion and subsequently leads to the death of the pest (Hameed et al., 2018; Raspor & Cingel, 2021).

Potato blight (*Phytophtora infestans*) is a serious fungal disease that affects the potato. All potato plants produce Temporin A, which is a protein resistant to fungal disease. Temporin A is a naturally occurring small antimicrobial peptide which improves resistance of the plant against bacterial wet rot as well as potato blight. Plant pathogens *P. erytroseptica* and *Erwinia carotovora* are majorly associated with this disease. Different research has shown that transgenic potato plants that express Temporin A can be effective against a variety of fungal pathogens.

Hydrogen peroxide production is yet another method for protecting plants from pathogens. The expression of the glucose oxidase gene from *Aspergillus niger* has been investigated in this regard. Resistance to potato blight (*P. infestans*) and bacterial rot caused by *Erwinia carotovora* is greater in potato plants that produce hydrogen peroxide. Transgenic plants carrying the ac2 gene from the amaranth *Amaranthus caudatus* can also be used to control fungal infections in potato plants (Liapkova et al., 2001). The protein produced by the ac2 gene is very similar to the cysteine/glycine-rich domains found in chitin binding proteins (Broekaert et al., 1992). This means that they can bind to chitin found in internal fungal cell walls, altering their polarity and preventing the fungus from growing (Raspar et al., 2021).

In addition to their resistant properties towards insects and microbes, studies have reported on the resistance ability of potato plants against various herbicides. For example, when the bar gene (PAT) of bacterium *Streptomyces hygroscopicus* was inserted into potato plants, it made them resistant against the herbicide phosphinothricin (Hameed et al., 2018).

In another study, when the gene glyceraldehyde-3-phosphate-dehydrogenase (GPD) from the oyster mushroom (*Pleurotussajor-caju*) was inserted into the potato plant genome, it resulted in potato plants that were resistant to increased soil salt levels. The influence of the protein was investigated by growing potato plants in soil containing sodium chloride. Unlike GPD-free potato plants, which died after a few days, a high tolerance towards salt was shown by the transgenic plants (Jeong et al., 2001; Patil et al., 2016).

8.6.7 *Secretion of human lactic β-casein*

Transgenic potato plants that produce human lactic β-casein have potential nutritional applications. Plant-produced human casein could be used in the future to make human milk proteins such as lysozyme and lactoferrin, or to make food for babies with preventive effects against various disorders related to the digestive system in children and/or with higher nutritional value (Bagri et al., 2018b; Patil et al., 2016; Chong et al., 1997).

8.6.8 *Secretion of human lactoferrin*

Human lactoferrin is a human glycoprotein. Human lactoferrin is inserted into the genome of potato plants in the same way as found primarily in mothers' milk, and to a lesser extent in bile and tears. It has a protective function in that it can bind iron, preventing access by bacteria. Because of the potential antimicrobial, nutritional, and immune regulatory activity of lactoferrin, transgenic potato plants containing a cDNA fragment encoding human lactoferrin have been developed (hLF). The biological properties of lactoferrin produced by transgenic potatoes have been confirmed in four different pathogenic bacteria of humans. The ability of GM potato plants to secrete human lactoferrin with bacteriostatic, bactericidal properties has been demonstrated (Hameed et al., 2018; Patil et al., 2016; Chong et al., 1997).

8.6.9 *Enhanced lipid synthesis in potato tubers*

Producing transgenic potato plants is also useful for studying various enzymatic pathways in different parts of plant bodies such as determining the factors that limit lipid synthesis in potato tubers. Potatoes have a high amount of starch, but they are lacking in lipids. Transgenic potato plants produce high levels of the enzyme acetyl-CoA carboxylase, which leads to increased fatty acid and triglyceride synthesis. As a result of the above findings, potato tubers could serve as lipid storage organs (Liu et al., 2017; Klaus et al., 2004).

8.6.10 *Production of edible vaccines*

Using the technique of genetic engineering, the expression of viral/bacterial antigens in the edible portion of plant cells is possible. A number of crops (such as maize, rice, potatoes, and soybeans) are being studied as potential carriers of edible vaccines against a variety of infections that includes the flu, *E. coli* toxins, rabies virus, *Helicobacter pylori*, and type B viral hepatitis (Kurup & Thomas, 2020; Zhang et al., 2017).

The benefits of GM potato are summarised in Figure 8.3.

8.7 Trade in the global market

The use of biotechnological applications in agriculture and medicine has resulted in an increase in GMOs. Societal benefits such as new drugs, improved plant varieties, and more nutritious and reliable food crops can be achieved by changing the characteristics of organisms. However, on the other hand, biotechnology is fraught with danger and uncertainty. Its potential effects on the environment, food safety, and human well-being are

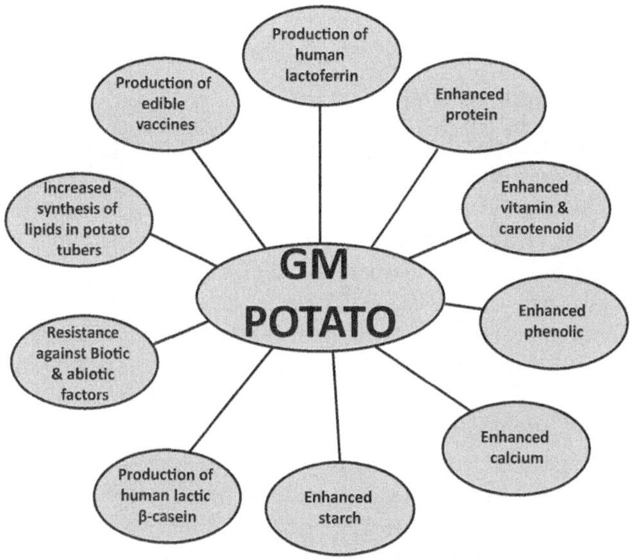

Figure 8.3 Benefits of genetically modified potato.
Source: Authors.

being actively debated at the international and national level. Many factors influence a country's position on GMOs, including awareness of policies, acceptance of risk levels, their capacity to conduct risk assessments in the region and implementation of appropriate legislation, as well as an awareness of the benefits gained from biotechnology. However, there is presently a stark difference between the prevalent international acceptance of the benefits of biotechnology in products related to the pharmaceutical industry and the general concern about its potential risks in farming and the production of food. Currently, the apparent advantages of GM crops include a reduction in the use of pesticides and insecticides, increased yield, and easier crop management. These advantage primarily profit agronomists and the agroindustry, allowing them to achieve higher yields at lower costs. However, the extended profits could be enhanced to include sustainable farming and improved food safety, which would benefit everyone, mainly developing countries. For instance, drought tolerance breeding could be extremely beneficial to tropical crops grown in hostile environments. Increasing the quantity of food yield per hectare may be a suitable approach to feed the rising population without using land from other practices such as animal grazing, forestry, or conservation.

Researchers have formed a GM rice strain to battle vitamin A deficiency, which is the foremost cause of blindness and impacts up to 250 million children worldwide (Zarrilli, 2005). Yet, according to Qaim (2009), because public opinion is generally negative, a couple of previously developed GM technologies for food crops were either never commercialized or were withdrawn from the market due to consumer acceptance and marketing issues (e.g. Bt and Roundup Ready).

8.8 Health and ethical issues

Although the GM potato has been associated with numerous benefits, it has also been associated with various risks to human health as well as to the plant itself. The process of genetic modification results in many changes in a plant's metabolic pathways and its own natural proteins, which can lead to the unexpected production of toxins or allergens in food – one of the most frequently cited risks of GE crops (Fagan et al., 2015). The majority of chemicals involved in crucial metabolic pathway are commonly shared by plants and animals and thus unlikely to be toxic. Hence, changes in molecules specific to plants, commonly called plant natural products, are the main source of perceived risks associated with plant compound changes.

The plant kingdom contains over 200,000 secondary metabolites (Springob & Kutchan, 2009). The amount of secondary metabolites produced by each crop species differs. For instance, the potato (*Solanum tuberosum*) is recognized for its diverse secondary metabolites, including sesquiterpenes, which provide resistance towards disease. These products serve as defensive agents by absorbing harmful UV radiation (Treutter, 2006), anti-nutrients, or by killing and/or preventing crop-damaging pathogens and insects (Dixon, 2004). Some of these compounds have been studied for their toxic properties, but others have not. When consumed in large quantities, these compounds can be harmful to individuals. One example is the consumption of green potato skin that is comprised of toxic products such as steroidal glycoalkaloids which cause digestive discomfort or vomiting and diarrhoea. Friedman (2006) reported that glycoalkaloids found in potatoes have been associated with negative as well as positive effects. The hazards related to these metabolites are well understood.

Another consideration is that food preparation can result in the secretion of these compounds. For instance, secretion of the carcinogen acrylamide is induced when *Solanum tuberosum* is cooked at high temperatures. Yet the risks of contamination from these modified crops are very low as potatoes reproduce by forming the tubers, in contrast to crops such as oil seed rape that produces pollen that can be taken off by the wind. Pollen from some GM crops, such as oilseed rape, has the potential to fertilize non-GM varieties. Hence, growers would need to cultivate these crops farther from other crops to reduce the risk. Pollen production is minimal in GM potatoes.

In industrial processes, potatoes containing amylopectin are used primarily to make starch and are generally avoided for human consumption. The GM potato EH92–527–1, which has a modified starch composition (a greater ratio of amylopectin : amylose), was examined for any toxic properties or allergy causing ability (Islam et al., 2020), yet human health was found to be unaffected.

GM potato has been associated with some ethical issues concerning producers, consumers, the environment and the agro-economic system. First, although the costs of fungicide are reduced and there is an increase in yield, GM seed potatoes are more expensive. Furthermore, the role of potato markets and the price squeeze that some farmers may be subjected to from agents and retailers may result in a decrease in production. Therefore, due to production costs, the price for the potato crop offered to farmers may be reduced. Secondly, consumers often doubt the safety and quality of the GM potato as they are generally wary of eating GM foods. Thirdly, the transformation method used when developing GM plants is associated with various risks and GM potato could have unintended negative consequences for non-target organisms (Hameed et al., 2018; Islam et al., 2020).

8.9 Current and future perspectives

In the agri-biotech industry, GM crops are widely accepted. These modified products provide a long-term foundation for ensuring food security for the world's growing population. The increasing rate of acceptance of this molecular technology by agriculturalists throughout the world underscores the successful completion of 25 years of commercial GM crop production (1996–2021). The use of sensitive and reliable approaches for labeling genetically altered organisms in both food and feedstuffs has become very important due to the expression of various traits in GM crops such as resistance towards herbicide, insecticide, tolerance to drought conditions.

GM foods as commodity crops have been a huge success for the past 25 years, raising significant profits for agriculturalists and offering various advantages. Presently, an increase in the production of modified crops has been reported since their introduction in 1996. Yet this rapid adoption of GM crops necessitates trait-specific assays that are precise and sensitive in detection of GMO in GM crops before their commercialization. It is necessary to devote time and effort to avoid potential risks by using such procedures as on-farm trials. An effective management framework with clear constraints should be used to measure the risks associated with the use of GM crops. Novel breeding tools, for example TALENs, ZFNs, CRISPR/Cas9, etc. have emerged recently, enabling the introduction of a site-specific genome to introduce significant agronomical traits in a wide range of crop plants.

8.10 Conclusions

Food security and safety research provide significant ways of meeting the increasing need for foodstuffs, particularly in countries short of food. Rapid advancements in the genetic engineering of plants has given novel tools for creating crops with enhanced quantity and better nutrition. In this milieu, *Solanum tuberosum* has tremendous potential to provide to aid food security by providing low-cost, high-energy food on a long-term basis. Several studies have shown that nutritional traits such as an increased content of protein and ascorbic acid (Bulley et al., 2012), beta-carotene level, triacylglycerol, tuber methionine, and amylose content can be incorporated into potatoes.

References

Ahmed, S., Zhou, X., Pang, Y., Jin, L., & Bao, J. (2018). Improving starch-related traits in potato crops: Achievements and future challenges. *Starch-Stärke*, 70(9–10), 1700113.

Akyol, H., Riciputi, Y., Capanoglu, E., Caboni, M. F., & Verardo, V. (2016). Phenolic compounds in the potato and its byproducts: An overview. *International Journal of Molecular Sciences*, 17(6), 835.

Andersson, M., Melander, M., Pojmark, P., Larsson, H., Bülow, L., & Hofvander, P. (2006). Targeted gene suppression by RNA interference: An efficient method for production of high-amylose potato lines. *Journal of Biotechnology*, 123(2), 137–148.

Bagri, D. S., Upadhyaya, D. C., Kumar, A., & Upadhyaya, C. P. (2018a). Overexpression of PDX-II gene in potato (*Solanum tuberosum* L.) leads to the enhanced accumulation of vitamin B6 in tuber tissues and tolerance to abiotic stresses. *Plant Science*, 272, 267–275.

Bagri, D. S., Upadhyaya, D. C., Jain, S. K., & Upadhyaya, C. P. (2018b). Biotechnological improvement of nutritional and therapeutic value of cultivated potato. *Frontiers in Bioscience*, 10, 217–228.

Broekaert, W. F., Marien, W., Terras, F. R., De Bolle, M. F., Proost, P., Van Damme, J., ... Rees, S. B. (1992). Antimicrobial peptides from *Amaranthus caudatus* seeds with sequence homology to the cysteine/glycine-rich domain of chitin-binding proteins. *Biochemistry*, 31(17), 4308–4314.

Brown, C. R. (2005). Antioxidants in potato. *American Journal of Potato Research*, 82 (2), 163–172.

Bulley, S., Wright, M., Rommens, C., Yan, H., Rassam, M., Lin-Wang, K., ... Laing, W. A. (2012). Enhancing ascorbate in fruits and tubers through over-expression of the l-galactose pathway gene GDP-l-galactose phosphorylase. *Plant Biotechnology Journal*, 10(4), 390–397.

Buziashvili, A., Cherednichenko, L., Kropyvko, S., Blume, Y. B., & Yemets, A. (2020). Obtaining transgenic potato plants expressing the human Lactoferrin gene and analysis of their resistance to phytopathogens. *Cytology and Genetics*, 54(3), 179–188.

Burlingame, B., Mouillé, B., & Charrondiere, R. (2009). Nutrients, bioactive non-nutrients and anti-nutrients in potatoes. *Journal of Food Composition and Analysis*, 22(6), 494–502.

Chakraborty, S., Chakraborty, N., Agrawal, L., Ghosh, S., Narula, K., Shekhar, S., ... Datta, A. (2010). Next-generation protein-rich potato expressing the seed protein gene AmA1 is a result of proteome rebalancing in transgenic tuber. *Proceedings of the National Academy of Sciences*, 107(41), 17533–17538.

Chakraborty, S., Chakraborty, N., & Datta, A. (2000). Increased nutritive value of transgenic potato by expressing a nonallergenic seed albumin gene from *Amaranthus hypochondriacus*. *Proceedings of the National Academy of Sciences*, 97(7), 3724–3729.

Charfeddine, M., Samet, M., Charfeddine, S., Bouaziz, D., & Gargouri Bouzid, R. (2019). Ectopic expression of StERF94 transcription factor in potato plants improved resistance to *Fusarium solani* infection. *Plant Molecular Biology Reporter*, 37(5), 450–463.

Che, Y., Zhang, N., Zhu, X., Li, S., Wang, S., & Si, H. (2020). Enhanced tolerance of the transgenic potato plants overexpressing Cu/Zn superoxide dismutase to low temperature. *Scientia Horticulturae*, 261, 108949.

Chong, D. K. X., Roberts, W., Arakawa, T., Illes, K., Bagi, G., Slattery, C. W., & Langridge, W. H. R. (1997). Expression of the human milk protein β-casein in transgenic potato plants. *Transgenic Research*, 6(4), 289–296.

Daniell, H. (2002). Molecular strategies for gene containment in transgenic crops. *Nature Biotechnology*, 20(6), 581–586.

Das Dangol, S., Barakate, A., Stephens, J., Çalıskan, M. E., & Bakhsh, A. (2019). Genome editing of potato using CRISPR technologies: Current development and future prospective. *Plant Cell, Tissue and Organ Culture (PCTOC)*, 139(2), 403–416.

DellaPenna, D., & Pogson, B. J. (2006). Vitamin synthesis in plants: Tocopherols and carotenoids. *Annual Review of Plant Biology*, 57, 711–738.

del Mar Martínez-Prada, M., Curtin, S. J., & Gutiérrez-González, J. J. (2021). Potato improvement through genetic engineering. *GM Crops & Food*, 12(1), 479–496.

Dixon, R. A. (2004). Phytoestrogens. *Annual Review of Plant Biology*, 55, 225–261.

Eggers, E. J., van der Burgt, A., van Heusden, S. A., de Vries, M. E., Visser, R. G., Bachem, C. W., & Lindhout, P. (2021). Neofunctionalisation of the Sli gene leads to self-compatibility and facilitates precision breeding in potato. *Nature Communications*, 12(1), 1–9.

El-Mounadi, K., Morales-Floriano, M. L., & Garcia-Ruiz, H. (2020). Principles, applications, and biosafety of plant genome editing using CRISPR-Cas9. *Frontiers in Plant Science*, 11, 56.

Ezekiel, R., Singh, N., Sharma, S., & Kaur, A. (2013). Beneficial phytochemicals in potato – a review. *Food Research International*, 50(2), 487–496.

Fagan, J., Antoniou, M., & Robinson, C. (2015). *GMO Myths and Truths*. Earth Open Source.

FAOSTAT (2020). *Statistical Yearbook: World Food and Agriculture 2020*. Food and Agriculture Organization of the United Nations. Available at: https://www.fao.org/3/cb1329en/CB1329EN.pdf

Forsyth, A., Weeks, T., Richael, C., & Duan, H. (2016). Transcription activator-like effector nucleases (TALEN)-mediated targeted DNA insertion in potato plants. *Frontiers in Plant Science*, 7, 1572.

Friedman, M. (2006). Potato glycoalkaloids and metabolites: Roles in the plant and in the diet. *Journal of Agricultural and Food Chemistry*, 54(23), 8655–8681.

Gaiero, P., Speranza, P., & de Jong, H. (2018). Introgressive hybridization in potato revealed by novel cytogenetic and genomic technologies. *American Journal of Potato Research*, 95(6), 607–621.

Getahun, B. B., Kassie, M. M., Visser, R. G., & van der Linden, C. G. (2020). Genetic diversity of potato cultivars for nitrogen use efficiency under contrasting nitrogen regimes. *Potato Research*, 63(2), 267–290.

González, M. N., Massa, G. A., Andersson, M., Turesson, H., Olsson, N., Fält, A. S., ... Feingold, S. E. (2020). Reduced enzymatic browning in potato tubers by specific editing of a polyphenol oxidase gene via ribonucleoprotein complexes delivery of the CRISPR/Cas9 system. *Frontiers in Plant Science*, 10, 1649.

Goutam, U., Thakur, K., Salaria, N., & Kukreja, S. (2018). Recent approaches for late blight disease management of potato caused by *Phytophthora infestans*. In *Fungi and their Role in Sustainable Development: Current Perspectives*, edited by P. Gehlot & J. Singh (Singapore: Springer), 311–325.

Hameed, A., Zaidi, S. S. E. A., Shakir, S., & Mansoor, S. (2018). Applications of new breeding technologies for potato improvement. *Frontiers in Plant Science*, 9, 925.

Hara-Skrzypiec, A., Śliwka, J., Jakuczun, H., & Zimnoch-Guzowska, E. (2018). QTL for tuber morphology traits in diploid potato. *Journal of Applied Genetics*, 59(2), 123–132.

Hofvander, P., Andersson, M., Larsson, C. T., & Larsson, H. (2004). Field performance and starch characteristics of high-amylose potatoes obtained by antisense gene targeting of two branching enzymes. *Plant Biotechnology Journal*, 2(4), 311–320.

Hussain, T., Aksoy, E., Çalışkan, M. E., & Bakhsh, A. (2019). Transgenic potato lines expressing hairpin RNAi construct of molting-associated EcR gene exhibit enhanced resistance against Colorado potato beetle (*Leptinotarsa decemlineata*, Say). *Transgenic Research*, 28(1), 151–164.

Islam, R., Parvin, A., Billah, M. M., Islam, M., Imran, M. A. S., Sarker, R. K., ... Abedin, M. Z. (2020). Assessment of the effects of genetically modified (GM) foods: A brief study on health and environmental concerns. *Journal of Material and Environmental Sciences*, 11(10), 1676–1688.

Jama-Rodzenska, A., Janik, G., Walczak, A., Adamczewska-Sowinska, K., & Sowinski, J. (2021). Tuber yield and water efficiency of early potato varieties (*Solanum tuberosum* L.) cultivated under various irrigation levels. *Scientific Reports*, 11(1), 1–12.

Jeong, M. J., Park, S. C., & Byun, M. O. (2001). Improvement of salt tolerance in transgenic potato plants by glyceraldehyde-3 phosphate dehydrogenase gene transfer. *Molecules & Cells*, 12(2), 185–189.

Kadam, U. S., Shelake, R. M., Chavhan, R. L., & Suprasanna, P. (2018). Concerns regarding 'off-target' activity of genome editing endonucleases. *Plant Physiology and Biochemistry*, 131, 22–30.

Karalis, D. T., Karalis, T., Karalis, S., & Kleisiari, A. S. (2020). Genetically modified products, perspectives and challenges. *Cureus*, 12(3).

Kaur, A., Sudhakara Reddy, M., Pati, P. K., & Kumar, A. (2020). Over-expression of osmotin (*OsmWS*) gene of *Withania somnifera* in potato cultivar 'Kufri Chipsona 1' imparts resistance to *Alternaria solani*. *Plant Cell, Tissue and Organ Culture (PCTOC)*, 142(1), 131–142.

Klaus, D., Ohlrogge, J. B., Neuhaus, H. E., & Dörmann, P. (2004). Increased fatty acid production in potato by engineering of acetyl-CoA carboxylase. *Planta*, 219(3), 389–396.

Kramkowska, M., Grzelak, T., & Czyzewska, K. (2013). Benefits and risks associated with genetically modified food products. *Annals of Agricultural and Environmental Medicine*, 20(3).

Kumar, K., Gambhir, G., Dass, A., Tripathi, A. K., Singh, A., Jha, A. K., … Rakshit, S. (2020). Genetically modified crops: Current status and future prospects. *Planta*, 251(4), 1–27.

Kumar, P., & Jander, G. (2017). Concurrent overexpression of *Arabidopsis thaliana* cystathionine γ-synthase and silencing of endogenous methionine γ-lyase enhance tuber methionine content in *Solanum tuberosum*. *Journal of Agricultural and Food Chemistry*, 65(13), 2737–2742.

Kurup, V. M., & Thomas, J. (2020). Edible vaccines: Promises and challenges. *Molecular Biotechnology*, 62(2), 79–90.

Kyriakidou, M., Achakkagari, S. R., Gálvez López, J. H., Zhu, X., Tang, C. Y., Tai, H. H., … Strömvik, M. V. (2020). Structural genome analysis in cultivated potato taxa. *Theoretical and Applied Genetics*, 133(3), 951–966.

Li, L., Yang, Y., Xu, Q., Owsiany, K., Welsch, R., Chitchumroonchokchai, C., … Thannhauser, T. W. (2012). The Or gene enhances carotenoid accumulation and stability during post-harvest storage of potato tubers. *Molecular Plant*, 5(2), 339–352.

Li, S., Zhang, N., Zhu, X., Ma, R., Yang, J., Tang, X., & Si, H. (2020). Enhanced drought tolerance with artificial microRNA-mediated StProDH1 gene silencing in potato. *Crop Science*, 60(3), 1462–1471.

Li, Y., Colleoni, C., Zhang, J., Liang, Q., Hu, Y., Ruess, H., … Huang, B. (2018). Genomic analyses yield markers for identifying agronomically important genes in potato. *Molecular Plant*, 11(3), 473–484.

Li, Y., Tang, W., Chen, J., Jia, R., Ma, L., Wang, S., … Ding, X. (2016). Development of marker-free transgenic potato tubers enriched in caffeoylquinic acids and flavonols. *Journal of Agricultural and Food Chemistry*, 64(14), 2932–2940.

Liapkova, N. S., Loskutova, N. A., Maĭsurian, A. N., Mazin, V. V., Korableva, N. P., Platonova, T. A., … Evsiunina, A. S. (2001). Isolation of genetically modified potato plant containing the gene of defensive peptide from Amaranthus. *Prikladnaia Biokhimiiai Mikrobiologiia*, 37(3), 349–354.

Liu, M. M., Li, Y. L., Li, G. C., Dong, T. T., Liu, S. Y., Pei, L. I. U., & Wang, Q. G. (2020). Overexpression of StCYS1 gene enhances tolerance to salt stress in the transgenic potato (*Solanum tuberosum* L.) plant. *Journal of Integrative Agriculture*, 19 (9), 2239–2246.

Liu, Q., Guo, Q., Akbar, S., Zhi, Y., El Tahchy, A., Mitchell, M., … Petrie, J. (2017). Genetic enhancement of oil content in potato tuber (*Solanum tuberosum* L.) through an integrated metabolic engineering strategy. *Plant Biotechnology Journal*, 15(1), 56–67.

Lombardo, L., & Grando, M. S. (2020). Genetically modified plants for nutritionally improved food: A promise kept? *Food Reviews International*, 36(1), 58–76.

Makhotenko, A. V., Khromov, A. V., Snigir, E. A., Makarova, S. S., Makarov, V. V., Suprunova, T. P., … Taliansky, M. E. (2019). Functional analysis of coilin in virus resistance and stress tolerance of potato *Solanum tuberosum* using CRISPR-Cas9 editing. *Doklady Biochemistry and Biophysics*, 484(1), 88–91.

Manrique-Carpintero, N. C., Coombs, J. J., Pham, G. M., Laimbeer, F. P. E., Braz, G. T., Jiang, J., … Douches, D. S. (2018). Genome reduction in tetraploid potato reveals genetic load, haplotype variation, and loci associated with agronomic traits. *Frontiers in Plant Science*, 9, 944.

Muñiz García, M. N., Cortelezzi, J. I., Fumagalli, M., & Capiati, D. A. (2018). Expression of the Arabidopsis ABF4 gene in potato increases tuber yield, improves tuber quality and enhances salt and drought tolerance. *Plant Molecular Biology*, 98(1), 137–152.

Nakayasu, M., Akiyama, R., Lee, H. J., Osakabe, K., Osakabe, Y., Watanabe, B., … Mizutani, M. (2018). Generation of α-solanine-free hairy roots of potato by CRISPR/Cas9 mediated genome editing of the St16DOX gene. *Plant Physiology and Biochemistry*, 131, 70–77.

Nie, L. Z., Yu, X. X., Ma, Y. H., Fang, Y. Y., Li, L. M., & Yu, Z. (2018). Enhanced drought and osmotic stress tolerance in transgenic potato plants expressing AtCDPK1, a calcium-dependent protein kinase. *Russian Journal of Plant Physiology*, 65(6), 865–873.

Park, S., Kang, T. S., Kim, C. K., Han, J. S., Kim, S., Smith, R. H., … Hirschi, K. D. (2005). Genetic manipulation for enhancing calcium content in potato tuber. *Journal of Agricultural and Food Chemistry*, 53(14), 5598–5603.

Parra-Galindo, M. A., Piñeros-Niño, C., Soto-Sedano, J. C., & Mosquera-Vasquez, T. (2019). Chromosomes I and X harbor consistent genetic factors associated with the anthocyanin variation in potato. *Agronomy*, 9(7), 366.

Patil, V. U., Singh, R., Vanishree, G., Dutt, S., Kawar, P. G., Bhardwaj, V., & Singh, B. P. (2016). Genetic engineering for enhanced nutritional quality in potato – a review. *Potato Journal*, 43(1), 1–21.

Pham, G. M., Hamilton, J. P., Wood, J. C., Burke, J. T., Zhao, H., Vaillancourt, B., … Buell, C. R. (2020). Construction of a chromosome-scale long-read reference genome assembly for potato. *Gigascience*, 9(9), giaa100.

Pribylova, R., Pavlik, I., & Bartos, M. (2006). Genetically modified potato plants in nutrition and prevention of diseases in humans and animals: A review. *Veterinarni Medicina-Praha-*, 51(5), 212.

Puchta, H. (2017). Applying CRISPR/Cas for genome engineering in plants: The best is yet to come. *Current Opinion in Plant Biology*, 36, 1–8.

Qi, X., Tang, X., Liu, W., Fu, X., Luo, H., Ghimire, S., … Si, H. (2020). A potato RING-finger protein gene StRFP2 is involved in drought tolerance. *Plant Physiology and Biochemistry*, 146, 438–446.

Raman, R. (2017). The impact of Genetically Modified (GM) crops in modern agriculture: A review. *GM Crops & Food*, 8(4), 195–208.

Raspor, M., & Cingel, A. (2021). Genetically modified potato for pest resistance: Thrift or threat? In *Solanum Tuberosum – A Promising Crop for Starvation Problem*, edited by M. Yildiz & Y. Ozgen (London: IntechOpen).

Shi, J., Wang, J., Wang, N., Zhou, H., Xu, Q., & Yan, G. (2019). Overexpression of StGA2ox1 gene increases the tolerance to abiotic stress in transgenic potato (*Solanum tuberosum* L.) plants. *Applied Biochemistry and Biotechnology*, 187(4), 1204–1219.

Singh, B., Kukreja, S., & Goutam, U. (2020). Impact of heat stress on potato (*Solanum tuberosum* L.): Present scenario and future opportunities. *The Journal of Horticultural Science and Biotechnology*, 95(4), 407–424.

Springob, K., & Kutchan, T. M. (2009). Introduction to the different classes of natural products. In *Plant-derived Natural Products*, edited by A. E. Osbourn & V. Lanzotti (New York: Springer), 3–50.

Tiwari, J. K., Plett, D., Garnett, T., Chakrabarti, S. K., & Singh, R. K. (2018). Integrated genomics, physiology and breeding approaches for improving nitrogen use efficiency in potato: Translating knowledge from other crops. *Functional Plant Biology*, 45(6), 587–605.

Treutter, D. (2006). Significance of flavonoids in plant resistance: A review. *Environmental Chemistry Letters*, 4(3), 147–157.

Upadhyaya, C. P., Bagri, D. S., & Upadhyaya, D. C. (2018). Genetically modified potato as a source of novel carbohydrates. In *Potato: From Incas to All Over the World*, edited by M. Yildiz (London: IntechOpen), 19–35.

van Esse, H. P., Reuber, T. L., & van der Does, D. (2020). Genetic modification to improve disease resistance in crops. *New Phytologist*, 225(1), 70–86.

Wang, L., Liu, Y., Li, D., Feng, S., Yang, J., Zhang, J., ... Gan, Y. (2019). Improving salt tolerance in potato through overexpression of AtHKT1 gene. *BMC Plant Biology*, 19(1), 1–15.

Weaver, C. M., Proulx, W. R., & Heaney, R. (1999). Choices for achieving adequate dietary calcium with a vegetarian diet. *The American Journal of Clinical Nutrition*, 70 (3), 543s–548s.

Yasumoto, S., Umemoto, N., Lee, H. J., Nakayasu, M., Sawai, S., Sakuma, T., ... Muranaka, T. (2019). Efficient genome engineering using platinum TALEN in potato. *Plant Biotechnology*, 36(3), 167–173.

Zarrilli, S. (2005). *International Trade in GMOs and GM Products: National and Multilateral Legal Frameworks.* Policy Issues in International Trade and Commodities, Study Series No. 29. https://unctad.org/system/files/official-document/itcdtab30_en.pdf

Zhang, C., Yang, Z., Tang, D., Zhu, Y., Wang, P., Li, D., ... Huang, S. (2021). Genome design of hybrid potato. *Cell*, 184(15), 3873–3883.

Zhang, H., Zhang, J., Lang, Z., Botella, J. R., & Zhu, J. K. (2017). Genome editing – principles and applications for functional genomics research and crop improvement. *Critical Reviews in Plant Sciences*, 36(4), 291–309.

9 Herbicide-tolerant alfalfa

Priya Sharma, Vishal Sharma, Shivali Panjgotra and Namrata Sharma

9.1 Introduction

Genetic engineering is one of the best tools of biotechnology which is used to produce transgenic plants with hand-picked characteristics. It facilitates the transfer of genes expressing a desired feature from one plant species to another which is not possible through classical plant breeding. Genetic engineering has been performed on various crop plants to improve their productivity, quality, as well as their resistance against herbicides, insecticides etc.

Alfalfa (*Medicago sativa* L.), also known as Lucerne or the Queen of Fodder, is a perennial forage crop or weed which is widely distributed in cool subtropical and warm temperate regions of the world. It is the most cultivated forage crop globally. Grown largely in the United States and Argentina, it is also cultivated significantly in Africa, Australia, the Middle East, China, Southern Europe, Chile, and Mexico. Its total worldwide production was approximately 32 million hectares in 2016 (NAFA, 2015), out of which, about 2–2.5 million hectares were produced by Asia. In India, it is cultivated on over 1 million hectares and considered to be the third most important forage crop after sorghum and berseem (Iqbal et al., 2018). Once cultivated, it shows regrowth for 6–7 years and acts as a perennial weed. Its distribution is widespread and it is also helpful in providing a habitat for wildlife (Putnam et al., 2001).

Alfalfa is considered to be 'green gold' because it is superior to the other forage crops. This plant has a high content of protein and energy with low fibre. It also stores starch and sugars in the roots and crown and uses them for regrowth. Farmers used this weed primarily for dairy cattle and secondarily for sheep, goats, horses, and beef cattle. Besides fodder, it is mostly harvested as hay (Putnam et al., 2001) and turned into alfalfa pellet and meal. People also eat it sprouted, using it in salads and sandwiches, and the dehydrated leaves are commercially used in supplementary dietary products such as tablets, powder, and in several herbal teas.

Unfortunately, weed competition affects the quality and quantity of alfalfa. Therefore, to improve its production, genetically engineered herbicide-tolerant

DOI: 10.4324/9781003278566-9

alfalfa has been produced. This means that herbicides can be used to kill the unwanted weeds from a field of alfalfa without the alfalfa itself being affected. This genetically modified (GM) alfalfa was developed by inserting a gene showing tolerance to herbicides (glyphosate) via *Agrobacterium tumefaciens* into the genome of conventional alfalfa. The first such GM glyphosate-tolerant alfalfa – Roundup (Glyphosate) Ready Alfalfa – was developed by FGI (Forage Genetics International) and Monsanto State University in 1998. This genetic engineering technique has also been widely used with canola, cotton, soybeans, and corn. These transgenic plants are identical to non-genetically modified plants but very reduced affinity for glyphosate herbicides (Padgette et al., 1996).

Glyphosate-tolerant alfalfa is commonly used by dairy farms because it is free from unwanted grasses and weeds. Dairy farms are totally dependent on such pure alfalfa crop fields, whereas horses and beef cattle depend on an alfalfa crop field mixed in with grasses and other annual weeds (Putnam, 2005). In the United Sates, about 40 percent of alfalfa is cultivated as pure field, 30 percent with a nurse crop, and the remaining 30 percent as a mixture of alfalfa and other grasses.

9.1.1 *Approval of GE alfalfa (herbicide-tolerant alfalfa)*

Genetically engineered herbicide-tolerant alfalfa was first deregulated by the USDA (United States Department of Agriculture) in June 2005. Deregulation means the removal of regulations or restrictions concerning the introduction and commercialization of a GE plant (Cowan & Alexander, 2013). In 2004, the Animal and Plant Health Inspection Service (APHIS) of USDA received a petition from MCFG (Monsanto Company and Forage Genetics International) to remove restrictions from two glyphosate (herbicide) tolerant alfalfa varieties. Thereafter, APHIS assessed the plant pest risk and compiled an Environment Assessment (EA) of the two herbicide-tolerant events of Roundup Ready alfalfa i.e. J101 and J163. The EA revealed that these two events did not pose plant pest risk (USDA/APHIS, 2010a). This result led to the deregulation of GE alfalfa lines in 2005 and meant that this variety could now be commercialized without needing any further approval from the USDA.

In 2006, the Centre of Food Safety took legal action against the Animal and Plant Health Inspection Service (APHIS)/USDA. The accusers claimed that the EA of GE herbicide-tolerant alfalfa was not clear and/or comprehensive and that an Environment Impact Statement (EIS) should be carried out in the proper way and in full depth. Subsequently, in February 2007, the US District Court of California (Northern) ruled that APHIS had not carried out a full, comprehensive assessment of the economic and environmental effects of this GE alfalfa as set out in the NEPA (National Environmental Policy Act). This ruling led to the dismissal of the deregulation of GE alfalfa. Farmers who had already planted their seedlings were allowed to harvest, use, and sell their crop. But new sale of seeds and plantings were no longer permitted.

Plantings of this GE alfalfa were suspended between 2007 and 2010 until in December 2010, a comprehensive EA was released by APHIS and on 27 January 2011, GE herbicide-tolerant alfalfa was again fully deregulated under the PPA (Plant Protection Act) of 2000. The adoption of this genetically engineered alfalfa in United States led to an increase in the productivity and quality of the plants. GE seeds have increased the plant yield compared with conventional alfalfa seeds by 10–20 percent, resulting in the further adoption of this plant in the near future. The timeline of the journey to approval of GE herbicide-tolerant alfalfa is summarised in Table 9.1.

9.2 Why are genetic modifications required?

Genetic modification is the process of altering the genome of an organism to improve certain features. This process had been performed for thousands of years through conventional techniques such as artificial selection and selective breeding, wherein humans identify the required trait in plants or animals and then take steps to perpetuate and enhance that trait in future generations. The downside is that this is limited to just naturally occurring traits/variations and takes considerable time. However, with new advancements in science, biotechnologists are now able to change a specific gene in a plant to improve, among other things, the quality, productivity, and efficiency of various crops or change other characteristics. This can be done by injecting the specific gene showing the desirable trait from one species to another. Soil and plant bacteria are mostly used for this genetic transformation.

Table 9.1 Chronological events in the approval of GM glyphosate-tolerant alfalfa

Year	*Description*
2004	The Animal and Plant Health Inspection Service (APHIS) of the USDA receive a petition from MCFG (Monsanto Company and Forage Genetics International) to remove the restriction or deregulate the two glyphosate-tolerant alfalfa lines, namely J101 and J163.
2005	Deregulation of GE herbicide-tolerant alfalfa is approved and given permission for fully fledged commercialization.
2006	The Centre for Food Safety in the United States prosecutes the APHIS/USDA, accusing them of not carrying out a comprehensive EIA as set out in the National Environmental Policy Act (NEPA).
2007	The United States District Court of California rules that the assessment report had not been carried out properly as required by NEPA and suspends the deregulation of GE herbicide-tolerant alfalfa lines.
2007–2010	Suspension of the plantings of GE glyphosate-tolerant alfalfa.
2010	APHIS/USDA release the Environment Impact Statement (EIS) for glyphosate-tolerant alfalfa.
2011	GM glyphosate-tolerant alfalfa is entirely deregulated as per the regulations of the Plant Protection Act 2000.

Source: Authors.

With respect to agriculture, farmers face many different challenges such as attack by disease causing insects and the presence of various competing weeds in the desired crop field. All these factors are responsible for reducing the crop yield and contributed to losses in income for farmers. Sometimes, these insects and pests are responsible for the complete destruction of the crop. As a result, farmers used various chemical-based pesticides and insecticides to protect their crop, ignoring the toxic side-effects these can cause. In case of alfalfa, the sprays of herbicides actually led to the destruction of desired crop plant and resulted in a complete loss of yield. In view of all these challenges faced by farmers, biologists turned to developing GM crops which are resistant to pests, disease causing insects, unwanted weeds, and thus eliminating the use of chemicals on the crop.

Another reason to adopt the GM crops is the stagnant yields of the major crops. Globally, the demand for food, grains, and milk and other dairy products is increasing with the growth in the world's population; a population that is expected to approximately double by 2050. To meet the demands of such a population, genetic modification in the crops is required to increase the yield or quality of the food because conventional breeding techniques cannot keep pace with what is required.

Additionally, GMOs are also helpful in the production of vaccines and medicines. Traditionally, these were extracted from animal parts, blood donors etc. But these medicines are known to possess several problems such as transmission of diseases and are unreliable. Vaccines from GMOs do not exhibit such problems and are also reliable and consistent in nature.

9.3 Genes involved in the development of agronomic trait

Glyphosate-tolerant alfalfa (Roundup Ready) is produced by inserting the cp4 epsps gene which is encoding the glyphosate-tolerant form of epsps from *Agrobacterium* sp. (Padgette et al., 1996). Glyphosate inhibits the epsps enzyme of the shikimate pathway in the plants and inhibits the production of the aromatic amino acids tyrosine, tryptophan, and phenylalanine. This leads to retarded growth of the plant and finally death. In GE glyphosate-tolerant alfalfa, the gene product of cp4 epsps is tolerant to glyphosate and is not affected by herbicides. This results in the continuous production of aromatic amino acids in GE herbicide-tolerant alfalfa. The complete description of the inserted gene is:

Gene: cp4 epsps
Gene Product: Herbicide-tolerant form of epsps (5-enolpyruvylshikimate-3-phosphate synthase) (enzyme used in Shikimate Pathway)
Trait: Herbicide (glyphosate) tolerant alfalfa
Source of the gene: *Agrobacterium tumefaciens* (Strain cp4)
Function: Shows high tolerance and very reduced affinity for glyphosate.

9.4 Mechanism of genetic modification leading to desired trait

In modern agriculture, herbicides are mostly used to destroy the weeds as this is a more energy-efficient method than manual/hand weeding. These herbicides are produced by the screening of a number of synthetic compounds (i.e. checking the toxicity level). Today, a number of phytotoxic synthetic herbicides are commercially available. Those herbicides that have both an absence of phytotoxicity and a lack of action/activity toward the desired crop have been selected for development. Nowadays, increasing concerns about phytotoxicity, the hazardous effect on the environment, plants, animals, and the presence of toxic synthetic compounds of herbicides in water and soil, have encouraged more detailed research into the production of new herbicides that are safe for plants, animals, for the environment and that are also biodegradable in nature.

Mostly herbicides perform their action by inactivating the target protein in plants, especially enzymes which are essential for performing vital functions such as photosynthesis, respiration, various biosynthetic pathways etc. Basically, two approaches have been used for the production of herbicide-tolerant plants (Botterman & Leemans, 1988):

1 Modification of the target site or protein or enzyme for the action of herbicide in the plant to make it insensitive to the herbicide or by the overproduction of the target protein or enzymes so that they can perform their normal function even in the presence of herbicide.
2 Introduction of a genetically engineered enzyme system in the plant to detoxify or degrade the herbicide before it can perform a function.

Glyphosate (Roundup) is a broad-spectrum and non-selective herbicide which is used worldwide to eliminate weeds. This herbicide inhibits the Shikimate pathway enzyme epsps (5-enolpryuvylshikimate-3-phosphate synthase) which performs a vital role in the pathway of aromatic amino acid biosynthesis. This epsps enzyme catalyzes the reaction of PEP (phospho-enol pyruvic acid) and S3P (shikimic acid-3-phosphate) to produce epsp, i.e. 5-enolpyruvylshikimate-3-phosphate (Schulz et al., 1985; Duke, 2017).

Weed competition decreases the yield of the desired crop, so the production of herbicide-tolerant alfalfa (Roundup Ready alfalfa) is essential. This is obtained by inserting the cp4 epsps coding sequence obtained from the soil bacteria *Agrobacterium* sp. strain cp4 into the genome of conventional alfalfa. The gene product of cp4 epsps of *Agrobacterium* sp. is similar to the epsps in plants with high tolerance and less affinity for glyphosate (Padgette et al., 1996). In plants, glyphosate acts by binding to epsps and blocking the Shikimate pathway and thus the biosynthesis of aromatic amino acids. In herbicide-tolerant alfalfa, however, by the activity or action of cp4 epsps enzyme, the biosynthesis of aromatic amino acids happens continuously even in the presence of glyphosate.

Agrobacterium tumefaciens is a highly causative soil bacteria for plants. It attacks the plant through wounds and causes crown gall disease. This bacterium is capable of transferring DNA between different kingdoms and therefore acts as a highly effective vector in the production of transgenic plants. *Agrobacterium* mediated genetic transformation requires the presence of two genetic components located on the bacterial Ti-plasmid: T-DNA (transfer-DNA having the gene of interest) and a virulence region having different *vir* genes. The product of these *vir* genes facilitates the movement of the T-DNA into the desired plant. The plasmid which is used in the production of glyphosate-tolerant alfalfa is PV-MSHT4. This vector possesses T-DNA having one copy of cp4 epsps flanked by right and left border sequences. This sequence having cp4 epsps was transferred into the genome of conventional alfalfa. This GE herbicide-tolerant plant produces cp4 epsps protein which is highly tolerant to glyphosate and also meets the demand for the production of aromatic amino acids.

9.5 Impact on modern agriculture – trade of GM crops

Genetic modification is an important application of biotechnology which deals with the manipulation of genetic material in different living organisms, thus permitting them to perform a desired function; and it has been happening for more than 10,000 years by traditional methods such as "artificial selection" and "selective breeding". The first such GM crop – "antibiotic resistant petunia and tobacco" – was successfully created in 1833. Since then, recent advancements have brought about a great revolution – GM crops.

In modern agriculture GM crops are employed mostly to help farmers and to prevent crop loss. The three most important features present in GM crops are:

a Resistance to insects
b Resistance to viruses
c Tolerance to herbicides

Crops which exhibit resistance to insects means that growers can apply fewer sprays of insecticide to protect the plants during the growing season. With herbicide-tolerant plants, farmers can efficiently control weeds without damaging the desired crop. Additionally, no tillage is required before planting the herbicide-tolerant crop, resulting in less labour work, lower soil erosion, and greater maintenance of soil fertility. Herbicide-tolerant GM crops account for 71 percent of all genetically engineered crops grown worldwide (James, 2008).

GM crops have proven to be beneficial with respect to the economy of the modern agricultural system. In total, the yield of GM crops has increased by more than 370 million tonnes worldwide (Raman, 2017). These GM crops have also recorded less ecological and environment impact. To meet the needs of our rapidly growing population, we need to accept GM crops. In current agriculture, the use of GM maize, wheat, rice, and soybean is increasing rapidly to fulfil the demands of this growing population.

Commercial agriculture faces many different challenges:

a a rapidly growing population
b attack by pests and various insects and diseases
c its burden on natural resources

All these challenges can be successfully overcome by adopting GM crops in modern agriculture. It is a solution that also benefits farmers worldwide.

9.6 Benefits of herbicide-tolerant alfalfa

Herbicide-tolerant (HT) alfalfa has many benefits, which are summarised below.

HT alfalfa is used by farmers for effective management of weed in alfalfa fields. Their unique characteristics allow for better or effective weed control throughout the growing season of the GE alfalfa crop. In the United States, 19 states recorded using herbicide-tolerant crops and all were using GE Alfalfa (Heap et al., 2010). Out of the 19, four states – Kansas, Ohio, Illinois, and Missouri – had more than 100 sites of herbicide-tolerant weeds with each state having more than 200,000 acres of land with alfalfa.

Use of GM HT alfalfa reduces crop damage. HT crops provide great benefits to the farmers in the production of a pure alfalfa crop, where weed control management is a big issue. Because glyphosate does not affect the alfalfa whether it is used pre or post planting, glyphosate herbicides can be used on alfalfa at much higher rates without damaging the crop.

Another benefit of HT alfalfa is safety. Glyphosate is among the safest herbicides used on farms because of its low toxicity; it becomes inactive when it combines with the particles of soil and ultimately becomes absorbed into the soil as inorganic compounds. Glyphosate which is not bounded to soil particles is easily degraded by microbes into carbon dioxide. With HT crops, farmers are able to use this herbicide to a greater extent to get better control of annual winter weeds such as garlic, wild onion, chickweeds, and various other perennial weeds such as dodder and nutsedge, and noxious weeds such as groundsel.

HT crops require reduced or no tillage before planting because of the efficient and flexible weed control that they provide. This reduced tillage brings many benefits to the environment as it leads to less soil erosion and run off and maintained soil fertility.

9.7 Trade in the global market

Genetically engineered plants have been commercially available in the USA for various major crops since 1996. Since that time, their adoption rate has increased rapidly. Nowadays, according to the Economic Research Service, USDA, in 2020 more than 90 percent of US crops such as cotton, corn, soybean, and alfalfa are produced by means of genetic engineering (USDA ERS, 2020). These genetically engineered crops are usually classified as insect-tolerant (Bt), herbicide-tolerant (HT) or can have a combination of both Bt and HT traits. Various other genetically engineered crops have also been developed by USDA. These exhibit, among other things, resistance against fungus, virus resistance, drought resistance, and enhanced vitamin, oil, and protein content. Of these types of genetically engineered crops, insect resistant and herbicide resistant are those most commonly used in the United States. Herbicide-tolerant varieties are most commonly used in the alfalfa crop.

Traditionally, alfalfa was utilized as a forage crop by farmers and it was planted locally or only a few kilometres away from where it was to be used. However, over the past two decades, trade in alfalfa has increased and at the international level its status has risen from a minor export to that of a major export, with trade approximately doubling in the last 16 years (Putnam et al., 2019). The demand of this crop has increased because of a number of factors: the limited production of forage, various water issues in Middle East and Asia, as well as the increasing the demand for milk and various other dairy products worldwide. Currently, the United States is one of the most hay exporting countries followed by Australia, Spain, Canada, and Italy. Various other countries are also involved in its export such as France, Germany, Mongolia, Romania, and Argentina but these countries make up less than 10–11 percent of the total world trade. Among the hay crops, alfalfa dominated in most of the countries. The greatest demand of this hay in Asia is led by Japan, followed by China and Korea. In 2017 Japan imported more than 2 million metric tons of this GE alfalfa and other grass hays compared with Korea's 212,000.

In recent years, China has appeared as the world's topmost importer of this alfalfa and represents the largest market for the United States (Wang & Yang, 2019). Imports to China increased from 19,601 metric tons in 2007 to 1.38 million metric tons in 2017. China imported 1.16 million tons of alfalfa hay in the initial months of 2021. The United States exhibited an export of nearly 1 million ton to China in the same period leading to 84.1 percent of China's total imports, with a value of $430 million (USDA FAS, 2021). Even so, China's import of this alfalfa is increasing significantly. This makes the US market more powerful as farming sectors in the USA are mostly dependent on the export markets. More exports to China lead to greater revenue for US farmers, helping to reduce its trade deficit with China (Jiang et al., 2010; Gale et al., 2015; USDA FAS, 2017).

The two main factors responsible for the increasing imports of this alfalfa in China are:

a) The rapidly increasing number of dairy farms
b) Less/limited growth of high-quality alfalfa in China

Dairy farms in China have grown rapidly from small/backyard (10 cows) to medium-sized farms (10–99 cows) and now to large farms with more than 100 cows. From 2008 to 2016, small/backyard farms in China decreased from 2.37 million to 1.29 million. The related increase in large commercial farm thus resulted in an increase in the demand for high-quality forage. Alfalfa is considered the "Queen of Forage Crops" because of its larger amounts of digestible protein, minerals, vitamins, and fibres as compared to other forage crops. Hence it is the preferred choice as feed for dairy cows and beef cows throughout the world; particularly for increasing the yield and quality of milk in the former. Although the dairy sector is new alternative in China, it gives a high yield and requires more use of alfalfa. This increase in demand by large dairy farms in China has thus led to an increase in alfalfa imports (Ministry of Agriculture in China, 2017).

The biggest challenge for Chinese importers is the introduction of genetically engineered alfalfa. Two genetically engineered alfalfa lines available commercially in United States are glyphosate-tolerant alfalfa (Roundup Ready) and HarvXtra (with less lignin content). However, China does not permit the import of these two genetically engineered traits of alfalfa. Therefore, only conventional alfalfa varieties which are free of these two traits are produced by US growers for export to China. In United States, the Export Authority assesses the Low Level Presence (LLP) of these two traits. China will reject the import of alfalfa even with very low concentrations of these traits or a LLP of only 0.01 percent concentration. This may change in years to come if China gives permission for GM traits.

In 2020, Japan was a major importer of alfalfa, with an import value of US $808.2 million. Its maximum import is of conventional alfalfa. Yet a petitioner in Japan applied for approval to import of GM HT alfalfa for use as a feed for dairy and beef cattle. Additionally, Japan gave permission of GE crop of up to 1 percent of imports and these crops have completed safety assessment (Ministry of Agriculture, Forestry, Fisheries announcement, 2002). This tolerance in Japan gives an allowance for those alfalfa producers that export their plants to Japan and who are concerned about trace levels of genetically engineered alfalfa in non-biotech shipments. Table 9.2 lists the top ten exporters/importers of alfalfa by value.

9.8 Health and ethical issues/risks

Genetically engineered crops have been developed to improve quality standards yet with same safety as traditional conventional crops. In the USA, all

Table 9.2 Top ten exporters and importers of Alfalfa along with export and import value in 2020

	Top exporter	*Export value ($ million)*	*Top importer*	*Import value ($ million)*
1.	United States	1,600	Japan	808.2
2.	Australia	371.7	China	617.6
3.	Spain	254.6	South Korea	324.0
4.	Canada	170.6	Unites Arab Emirates	313.3
5.	Italy	148.5	Saudi Arabia	256.0
6.	France	88.9	United States	111.0
7.	Netherlands	59.2	Taiwan	87.4
8.	Armenia	50.9	Germany	58.2
9.	South Africa	43.7	Netherlands	56.2
10.	Germany	34.3	Switzerland	47.7

Source: Authors.

these genetically engineered crops and their products are assessed by Food and Drug Administration (FDA) before they are sold into the market. Food products from GE crops have been shown to have the same chemical composition as foods from traditional crops, albeit with different or enhanced nutritive compositions.

The introduction of genetically engineered herbicide-tolerant alfalfa and its effects on the environment and human health have been widely discussed in the United States. It has been proven that the introduction of this herbicide-tolerant alfalfa has doubled the yield as well as improving the quality of the product. Besides its benefits, its negative effects on humans, animals and on the environment are also elaborated below.

9.8.1 *Effect on human health*

A small amount (~2.5 percent) of alfalfa seeds are used for human consumption as dietary supplements, sprouts, and in various herbal teas. Seeds of this (normal) alfalfa grown for sprouts are subject to more restrictions throughout the season concerning the application of chemicals. Chemicals which are applied to this plant must be evaluated properly. In genetically engineered alfalfa, the gene cp4 epsps is transferred into the genome of alfalfa via *Agrobacterium tumefaciens*. It is therefore necessary to evaluate the risks or safety of this genetically engineered plant on human health, particularly that:

1 The protein which is produced by the cp4 epsps gene in GE alfalfa is functionally and biochemically equivalent to the naturally occurring epsps protein in the crop.

2 This cp4 epsps protein is easily susceptible to the proteolytic digestion in stomach of humans by the action of gastric juice.
3 *Agrobacterium tumefaciens* which is used for the transformation experiment is not known for its toxicity to humans and animals. The cp4 epsps gene of this soil bacteria acts as a genetic donor for many other herbicide-tolerant crops and these crops have been consumed by humans and animals without any toxicity since 1996. The product of this gene shows no similarity to any protein which can cause adverse health effects on animals and humans.
4 The EPA also suggests that genetically engineered alfalfa having cp4 epsps poses negligible risk to animal and humans (ASSAF, 2010; NRC, 2004).

9.8.2 Impacts on the environment

The United States Department of Agriculture (USDA) has performed various field tests to determine the safety of GE herbicide-tolerant alfalfa. The APHIS of the USDA has also assessed the plant pest risk and compiled an environmental assessment of the two herbicide-tolerant events of alfalfa, i.e. J101 and J163. The environment assessment, and the second more thorough one, revealed that these two lines did not pose plant pest risk. As mentioned previously, this result led to the deregulation of GE alfalfa lines in 2005 and again in 2010 once they had been reassessed after a short spell of re-regulation from 2007 to 2010.

Yet, the deregulation of this GE alfalfa has resulted in an increase in the use of glyphosate-based herbicides and to their greater exposure by other crop plants and various aquatic and terrestrial plants. Based on their report, the US EPA (1993, 2006) considers glyphosate to be a low toxic and ecologically low risk herbicide (Cerdeira & Duke, 2006). Yet, there remain many issues to be addressed and discussed.

a Glyphosate acts as non-selective herbicide and therefore removal of restrictions from glyphosate-tolerant alfalfa would adversely affect non-target plants because of its increased use. These plants when exposed to herbicides experience high rates of toxicity. Also, if these plants use the Shikimate pathway for the production of aromatic amino acids, then glyphosate would block this pathway, leading to the death of plants (USDA-FS, 2003).
b Excessive use of herbicides in GE alfalfa crop fields can cause the crop to be more susceptible to diseases.
c Some research has also revealed that glyphosate-based herbicides might be responsible for the interference with the translocation of phloem as well as the movement of various important minerals like manganese from the soil via their binding with the glyphosate either in roots or in the soil. As a result, the quality of the seed and the growth of the plant may be affected.

d The alfalfa plant forms a symbiotic relationship with *Sino rhizobium meliloti* – a nitrogen-fixing bacterium – and plays an important role in the conversion of atmospheric nitrogen into fixed nitrogen in the soil. When a GE herbicide-tolerant alfalfa plant is sprayed with glyphosate, its nitrogen fixation property is not altered and, overall, shows no negative impact on the yield of alfalfa and nitrogen fixation.

e GT alfalfa has been shown to have a negative impact on local and regional biodiversity by reducing the number of insects, birds and small mammals. However, the rate and extent of this impact on biodiversity depends on the spatial and geographical region of the crop field (Castillo et al., 1999)

f HT alfalfa shifts the mode of agriculture from conventional plough-based to less or no tillage production systems which has resulted in lower greenhouse gas emissions owing to the reduced use of tractors (Brookes & Barfoot, 2017).

9.9 Current status and future prospects

Doubts about the safety of GM crops in the agricultural industry could mean that their introduction would be difficult in poor and developing countries. Yet, in spite of these controversies, their adoption has increased rapidly because of the increased productivity, high food safety, less environmental impact, and great profit to the world's economy. The introduction of GM crops has increased crop yield of the crop by 22 percent, which results in a 68 percent increase in farmers' profits (Klümper & Qaim, 2014).

With respect to alfalfa, weeds compete with desired crop plants for essential nutrients, sunlight, water, and space. This competition between the intended crop and unwanted weed results in a loss of yield. To prevent this yield loss, active weed management was carried out by the use of herbicides. However, the spraying of non-selective herbicides also led to the death of the alfalfa crop. Therefore, development of a HT alfalfa variety became essential. Because of several benefits of this GM HT crop, its current importance and the demand for it keep on rising rapidly. Livestock is also main factor as to why increasing importance is being placed on this alfalfa hay. There is immense pressure on the livestock market to fulfil the global demand for high value animal protein and this will, of course, significantly increase the demand for GM Alfalfa and its high yield. Globally, the market of GM HT crops is projected to register a Compound Annual Growth Rate of 6.9 percent during 2021–2026, although from 2020 onwards it has been adversely affected by the Covid-19 pandemic and the subsequent disruption to market supply chains. That said, the future prospects of GM crops for meeting the need of the ever-growing global population in a responsible and sustainable manner remain extremely promising.

9.10 Conclusion

Adoption of GM HT alfalfa contributes to maximum profits for farmers by increasing yield, giving better weed control management, being energy efficient, reducing injury to the target crop, and low weed management costs. Its introduction also reduces the environment impact of weed management. Because of all these unique characteristics, this GM HT crop is highly preferable to the farmers. This HT technology also shifted the mode of agriculture from highly plough-based to one of less or no tillage in the crop field before planting. This less or no tillage agricultural system results in reduced emissions of greenhouse gases because of the lower use of tractors and also helps to preserve the fertility of the soil.

References

ASSAF (Academy of Science of South Africa) (2010). *Workshop Proceedings Report GMOs for African Agriculture: Challenges and Opportunities*. Available at: www.assaf.org.za

Botterman, J., & Leemans, J. (1988). Engineering of herbicide resistance in plants. *Biotechnology and Genetic Engineering, Reviews*, 6(1), 321–340.

Brookes, G., & Barfoot, P. (2017). Environmental impact of genetically modified (GM) crop use 1995–2015: Impact on pesticide use and carbon emission. *GM Crops & Food*, 8(2). doi:10.1080/21645698.2017.1309490.

Castillo, M., Flores, M., Mavingui, P., Romero, E. M., Palacios, R., & Hernandez, G. (1999). Increase in alfalfa nodulation, nitrogen fixation, and plant growth by specific DNA amplification in *Sinorhizobium meliloti*. *Applied and Environmental Microbiology*, 65, 2716–2722.

Cerdeira, A. L., & Duke, S. O. (2006). The current status and environmental impacts of glyphosate-resistant crops: A review. *Journal of Environmental Quality*, 35, 1633–1658.

Cowan, T., & Alexander, K. (2013). *Deregulating Genetically Engineered Alfalfa and Sugar Beets: Legal and Administrative Responses*. Congressional Research Service Report for Congress, R41395. Available at: https://www.fas.org/sgp/crs/misc/R41395.pdf

Duke, S. O. (2017). The history and current status of glyphosate. *Pest Management Science*, 74(5), 1027–1037.

Gale, F., Hansen. J., & Jewson, M. (2015). *China's Growing Demand for Agricultural Imports*. EIB-136, US Department of Agriculture, Economic Research Service. Available at: https://www.ers.usda.gov/webdocs/publications/43939/eib-136.pdf

Heap, I., Glick, H., Glasgow, L., & Beckie, H. (2010). International Survey of Herbicide Resistant Weeds Database. Available at: https://www.weedscience.org/Home.aspx

Iqbal, S., Ahmad, M., Zaffar, G., & Dar, E.A. (2018). Alfalfa (*Medicago sativa* L): A valuable source of forage/fodder crop under cold arid conditions of Ladakh (Jammu and Kashmir). *Research Journal of Chemical and Environmental Science*, 6(4), 74–84.

James, C. (2008). Global status of commercialized biotech/GM crops. ISAAA Brief No. 39. Available at: https://www.isaaa.org/resources/publications/briefs/39/download/isaaa-brief-39-2008.pdf

Jiang, Y. Z., Zhang, J. B., Nan, Z. B., Wang, L. J., Wang, Q., Parsons, R., & Zhang, G. (2010). China's dairy market: Trends, disparities, and implications for trade. *China Agricultural Economic Review*, 2, 356–371.

Klümper, W., & Qaim, M. (2014). A meta-analysis of the impacts of genetically modified crops. *PLoS One*, 9(11), e 111629.

Ministry of Agriculture of China (2017). *National Alfalfa Industry Development Plan 2016–2020.*

NAFA (2015). *Best Management Practices for Roundup Ready Alfalfa Seed Production.* NAFA Coexistence Documents, February.

NRC (National Research Council) (2004). *Safety of Genetically Engineered Foods: Approaches to Assessing Unintended Health Effects* (Washington, DC: National Academies Press).

Padgette, S. R., Re, D. B., Barry, G. F., Eichholtz, D. E., Delannay, X., Fuchs, R. L., … & Fraley, R. T. (1996). New weed control opportunities: Development of soybeans with a Roundup Ready gene. In *Herbicide-Resistant Crops*, edited by S. O. Duke (Boca Raton, FL: CRC Press), 53–84.

Putnam, D. (2005). Market sensitivity and methods to ensure tolerance of biotech and non- biotech alfalfa production systems. In *Proceedings of the35th California Alfalfa & Forage Symposium, Visalia, CA, 12–14 December, 2005.* UC Cooperative Extension, Agronomy Research and Extension Center, Plant Sciences Department, University of California, Davis, CA.

Putnam, D., Russell, M., Orloff, J., Kuhn, J., Fitzhugh, L., Godfrey, L., … Long, R. (2001). *Alfalfa, Wildlife, and the Environment: The Importance and Benefits of Alfalfa in the 21st Century.* California Alfalfa & Forage Association, Novato, CA. Available at: https://agric.ucdavis.edu/files/242006.pdf

Putnam, D. H., Matthews, W., Hanon, T., & Sumner, D. (2019, January 22). Long-term hay exports have increased dramatically, but are dampened by recent trade disputes. *Alfalfa & Forage News.* Available at: https://ucanr.edu/blogs/blogcore/postdetail.cfm?postnum=29204

Raman, R. (2017). The impact of genetically modified (GM) crops in modern agriculture: A review. *GM Crops and Food*, 8(4), 195–208.

Schulz, A., Krüper, A., & Amrhein, N. (1985). Differential sensitivity of bacterial 5-enolpyruvylshikimate-3-phosphate synthase to the herbicide glyphosate. *FEMS Microbiology Letters*, 28, 297–301.

US Department of Agriculture (USDA), Animal and Plant Health Inspection Service (APHIS) (2010). *Glyphosate-Tolerant Alfalfa Events J101 and J163: Request for Non-regulated Status. Final Environmental Impact Statement – December 2010.*

US EPA (United States Environmental Protection Agency) (1993). *Reregistration Eligibility Decision (RED): Glyphosate.* Technical Report, Office of Prevention, Pesticides and Toxic Substances. EPA 738-R-93-014.

US EPA (United States Environmental Protection Agency) (2006). *Glyphosate – Human Health Risk Assessment for Proposed Use on Indian Mulberry and Amended Use on Pea, Dry.* Office of Prevention, Pesticides, and Toxic Substances. PC Code: 417300, Petition No: 5E6987, DP Num: 321992, Decision No. 360557.

USDA ERS (2020, November 17). Adoption of Genetically Engineered Crops in the U.S. Dataset, USDA, Economic Research Service. Available at: https://www.ers.usda.gov/data-products/adoption-of-genetically-engineered-crops-in-the-us.aspx

USDA FAS (United States Department of Agriculture Foreign Agricultural Service) (2017). Spain: Fodder demand in the Middle East drives Spanish alfalfa area growth. June 16. Available at: https://www.fas.usda.gov/data/spain-fodder-demand-middle-east-drives-spanish-alfalfa-area-growth

USDA FAS (United States Department of Agriculture Foreign Agricultural Service) (2021). *Market Overview – Alfalfa Hay and Other Forages*. Available at: https://apps.fas.usda.gov/newgainapi/api/Report/DownloadReportByFileName?fileName=Market%20Overview%20-%20Alfalfa%20Hay%20and%20Other%20Forages%20_Beijing%20ATO_China%20-%20People%27s%20Republic%20of_11-28-2021.pdf

USDA FS (United States Department of Agriculture, Forest Service) (2003). *Glyphosate – Human Health and Ecological Risk Assessment: Final Report*. Technical report, USDA, Forest Service. Available at: https://www.fs.fed.us/r5/hfqlg/publications/herbicide_info/2003_glyphosate.pdf

Wang, Q., & Yang, Z. (2019). China's alfalfa market and imports: Development, trends, and potential impacts of the US–China trade dispute and retaliations. *Journal of Integrative Agriculture*, 19(4), 1149–1158.

10 Rainbow papaya

Naveet Kaushal, Nitesh Sood, Mandeep Kaur and Talwinder Singh

10.1 Introduction

Papaya, *Carica papaya*, belongs to *Caricaceae* family and is the most financially significant member in the family (Carvalho & Renner, 2012). It is the world's third most farmed tropical crop. Plantations have the potential to produce fruit over a period of 9 months, with a tree density of 1,500–2,500 trees per hectare considered financially viable to produce an average yield of between 125,000 and 300,000 pounds for each acre (Gonsalves, 1998). Among the world's fruits, papaya is the torchbearer for its retinol, ascorbic acid, potassium, folate, niacin, riboflavin, iron, fibre, and protein degrading enzymes such as Papain and Chymopapain (Huerta-Ocampo et al., 2012). Papaya (*Carica papaya* L.) is an aggressive, short-lived, tropical plant, which is nowadays more cultivated for its papain, pectin, and antibacterial substances (Niklas & Marler, 2007). In modern times, it is farmed extensively in tropical and subtropical lowland areas in India, Brazil, Peru, USA, Ethiopia, Mexico, Indonesia, and China (Evans & Ballen, 2012).

Hallé et al. (1978) conferred with Corner's reference model which explains a monopodial, solitary, orthotropic, and non-branching trunk with the flowers occurring mainly along a single meristem, as well as some additional inflorescences which can be responsible for its undefined growth. It has huge palmate-shaped vegetation. The female flowers develop at the axils of the leaf petiole, which then bear fruits; their male counterparts construct a spike. Papaya has three types of flower: female, male, and hermaphrodite. In their natural habitat, papaya is most frequently diclinous (dioecy), whereas farmed papaya trees are often dioecious or hermaphrodites (Carvalho & Renner, 2012; Chávez-Pesqueira et al., 2014). Cultivars can interbreed, creating descendants with the same personality as the mother plant (Manshardt, 1992). Nearly half the trees of the gynodioecious farmed papaya plantation are hermaphrodites, with the remaining 25 percent female; dioecious plants are also found (VanBuren et al., 2015).

There are several observable differences in morphology and reproduction between the three sexes. The female flower has an ovary that emerges in the plant stem through the axils of the petioles. This ovary ought to primarily

DOI: 10.4324/9781003278566-10

obtain the pollen of an alternative plant (male or hermaphrodite), since the stamens are not the morphological part of female florae, consequently both fertilized to form fruits bearing seeds in it. The same as female flowers, they blossom on the pedicels, sub peduncular with some secondary flowers. Hermaphrodite plants are the most favored by commercial growers owing to their reliable yields via self-pollination. However, elevated temperatures and water stress can consequently transform a hermaphrodite to a male (Gonsalves, 1998; Ming et al., 2007). The visible differences in male flowers are in their being branching, drooping, peduncular, with multi-inflorescences. Aborted pistils inhabit the interior of the flowers through a sex-linked region that functions like an XY chromosome in the stamens bearing a high pollen burden (Carvalho & Renner, 2012). Sex is governed by unique Y chromosome regions in males and hermaphrodites: Yh in hermaphrodites and Y in males (VanBuren et al., 2015).

In tropical climates, female plants produce flowers and fruits throughout the year; nevertheless, despite uninterrupted blooming, fruit set is limited during arid seasons in subtropical areas (Gonsalves, 1998). Freshly consumed mature fruits can be processed into jam, jelly, marmalade, and confectionery. Fruits that are "green" or immature can be used as vegetables. Papain, a cysteine endopeptidase, is a protein commonly found in the latex of many plants, including papaya (Endo et al., 2002). Latex is a complex emulsion of chemicals possessing a vital proteolytic action which is exploited for its antimicrobial character. It is a major allergen to humans owing to its connection with cysteine proteases. It is great for tenderising red meat, brewing beer, and treating warts and marks on the skin (Ming et al., 2008).

10.2 Why genetic modifications are required

Genetic modifications possibly offer a safe, significant means to expand food security and progress sustainable agriculture. *C. papaya* is still the only species in the genus *Caricain* which further belongs to the *Caricaceae* family. Globally it is greatly affected by the papaya ring spot virus (PRSV), which only can be mitigated with genetic transformation technologies. In Hawaii, USA two PRSV virus-resistant genotypes – "SunUp" (red fleshed) and "Rainbow" (yellow fleshed) – have been successfully developed (Gonsalves, 1998). The development of these varieties was a boon to scientists and to the growers who were suffering from the effects of the virus before the launch. This virus transmits in seconds to the healthy plants, with onset of morphological changes occurring three weeks after infection. Young seedlings, unable to bear the shock, die rapidly and never bear fruit in their life span. However, in older trees it causes chlorosis leading them to produce insignificant sized fruits and ultimately dooming them to a slow death. It is evident from the past that this demon has grasped entire plantations in the Puna region which has destroyed the morale and the returns for the growers.

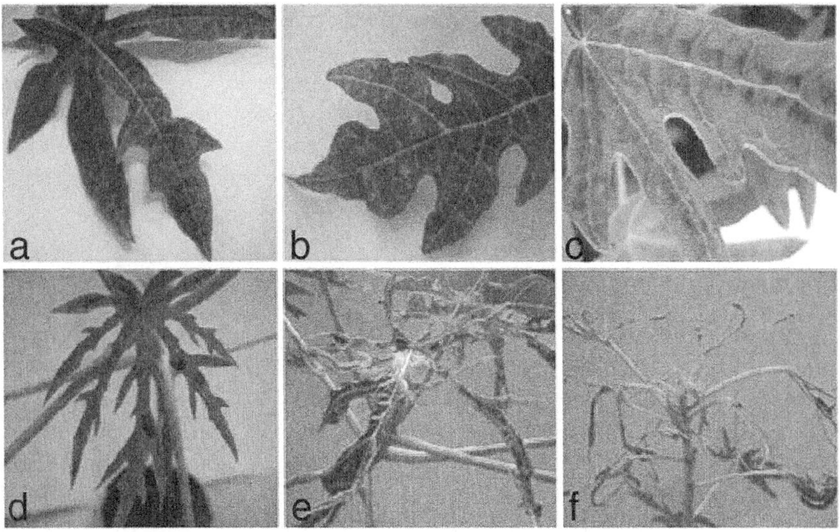

Figure 10.1 Papaya ring spot virus: (a) vein clearing; (b) mild mosaic; (c) clear intense color zones distributed on the adaxial leaf surface; (d) swelling and intense color zones on the adaxial leaf surface; (e) slight deformation on young leaves; (f) leaf distortion.

Genetic transformation was the need of the hour because conventional breeding programs could only offer minute improvisations in the texture of the fruit and its sugar content. Ethylene, a phyto-hormone, is responsible for the major textural changes during ripening, making the fruit more perishable after harvesting. This softening is a multifaceted, regulated process caused by the action of membrane perforating enzymes, like poly-galacturonase and beta-galactosidase. These notorious enzymes are now targeted via the genetic engineering procedure for reducing the perishability of the fruit and thus reduce losses.

In some nations, the regulation of cutting down the infected trees after the onset of the first symptom of PRSV is a major hurdle in the success of traditional papaya breeding programs, and hence molecular markers can play a crucial role as saviors for these programs. The genus *Vasconcelleais* is the nearest relative of *C. papaya* and is majorly used in-cross breeding programs (Chávez-Pesqueira et al., 2014). Specifically, papaya is a cross-pollinated plant which has failed all the attempts to achieve intra-breeding success and to produce an ultra-breeding line via a combination of gynodioecious and dioecious lines. This property is mainly exploited by breeding programs to glue together the concerned genetic characteristics of gynodioecious and dioecious lines simultaneously (Aquilizen, 1987).

The sequencing of the papaya genome began in 2004 via an integrated multi-institutional partnership at the University of Hawaii, and it was finished in 2007. Papaya is a eukaryotic plant with nine sets of chromosomes and a genome measuring 372 Mbp/1C (Bennett & Leitch, 2005). Within the genus C. *papaya*, there are numerous phenotypic differences; it has been crossed with various *Vasconcellea* species – including C. *papaya* with *Vasconcellea cauliflora*, C. *papaya* with *Vasconcellea quercifolia*, and C. *papaya* with *Vasconcellea pubescens* – to produce resistant inter-specific hybrids (Drew, 2005). The purpose of the genome sequencing study was to map and clone papaya genes that would actually strengthen monetary gains and crop productivity.

10.2.1 Plant tissue culture

Amongst the several transformation techniques, the main one is plant tissue culture. Immature zygotic embryos are the most successful transformers of papaya plants (Fitch & Manshardt, 1990). With a biolistics system, somatic embryos conferred 41 percent success against kanamycin resistance in a time span of six months (Ming et al., 2008a). However, the main drawback of the biolistics transformation technology in papaya is that it is not applicable for the superior females or hermaphrodites but only to a progeny of the plants derived from open pollination. This result suggests the absence of barriers to self-fertilization (Drew, 2005).

10.3 Genes involves in agronomic trait development

10.3.1 Tolerance to biotic and abiotic stress

Adaptation to biotic and abiotic stresses in the wild has always been a major concern for scientists (Agarwal et al., 2011). Transcriptional level gene regulation has been a major draw for researchers. The responsible genes are called transcriptional factors (TFs) and these proteins have the capability of activating or limiting downstream target gene transcription. For the aforementioned transcription events, sequence-specific binding to target gene promoters is required (Tennant et al., 2001). The WRKY TFs perform a dual role in plant responses to biotic and abiotic stressors (Agarwal et al., 2011). Cold, dryness, salt, wounding, and disease elicit a polygenic cooperative defence response in flowers, which is analogous to the floral response to abiotic and biotic stressors. To date, however, the application of DNA polymorphism sequencing methods and the identification of vital parameters in genetic mapping for papaya progeny is scarce. Moving in this direction, Amplified Fragment Length Polymorphisms (AFLP) have been used to create the first high-density linkage map of 1498 base pairs (Ma et al., 2004). This highly dense map was generated via 706 Simple Sequence Repeat (SSR) markers (Chen et al., 2007). These repeats were further used to strengthen the map resolution by using 14 quantitative trait loci (QTL) pertaining to fruit quality that were revealed by applying 277 AFLP and 712 SSR markers. So far, 21 QTLs for fruit quality features have been discovered, allowing for the isolation of key genes and the advancement of marker-assisted selection strategies. SNPs and In-Dels that cause phonotypical changes have been identified using Next Generation Sequencing (NGS) to measure the genetic map of fruit quality (Nantawan, et al., 2019; Luo et al., 2016; Martínez-García, 2013).

10.4 Mechanisms of genetic modification leading to desired traits

10.4.1 Spread of virus and transformation of papaya

Before the several failed attempts to develop transgenic papaya, in 1987 Cornell University isolated the virus's coat protein as a source of the desired gene. This intracellular obligate parasite exploits wounds, either manually or via vector-mediated processes, to infect the host cell. There are two types of infestation: short distance (cell to cell movement) and long distance (cell to cell movement). Most specific reactions are synchronised with internal parasite motions. PRSV has a limited host range, only affecting the *Caricaceae*, *Chenopodiaceae*, and *Cucurbitaceae* families. C. *papaya*, *Cucurbita pepo*, and *Cucumis metuliferus* are PRSV's propagation hosts. *Chenopodium quinoa* and *Chenopodium amaranticolor* are the PRSV lesion assay hosts (Tripathi et al., 2008). There are two strains of PRSV which can be differentiated on the basis of their host range (Gonsalves & Ishii, 1980). PRSV type P (PRSV-P) could indeed infect papayas, however PRSV type W (PRSV-W) only can infect cucumbers. Post-transcriptional gene silencing (PTGS) is now commonly used in papaya agriculture to control PRSV. PRSV vectors that feed on infected papaya plants and then feed on healthy papaya plants require CP and HC-Pro to transmit the virus. R1 plants of the transgenic papaya line 55–1 aid in the expression of the single coat protein (CP) gene, resulting in the development of two lines. Because it is an F1 hybrid of SunUp and the non-transgenic papaya cultivar Kapoho, UH SunUp (SunUp) is homozygous for the CP gene insertion while UH Rainbow (Rainbow) is hemizygous for the CP gene. The resistance of the transgenic papaya is RNA-mediated via post-transcriptional gene silencing, according to RNA analyses and nuclear run-on tests (Tennant et al., 2001).

10.4.2 Characteristics of papaya hybrids

'UH Rainbow' papayas differ from inbred papayas. Inbred lines have a 2:1 sex-segregation ratio, whereas hybrids have a 1:1 sex-segregation ratio (one hermaphrodite to one female plant). As a result, farmers will need to leave five seedlings per hole (rather than three) to assure a 97 percent chance of one hermaphrodite per hole at flowering thinning time. Plants grown from F1 hybrid seeds do not reproduce true, thus farmers must acquire new F1 seeds for each planting. Post-transcriptional gene silencing (PTGS), a host defence response to foreign RNA, provides the basis for papaya's RNA-mediated resistance (Kung, 2015). Sequence similarity between a transgene and the matching viral genome is required for PTGS-mediated transgenic resistance (Giladi et al., 1995).

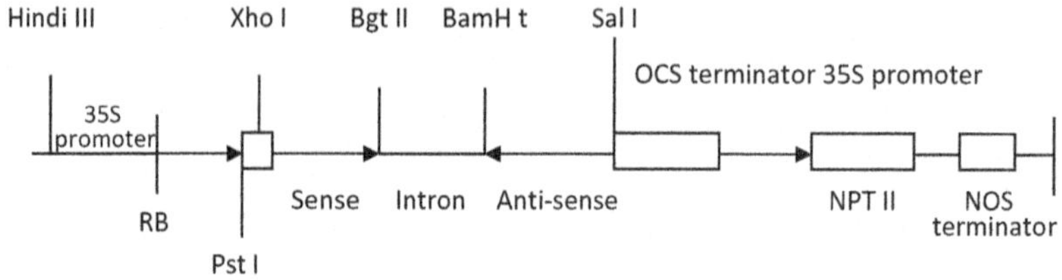

Figure 10.2 Components of constructed RNAi vector pCAMBIA 2300–35S-OCS.35S promoter = *Cauliflower Mosaic Virus* promoter; NOS terminator = nopaline synthase gene terminator; OCS terminator = octopine synthase teminator; NPTII = neomycin phosphotransferase gene; Sense and Anti-sense = conserved 544 bp fragment of CP gene and inverted repeat sequence.

Source: Authors.

10.5 Impact on modern agriculture – trade of GM crops

GM crops are a boon to farmers and the environment. All the world has embraced GM crops. In 2013, transgenic soybeans made up 93 percent of all US agricultural soybean output. Crop yields have increased by 21 percent and, thanks to GM technology, not only has the yield increased but effective pest control with less crop losses have also been reported. Moreover, the use of pesticides, and thus their cost, has reduced by almost 40 percent. It is a known fact that the GM seeds are more expensive, but these extra charges are counterbalanced and even rewarded via savings in chemical and mechanical pest control (Klümper & Qaim, 2014). The monetary benefits to the growers of the Rainbow papaya are almost 69 percent, which is at par with the conventional cultivars and thus ensuring the returns to the stake holders (Klümper & Qaim, 2014). Organic farmers believe that the use of biotechnology to eliminate the use of chemicals in agriculture and generate more sustainable and ecologically friendly crops is the greatest outcome of GMOs.

10.6 Benefits to society

The risk associated with introducing a novel GM crop means that these foods are tested more carefully than any other foods. And yet, they have repeatedly stood up to this stringent testing (Tripathi et al., 2011). The gene inserted to the papaya represents a minute fraction of the virus genome which is expertly attenuated before the ligation or, alternatively, the final product is efficiently transformed, ensuring the health safety of the consumer (Voosen, 2011). Even after 15 years of introduction, no adverse health effects have ever been reported. Consumers who really are worried about the consequences of pesticides upon land, farmers, and even themselves, should accept the Rainbow papaya considering it has mitigated the need for the chemicals. Rainbow papaya has proved to be savior for its industry. Puna's papaya industry might have died if the Rainbow papaya had not been developed. Previous solutions of spraying, chopping down sick trees, and evacuating contaminated farms were all unsuccessful –the parasite quickly spread and wiped out the vegetation. It would be a sad irony if, after developing a safe method to protect the Puna papaya industry, we let our misplaced worries of biotechnology do what the virus could not do – destroy the industry.

10.7 Trade in the global market

Amongst the 60 nations in global production, India, a developing economy, is also a main grower of papaya. Worldwide production in 2019/20 was nearly 13.9 Mt, hence showing its good acceptability in the global market. Asia is also a main grower of this fruit crop, holding a 57 percent share of global production (FAO STAT, 2021).

The genetic engineered papaya lines Sun Up and Rainbow were commercialized in 1998 and in 2006 Huanong 1 was made available to the general masses in China. It was presented via a single proprietor of Chinese origin through a Chinese seed company in the shape of micro-propagated seedlings. Notably, numerous other Asian countries are also in this race of flourishing ground trials of PRSV-resistant varieties that still, to date, have not received a nod from nations like Jamaica, Taiwan, and Thailand. The majority of the previous English colonies are in the preliminary phases of testing and hence are still skeptical contributors in the present global trade (Mendoza et al., 2008). Even after the generation of a line, several other concerns such as acceptability by the masses, regulations, and intellectual property rights continue to play a big role in deciding their future fate as world transformers.

Another significant issue encountered by papaya growers are the post-harvest losses from insect pest fungal malformations, metabolic changes, and above all the fruit's perishability. Post-harvest management, which is the most crucial aspect, is solely dependent upon the availability of electricity and the growers' connectivity to urban areas, both of which are heavily compromised in the developing nations. These foresaid factors are responsible for the majority of losses in the underdeveloped and developing nations of the world. Many growers can only find a market for their produce if they personally bear the cost of freight and packing. India is the leading papaya producer, and is also applying its hands to transgenic papaya production, contributing nearly 40 percent of the global papaya yield since 2017 (Evans & Ballen, 2012, updated 2018).

10.8 Health and ethical issues

10.8.1 Disruption to the food chain

Whenever a transgenic plant is introduced to the commercial market, a main issue is the development of minor pests which have been neglected while dealing with the major pests of the same variety. As already mentioned, though minor, if ignored these pests have great potential to affect the main predators of the food chain. Consequently, this will disrupt the ecological balance and adds a potential new living entity to the transgenic game (Snow & Palma, 1997). In fact, insects and weeds will almost certainly evolve ways to counteract our sophisticated design of transgenic crops in a few years as a result of human-made stresses in their environments (Bawa & Anilakumar, 2013).

10.8.2 Resistance to antibiotics

Humans, the most intelligent creation of the Almighty, have always been tricked by Mother Nature since the inception of Earth. In the contemporary era, the genetic engineered transformants are the 'lining systems' of the desired traits, but the hidden pressure caused by them on the ecosystem can prove to be a big cost for their developers. Apart from the main desired characteristics, a number of unpreventable characteristics like weed resistance and tolerance to antibiotics have always been a question of worry. For example, in direct transfer systems, a microbe capable of holding antibiotic resistance genes is also a potent virulent agent. Thus, because of the risk of gene shuffling, there has always been considerable worry about the safety of GE foods and ultimately their effect on the health of consumers (Tabashnik, 1994).

10.9 Current status and future prospects

Until now, certain nations have feared to give the green light for growing transformed crops on their land. However, investigations into the standardization of Ceruloplasmin (CP) transgenic resistance and fruitful field trials with controlled parameters have, in the last 20 years, become more accepted by governments. The perseverance and hard work put into researching the intricate network of the associated living entities and the continuous ground trials stretching for more than a decade have facilitated the generation of mono, di, and ultra resistance against PRSV. These outcomes are in agreement with genome sequences of papaya from the University of Hawaii (Ming et al., 2008; Yu et al., 2009). Quantitative trait loci along with marker-assisted selection have made possible future exhaustive searches targeted on the dietary aspects of papaya (Cai et al., 1999). *Vasconcellea* ferals, tolerant to a plethora of diseases, will certainly become a focus of research with the hope of transferring this tolerance to other, commercial varieties. For example, a survey conducted from 2006 to 2008 at the plantations of papaya at diverse sites in India, discovered that maximum leaf curl disease occurrence (23.50 percent) was recorded in Lucknow district.

10.10 Conclusions

Nuclear alterations have enabled the human race to take control of today's agriculture produce. Previously grown varieties were really hard to cultivate and harvest but owing to scientific advancements, present varieties are easy to grow and reap and are more nutritious. In reality, rejection of the genetic modification of foods will ultimately mean the rejection of agriculture (allianceforscience.cornell.edu.). On the opposite side, these transgenic plants are now open heartedly in the local area (hawaiipapaya.com).

References

Agarwal, P., Reddy, M. P., & Chikara, J. (2011). WRKY: Its structure, evolutionary relationship, DNA-binding selectivity, role in stress tolerance and development of plants. *Molecular Biology and Cell Reproduction*, 38, 3883–3889.

Aquilizen, F. A. (1987). Breeding systems for fixing stable papaya inbred lines with breeding potential for hybrid variety production. In *The Breeding of Horticultural Crops*. Food and Fertilizer Technology Center for the Asian and Pacific Region, Taipei, 101–106.

Bawa, A. S., & Anilakumar, K. R. (2013). Genetically modified foods: Safety, risks and public concerns – a review. *Journal of Food Science and Technology*, 50(6), 1035–1046.

Bennett, M. D., & Leitch, I. J. (2005). Nuclear DNA amounts in angiosperms: Progress, problems and prospects. *Annals of Botany*, 95, 45–90.

Cai, W., Gonsalves, C., Tennant, P., Fermin, G., Souza, M., Sarindu, N., Jan, F., Zhu, H., & Gonsalves, D. (1999). A protocol for efficient transformation and regeneration of *Carica papaya* L. *In Vitro Cell and Plant Development Biology*, 35, 61–69.

Carvalho, F. A., & Renner, S. S. (2012). A dated phylogeny of the papaya family (*Caricaceae*) reveals the crop's closest relatives and the family's biogeographic history. *Molecular Phylogenetics and Evolution*, 65, 46–53.

Chávez-Pesqueira, M., & Núñez-Farfán, J. (2016). Genetic diversity and structure of wild populations of *Carica papaya* in Northern Mesoamerica inferred by nuclear microsatellites and chloroplast markers. *Annals of Botany*, 118, 1293–1306.

Chávez-Pesqueira, M., Suárez-Montes, P., Castillo, G., & Núñez-Farfán, J. (2014). Habitat fragmentation threatens wild populations of *Carica papaya* (*Caricaceae*) in a lowland rainforest. *American Journal of Botany*, 101, 1092–1101.

Chen, C., Yu, Q., Hou, S., Li, Y., Eustice, M., Skelton, R. L., ... Ming, R. (2007). Construction of a sequence-tagged high-density genetic map of papaya for comparative structural and evolutionary genomics in Brassicales. *Genetics*, 177, 2481–2491.

Chiang, C.H., Lee, C.Y., Wang, C. H., Jan, F.-J., Lin, S.-S., Chen, T.-C., ...Yeh, S.-D. (2007). Genetic analysis of an attenuated *Papaya ringspot virus* strain applied for cross-protection. *European Journal of Plant Pathology*, 118(4), 333–348.

Drew, R. A. (2005). *Development of New Papaya Varieties for Southeast and Central Queensland*, Report No. FR02024, Project No. FR02024. Horticulture Australia Limited, Sydney, Australia.

Endo, S., Kasahara, T., Sugita, K., & Ebinuma, H. (2002). A new GSM-MAT vector containing both ipt and iaaM/H genes can produce marker free transgenic tobacco plants with high frequency. *Plant Cell Reproduction*, 20, 923–958.

Evans, E. A., & Ballen, F. H. (2012, updated 2018). *An Overview of Global Papaya Production, Trade, and Consumption*. Publication FE913 (Gainesville, FL: University of Florida). https://edis.ifas.ufl.edu/pdf/FE/FE913/FE913-4432060.pdf

FAO STAT (2021). Production quantities of papayas by country, average 2019–2020. Crops and Livestock Products. http://www.fao.org/faostat/en/#data/QCL/visualize

Fitch, M. M. M., & Manshardt, R. M. (1990). Somatic embryogenesis and plant regeneration from immature zygotic embryos of papaya (*Carica papaya* L.). *Plant Cell Reproduction*, 9, 320–324.

Giladi, E., Raz, E., Karmeli, F., Okon, E., & Rachmilewitz, D. (1995). Transforming growth factor-beta gene therapy ameliorates experimental colitis in rats. *European Journal of Gastroenterology & Hepatology*, 7, 341–347.

Gonsalves, D. (1998). Control of papaya ringspot virus in papaya: a case study. *Annual Review of Phytopathology*, 36, 415–437.

Gonsalves, D., & Ishii, M. (1980). Purification and serology of papaya ringspot virus. *Phytopathology*, 70(11), 1028–1032.

Hallé, F., Oldeman, R. A. A., & Tomlinson, P. B. (1978). Papaya. In *Tropical Trees and Forests – An Architectural Analysis* (Berlin: Springer).

Huerta-Ocampo, J. Á., Osuna-Castro, J. A., Lino-López, G. J., Barrera-Pacheco, A., Mendoza-Hernández, G., & De León-Rodríguez, A. (2012). Proteomic analysis of differentially accumulated proteins during ripening and in response to 1-MCP in papaya fruit. *Journal of Proteomics*, 77, 2160–2169.

Klümper, W., & Qaim, M. (2014). A meta-analysis of the impacts of genetically modified crops. *Journal of Agromedicine*, 12(1), 27–37. www.bit.ly/metaanalysisgmo

Kung, Y.-J., You, B.-J., Raja, J. A. J., Chen, K.-C., Huang, C.-H., Bau, H.-J., ... Yeh, S.-D. (2015). Nucleotide sequence-homology-independent breakdown of transgenic resistance by more virulent virus strains and a potential solution. *Scientific Reports*, 5, 98–104.

Luo, C., Shu, B., Yao, Q., Wu, H., Xu, W., & Wang, S. (2016). Construction of a high-density genetic map based on large-scale marker development in mango using specific-locus amplified fragment sequencing (SLAF-seq). *Frontiers in Plant Science*, 7, 1–8.

Ma, H., Moore, P. H., Liu, Z., Kim, M. S., Yu, Q., Fitch, M. M. M., ... Ming, R. (2004). High-density linkage mapping revealed suppression of recombination at the sex determination locus in papaya. *Genetics*, 166, 419–436.

Manshardt, R. (1992). Papaya. In *Biotechnology of Perennial Fruit Crops*, edited by F. A. Hammerschlag & R. E. Litz (Oxford: Cambridge University Press), 489–511.

Martínez-García, P. J., Parfitt, D. E., Ogundiwin, E. A., Fass, J., Chan, H. M., Ahmad, R., ... Crisosto, C. H. (2013). High density SNP mapping and QTL analysis for fruit quality characteristics in peach (*Prunus persica* L.). *Tree Genetic Genomes*, 9, 19–36.

Mendoza, T., Laurena, C., & Botella, J. (2008) Recent advances in the development of transgenic papaya technology. *Biotechnology Annual Review*, 14, 423–462.

Ming, R., Hou, S., Feng, Y., Yu, Q., Dionne-Laporte, A., Saw, J. H., Senin, P., ... Alam, M. (2008). The draft genome of the transgenic tropical fruit tree papaya (*Carica papaya* Linnaeus). *Nature*, 452, 991–996.

Ming, R., Yu, Q., & Moore, P. H. (2007). Sex determination in papaya. *Seminars in Cell & Developmental Biology*, 18(3), 401–408.

Nantawan, U., Kanchana-Udomkan, C., Bar, I., & Ford, R. (2019) Linkage mapping and quantitative trait loci analysis of sweetness and other fruit quality traits in papaya. *Plant Biology*, 19, 1–11.

Niklas, K. J., & Marler, T. E. (2007). *Carica papaya* (Caricaceae): A case study into the effects of domestication on plant vegetative growth and reproduction. *Journal of Applied Botany*, 94, 999–1002.

Snow, A. A., & Palma, P. M. (1997). Commercialization of transgenic plants: Potential ecological risks. *Bioscience*, 47(2), 86–96.

Tabashnik, B. E. (1994). Evolution of resistance to *b*acillus *t*huringiensis. *Annual Reviews of Entomology*, 39, 47–79.

Tennant, P., Fermin, G., Fitch, M. M., Manshardt, R. M., Slightom, J. L., & Gonsalves, D. (2001). Papaya ringspot virus resistance of transgenic Rainbow and SunUp is affected by gene dosage, plant development, and coat protein homology. *European Journal of Plant Pathology*, 107, 645–653.

Tripathi, S., Suzuki, J. Y., Carr, J. B., McQuate, G. T., Ferreira, S. A., Manshardt, R. M., ... Gonsalves, D. (2011). Nutritional composition of Rainbow papaya, the first commercialized transgenic fruit crop. *Journal of Food Composition and Analysis*, 24, 140–147.

Tripathi, S., Suzuki, J. Y., Ferreira, S. A., & Gonsalves, D. (2008). Papaya ringspot virus-P: Characteristics, pathogenicity, sequence variability and control. *Molecular Plant Pathology*, 9, 269–280.

VanBuren, R., Zeng, F., Chen, C., Zhang, J., Wai, C. M., Han, J., ... Ming, R. (2015). Origin and domestication of papaya Y^h chromosome. *Genome Research*, 25, 524–533.

Voosen, P. (2011). Crop savior blazes biotech trail, but few scientists or companies are willing to follow. *New York Times*, September 21. www.bit.ly/cropsavior

Yu, Q., Tong, E., Skelton, R. L., Bowers, J. E., Jones, M. R., Murray, J. E., ... Ming, R. (2009). A physical map of the papaya genome with integrated genetic map and genome sequence. *Biomolecular Genomics*, 10, 371–375.

11 Genetically modified canola

Rosy Bansal and Jasmeet Kour

11.1 Introduction

The genetic machinery of living organisms such as animals, plants, and microbes can be modified by gene technology. This is the process of combining the genes from different organisms via recombinant DNA technology, and the subsequent organism is then considered to be a 'genetically modified' (GM), 'genetically engineered', or 'transgenic' organism. Many common commercially produced transgenic crops like soybeans, corn, cotton, and canola are herbicide and pesticide resistant. Sweet potato and rice are genetically engineered to enhance their iron and vitamin content in order to curb chronic malnutrition in Asia. GM crops capable of surviving extreme weather conditions are also being tested in the field for economic reasons. Other innovations include: fruits such as bananas that have the potential to immunise against communicable ailments like hepatitis B; sea animals that mature sooner; nuts, fruits, and similar trees that can be harvested years earlier; and flora used in making new plastic materials with exclusive qualities.

Canola is a member of the *Brassicaceae* family and is a flowering plant with vivid yellow flowers. *Brassica napus*, *Brassica rapa*, and *Brassica juncea* are all members of this family. *B. napus*, sometimes known as rapeseed, is a plant native to the Mediterranean region and Northern Europe. Rapeseed was discovered to be a high-erucic acid crop in about 2000 BC, with the oil containing more than 40 percent erucic acid. Because of concerns about erucic acid concentration based on animal research, rapeseed oil with its higher erucic acid content was produced in modest volumes in Northern USA for only industrial, non-food purposes. Canadian scientists, on the other hand, succeeded in increasing the value of earlier rapeseed cultivars via classical plant breeding in 1976, resulting in the development of economically consumable canola cultivars.

A gene from *Bacillus thuringiensis* subsp. kurstaki has been used to generate several lines of GM cotton plants that provide improved protection against key lepidopteran pests. In terms of macronutrients and gossypol, the lines were said to be "essentially identical" to parent lines. Aflatoxin and

DOI: 10.4324/9781003278566-11

cyclopropenoid fatty acid levels were lower than in ordinary seeds. However, it was unclear whether the GM and non-GM lines were similar due to the use of erroneous statistics, especially when environmental pressures could have unpredictable impacts on anti-nutrient/toxin levels.

Canola has grown to be one of the world's most important oilseed crops in the last 40 years. Palm and soybean oil are followed by canola oil which stands as the third vegetable oil crop by volume. According to the USDA (European Commission, 2017), out of various oilseeds, the production of soyabeans was reported as the highest in the world from 2012–2016 followed by rapeseed or canola. In addition to this, as per the statistic for rapeseed production globally, in 2020/2021, it was Canada which was reportedly acclaimed as a pioneer in canola production worldwide with a production volume of 19.5 million metric tons along with the worldwide production of canola oil estimated at 29.2 million metric tons in 2020/2021 (Shahbandeh, 2022).

GM technology has a bright future in meeting future world requirements for food, feed, and fibre in a sustainable and responsible manner. However, GM crops are simply one part of the solution. We need all of the resources at our disposal, including conventional and organic food production techniques, to fulfil the targeted yields, nutritional quality, and the sustainable production required in the future.

As the United Nations Food and Agricultural Organization (FAO) reported, 868 million people worldwide were hungry or malnourished in 2012. Despite the fact that these numbers have decreased since the year 1990, there is still much to be done. Consider the following: the number of people who have died as a result of starvation and malnutrition, a preventable disease, is greater than the combined death tolls from AIDS, malaria, and tuberculosis. Canola is a popular oilseed that is grown in many places of the world. For agronomic reasons, it's critical to produce canola that contains a large oil content. For example, in Iran, because a substantial portion of the vegetable oil for human use is imported, oilseed culture and management are crucial for increasing productivity. The Foreign Agricultural Service (USDA) has reported that in 2020/21, the area, yield, and production of rapeseed (canola) in the world is estimated at 34.86 million hectares, 2.11 metric tons per hectare, and 73.59 million metric tons, respectively. Apart from this, the area, yield, and production of rapeseed (canola) in India is estimated at 6.70 million hectares, 1.27 metric tons per hectare, and 8.50 million metric tons, respectively (Foreign Agricultural Service/USDA, 2022). There is increased interest in the use of seeds with high oil content for health, industrial, and pharmacological purposes, with an average oil content of 40 percent on the dry seed weight basis.

Plant oils play a significant role as additions to food (cooking oils and salad dressings, margarine, and mostly processed foods) and for commercial purposes (soaps and detergents, cosmetics items, paints and lubricants), and also provide direct nutritional benefits to the human diet. Plant based oils are mainly derived from the seeds of the plant and the types of fatty acids produced determine the physical qualities of the oil. That is why some plant-based oils are more suitable for certain purposes than others, such as cooking with peanut oil versus cosmetics and industrial lubricants with jojoba oil. Researchers are able to design a plant-based oil to be more suited for a given usage by utilising genetic engineering techniques to modify the genetic makeup of the fatty acids present in the plant.

Canola plants are between 2 and 4 ft tall with a stem in middle that branches out. The yellow flowers of canola have four petals. The seed pods range in length from 1 to 1.5 inches and are about 1/8inch wide. The oil-rich seeds – which resemble those of mustard in appearance – are picked from these pods and then the vegetable oil is produced by either mechanical extraction or a solvent extraction process. The 'leftovers' from this process are known as rapeseed or canola meal, which is a high-protein 'oil meal' that is frequently used as cattle fodder or as a part of livestock feed. The production of canola oil involves a number of procedures. As per the Canola Council of Canada, these are:

a **Cleaning the seeds:** To remove contaminants such as plant stems and dirt, canola seeds are separated and cleaned.
b **Seed flaking and conditioning:** Seeds are pre-heated to around 95°F (35°C) before being 'flaked' by roller mills, which burst the seed's cell wall.
c **Cooking seeds:** A series of steam-heated ovens fry the seed flakes. This heating process usually takes 15–20 minutes at 176–221°F (80°–105°C).
d **Pressing:** In a succession of screw presses or expellers, the cooked canola seed flakes are crushed. With this technique 50–60 percent of the oil from the flakes can be removed, leaving the remaining oil to be recovered through other methods.
e **Solvent extraction method:** To obtain the residual oil, the remaining seed flakes, which contain approximately 18–20 percent oil, are further broken down with hexane.
f **Desolventizing:** Hexane is then removed from the canola meal a third time by steaming it at 95–115°C.
g **The oil is processed:** Steam distillation, phosphoric acid exposure, and filtering via acid-activated clays are all used to purify the extracted oil.

The Oil Stability Index (OSI; a method that determines the relative resistance of fats or oils to oxidation) of classic canola is 10 hours, compared to 7 hours for soybean salad oil. The OSI in some high oleic canola oils is 13–22 hours (Scarth & McVetty, 1999).

11.2 Why are genetic modifications required?

GM technologies give living organisms new features, or phenotypes. The new characteristics are obtained by using DNA modification procedures to alter the organism's genetic make-up. DNA is found in every cell of an organism. A chain of fragments called genes makes up DNA strands. Each gene carries instructions for making a specific protein, and the production of various combinations of proteins is the focus of research which determines an organism's characteristics and qualities. Genetic modification is likely to alter the production of specific proteins or to create entirely new proteins by modifying the underlying DNA in a gene, which can change the organism's features. This can be performed either by removing the DNA segment entirely or by transferring an anticipated gene of one organism to the DNA of another organism.

At least 18 crops have been tested in the field with transgenic plants expressing Bt (*Bacillus thuringiensis*) toxins that provide insect resistance. However, only potato, cotton, and corn are commercially accessible. Since 1996, pest-resistant transgenic crops have been grown in broad regions across the United States. Bt corn and Bt cotton were planted on around 13 million hectares in 2001, largely in the United States and Canada. Bt crops are also grown in China, India, South Africa, and Argentina, with China seeing the most growth. As a result, some commercial experience with insect-resistant transgenic crops is being gained (Zhang et al., 2016).

Canola is the largest oilseed crop in China and comprises 20 percent of world production (Hu et al., 2017). USDA data reported that Canola production in 2014/2015 in China reached 14.7 million tons (via http://apps.fas.usda.gov/psdonline/). Herbicide-resistant GM canola is also currently licensed for commercial production in Australia. Yet, due to marketing concerns, South Australia and Tasmania have passed legislation prohibiting the cultivation of GM canola.

Problems connected to the insect populations acquiring tolerance to pesticides can be predicted based on previous expertise with traditional plant breeding as well as field and laboratory molecular breeding research. Pests have always been able to overcome the resistance that traditional breeding programs have introduced into crops and some Lepidopteran species have adapted to Bt toxins generated by Bt biopesticide sprays. According to the current data, resistance to GM Bt crops is therefore expected. Furthermore, there have been reports in the field that secondary pest outbreaks will emerge.

Herbicide-resistant rapeseed cultivars are far more common than canola. The canola oil consumed mainly in the United States hails from rapeseed plants farmed in Western Canada, with the exception of a small region in the upper Midwest. Since the mid-1990s, Canadian regulatory bodies have approved a number of GE herbicide-resistant rapeseed types, which now represent more than 60 percent of the country's total rapeseed acreage (most of these varieties have also been accepted for planting and food usage in the United States) (Smyth, 2014).

Pesticide-tolerant crops raise the question of the usage of agricultural herbicides to a certain extent. Weeds have for several decades been controlled by farmers with the spraying of post-emergent herbicides straight over the crops because of the increasing utilization of no-till agricultural techniques to reduce soil erosion. Many agronomists employ a combination of herbicides to suppress weeds as most of the herbicides have a small range of plants which they can kill (otherwise the crop plants would be killed too). Weeds competing for soil nutrients, water, and sunlight can drastically affect agricultural yields when they grow in the same field as crop plants.

Herbicide-resistant varieties of rapeseed require fewer herbicide applications owing to the elimination of numerous herbicide treatments in favour of a single herbicide. However, because different herbicides are sprayed at different rates, it's unclear whether herbicide-resistant varieties actually require a lower amount of herbicides in terms of pounds applied. That said, according to some reports, the herbicides used on genetically altered rapeseed are less toxic than the herbicides they replace.

11.3 Genes involved in agronomic trait development

Transgenic GM canola seeds with very high oil content have been developed by a group of researchers. According to Ton et al. (2020), seed-specific overexpression of the two genes BnLEC1 and Canola BnL1L, which is being controlled by storage protein 2S-1 promoter of truncated canola, also known as the napin A promoter, greatly improves the oil content of the transgenic plant to an adequate level with no discernible detrimental impacts on other main agronomic parameters.

Seed oil content is a complicated quantitative variable regulated by the environment and is under the control of a large number of genes; increasing the oil content of seed is the main aim of breeding *B. napus* (Tan et al., 2011). The embryo genotype, maternal genotype, and cytoplasm of seed influence the oil content and have a role in determining seed oil content. Genetic components from genome cytoplasm (mitochondria as well as plastid genomes) govern cytoplasmic effects, which are normally inherited from the mother. In spite of the fact that cytoplasmic effects in agricultural crops have been known for decades, with the exception of those governing cytoplasmic male sterility, the molecular mechanisms by which agronomic traits are determined by them are mostly unknown, and hence has become the focus of intense research (Liu et al., 2019).

In 2017 Villanueva-Mejia and Alvarez developed canola transgenics with altered fatty acid compositions by the use of the *B. napus* variety "CY2" as the transgenic recipient of BnFAE1 – a fragment involved in the production of fatty acids with very long chain. The hypocotyls were co-cultured with *Agrobacterium tumefaciens* EHA105 to introduce the genetic make-up in canola cells after placing a fragment of BnFAE1 driven by napin A promoters. The researchers constructed the seed-specific knockdown of BnFAE1, dramatically modifying the composition of fatty acids due to the high erucic acid content of CY2 of about 40 percent and low oleic acid of around 20 percent. They showed that in F1 hybrid seeds resulting from crossings between BnFAE1-Ri lines and high erucic acid varieties, the RNAi construct of BnFAE1 could efficiently affect the levels of mRNA in the BnFAE1 gene. At the conclusion of their research, Villanueva-Mejia and Alvarez had obtained canola transgenic lines with significantly lower erucic acid levels (less than 3 percent).

11.4 Mechanism of genetic modification leading to desired trait

The era of scientific crop development began around 1900, at which time the significance of Gregor Mendel's nineteenth-century work on pea trait inheritance became generally understood. Since then, a variety of approaches for increasing agricultural yields, quality, and resistance to disease, insects, and environmental stress have been developed.

11.4.1 Development of new gene combinations

Most of the plant breeding operations rely on manual cross-pollination between genetically diverse plants. Development of hybrid varieties by crossing two parental strains to produce offspring with greater vigour and induced mutations to create desirable varieties are two other approaches included in the traditional plant breeding toolbox. The transmission of only the required gene or genes to the recipient plant means that GM technology is significantly more exact. GM canola can be made by scientists by first inserting the gene(s) coding for certain features into a plant cell and then regenerating the plant via tissue culture (Zhang et al., 2016).

11.4.2 Methods of gene transfer

11.4.2.1 Direct gene transfer

Microparticle bombardment is the most common technique for introducing exogenous DNA. Sanford developed the process in the late 1980s. Naked, designed DNA is coated on gold or tungsten microparticles, which are then driven by pressurised helium into specified tissues, such as embryonic tissues from the seed or meristems. Other methods of DNA delivery include electroporation (moving negatively charged DNA down an electric potential gradient) into protoplasts, microinjection, chloroplast transformation, silicon-carbide slivers, and by means of mesoporous silica nanoparticles.

11.4.2.2 Indirect transfer using bacterial vehicle

The usage of *Agrobacterium tumefaciens* to introduce exogenous genes into plant cells ushered in a new era. Plants are infected by the soil bacterium *A. tumefaciens*, which causes a gall at the crown. The bacteria modify the plant's genome, inducing cell multiplication and allowing the plant to create modified amino acids as a specialised food source for themselves. The bacteria have a tumour-inducing plasmid (the "Ti-plasmid") that allows them to insert genes; researchers hijack the plasmid by adding "designer genes" into the Ti-T-DNA plasmid's (transfer-DNA) portion.

The inserted DNA fragment contains one or more genes, which contain the DNA sequence information encoding specific proteins along with DNA segments that regulate the production of the proteins. The inserted fragment also sometimes contains a marker gene to easily identify plants that have incorporated the transferred genes, also known as transgenes, into their chromosomes. As mentioned above, transgenes can be entrenched into the target plant in a number of ways. However, transgene insertion is a time-consuming process, with only a small percentage of plant cells or tissues successfully incorporating the foreign gene. Various strategies are used to identify the small percentage of cells/tissues that have actually been transformed.

The next step is then to develop those cells or tissues, in which the foreign gene has been successfully incorporated, into whole plants capable of producing seed. This is carried out through a process called tissue culture; that is, growing plants on agar or a similar medium in the presence of plant nutrients and hormones under controlled environmental conditions (Belide et al., 2013).

11.5 Impact on modern agriculture – trade of GM crops

The introduction of GM crops to the market in 1996 is synonymous with a growing interest in how agricultural practices may indirectly harm the environment (Fontes et al., 2002). In many regions of the world, discussions regarding GM pest-resistant crops have raised concerns about their possible implications on biodiversity, particularly on non-target creatures such as insect herbivores and natural enemies, as well as soil bacteria. The possibility of insect pest resistance development is also a matter of concern since it could jeopardise vital natural resources like the bacteria *Bacillus thuringiensis* Berliner – a natural microbial pesticide. Another concern usually raised is the possible hazard posed by pest-resistant GM crops in that the inserted gene could be transferred from the GM crops to their wild or weedy cousins, and the repercussions of such a transfer. These are the questions that have been the focus of research for the last few years. Yet, in recent years the introduction of GM crops using modern biotechnology is being hailed as a major agricultural technical achievement on par with the green revolution of the early 1970s. With the expansion in global population, one of the best techniques to deal with pest management paving the way to food security, agricultural sustainability, and environmental protection has been proven in the form of these genetically engineered crops

Parallel to this high anticipation, the use of genetic engineering to molecularly develop cultivated plants has sparked a worldwide discussion concerning their environmental impact. The information gathered has been utilised for assessing the environmental risk GM crops, which is a critical step in the global biosafety governing process (Anderson, 2019).

Globally, pest-related agricultural losses – crop diseases and epidemics as well as the insect problem – are linked with a 20–40 percent loss of crops each year in terms of dollars, costing the agriculture business more than US $290 million every year. Various crop diseases and pest infestation are expected to spread poleward (2.7 kilometres annually), as understood by occurrences of coffee leaf rust and wheat rust in Central America (Bebber et al., 2013). These occurrences have generally been ascribed to a combination of globalization and climate change, which has given rise to a greater plant, insect, and disease mobility, as well as an increase in disease vectors. GM crops have mostly succeeded in reducing the aforementioned key agricultural difficulties while also bringing several benefits to growers around the world. Over the 17-year period from 1996–2013, they created 117.6 billion dollars in international farm revenue. From the years 2010–2012, worldwide annual net income scaled by 34.3 percent. Furthermore, GM crops lowered pesticide (active ingredient) use by 37 percent and environmental influence (insecticide and herbicide) by 18 percent while increasing global output by 22 percent. More than 300 million acres of traditional crops would have been required to satisfy the same yield demands, significantly exacerbating agriculture's current environmental and socioeconomic challenges (Oliver, 2014).

Canola is planted as a rotation crop in Australia, giving agriculturalists a viable alternative and the benefits from ongoing cereal yield stages and the weed/pest mechanisms that go with it. Broad-leafed weed and cereal root disease control, as well as improved subsequent cereal crop growth, are other advantages. It is the most widely produced rotation crop in Western Australia, covering 400,000 to 800,000 acres of farmland and being the most effective of four rotation crops (oats, lupin, canola, and field peas). Between 2002 and 2007, Western Australia alone produced 440 million tonnes of canola worth Aus$200 million. Nonetheless, canola is known to be high-risk crop, especially in terms of blackleg disease caused by the fungus *Leptosphaeria maculans* and weeds such as charlock (*Sinapis arvensis*), wild radish (*Raphanus raphanistrum* L.), and Buchan (*Hirschfeldia incana*) which intensify anti-nutritional content and composition (James, 2011).

The two herbicide-resistant GM canola varieties – Roundup Ready (Monsanto) and InVigor (Bayer CropScience) – were introduced into Australia in 2008–09. Roundup Ready is glyphosate resistant because it has gene variations with transformed 5-enolpyruvylshikimate-3-phosphate synthase (EPSP) and a glyphosate oxidoreductase gene. It received OGTR (Office of Gene Technology Regulator) approval after experiments revealed that it had a lower environmental effect than canola varieties which were tolerant to triazine (43 percent), and remains the only OGTR-approved GM canola to date (ACIL Tasman, 2007). Agriculturalists have benefited from the canola variety Roundup Ready since it has helped to reduce weeds that were previously problematic to control. GM canola harvest area (hectares) increased to 14 percent in 2014, up from 4 percent in 2009, showing an almost three-fold growth and Australia's increasing contribution to biotechnology produce acreage (Fisher & Tozer, 2009).

The 'Oscar' cultivar, as well as leading commercial varieties like AV Jade, AV Spectrum, AV Zircon, and the high oleic content of the Monola variety, were used as transformation recipients. The binary vector GA7 mod-B was used to alter B. napus cotyledonary petioles with just minor changes. Disease-ridden cotyledonary petioles were grown on MS media as previously defined but with the addition of 5 mg L1 PPT for selection and biweekly subcultures. To extend the shoots for an additional two weeks, shoots of the resistant callus were moved to shooting media (medium supplemented with 0.1 mg L1 gibberellic acid + 3 mg L1 silver nitrate + 250 mg L1 cefotaxime + 5 mg L1 PPT). One or two leaves from healthy young branches were taken and placed in rooting media (1/2 MS with 1 mg L1 NAA + 20 mg L1 ADS + 3 mg/mg L1 $AgNO_3$ + 250 mg L1 cefotaxime). To validate the T-DNA insertion, DNA was extracted from the leaf using an isolation kit of plant DNA (Bioline, Alexandria, NSW, Australia). The lines that had been confirmed were transplanted into pots and grown in a glasshouse (Belide et al., 2013).

The most recent triumph in canola enhancement is the approval of DHA canola for commercial release, which influences natural resources of fatty acid production while relying heavily on modern genetic and gene technology (Petrie et al., 2020). Glucosinolates in rapeseed meal have high toxicity and palatability concerns and have thus been avoided for a long time, yet now they have been shown to include a number of bioactivities that enhance plant defence against non-adapted diseases and insect pests, treatment of tumours, and the nutritive value of canola meal. Glucosinolate production, along with its functions in physiological and biological developments in *Brassica napus*, have been studied using genomic and transcriptome techniques (Lu et al., 2019). Pathogen resistance and herbicide tolerance improvement require ongoing attention because disease causing agents and weeds can develop resistance after the introduction of pathogen-resistant or herbicide-resistant varieties, with the rate of resistance varying depending on how those varieties are managed and rotated in conjunction with other management practices (Hasan et al., 2021).

11.5.1 GE canola approval

In the USA GM crops must meet state and federal criteria before they can be sold commercially. The United States Department of Agriculture (USDA), the United States Environmental Protection Agency (EPA), and the United States Food and Drug Administration (FDA) are all involved in the Coordinated Framework for Biotechnology Regulation. The Animal and

Plant Health Inspection Service (APHIS) of the USDA is in charge of overseeing field testing of agricultural biotechnology products. Such items, that comprise certain genetically altered plants, microorganisms, and invertebrates, are classified as "regulated articles" and are subject to either a notice or permit process (Fernandez-Cornejo et al., 2014). To allow technology providers to undertake field testing, APHIS gives approvals of field proclamations of those organisms (primarily genetically engineered plants) which are classified as "regulated articles."

Genetically engineered plants which meet six particular conditions (specified according to the regulations) go through a shortened administrative process called notification. As part of a notification, applicants must give evidence on the kind of the plant and genes being introduced, the particular genetic alterations, the size and scope of the introduction, and the origin of destinations for movement or the location of a field trial. An APHIS authorization is necessary for genetically engineered plants that do not to meet the standards for notification. This procedure entails a more thorough examination. Here, certified applicants must also provide information on the performance of the test, together with specified methods for minimising the potential harm to other plants and ensure that the organisms being tested are contained and do not persist once the field test is completed.

In response to a petition from AgrEvo, the first GM herbicide-tolerant (GEHT) canola plant was approved in January 1998. It was resistant to phosphinothricin, a (glufosinate ammonium) herbicide. APHIS deregulated the first glyphosate-tolerant canola cultivar in January 1999. Between 1999 and 2002, several glufosinate-tolerant cultivars were deregulated (USDA and APHIS, 2010, 2015). A list of GE canola/rapeseed deregulations is shown in Table 11.1.

Table 11.1 GE canola/rapeseed traits and deregulation dates

S.No.	*Company*	*Trait*	*Deregulation date*
1.	Calgene	Oil profile altered	31 Oct 1994
2.	AgrEvo	Glufosinate tolerant	29 Jan 1998
3.	Monsanto	Glyphosate tolerant	27 Jan 1999
4.	AgrEvo	Glufosinate tolerant and pollination control	22 March 1999
5.	Aventis	Glufosinate tolerant	23 Dec 2002
6.	Aventis	Glufosinate tolerant and pollination control	23 Dec 2002
7.	Monsanto	Glyphosate tolerant	2 Jan 2003
8.	Pioneer	Glyphosate tolerant	18 July 2013
9.	Monsanto	Glyphosate tolerant	25 Sep 2013

Source: Authors.

11.6 Benefits to society

In 2013 Lin et al. found evidence that monounsaturated fatty acids (MUFAs) had a more beneficial effect on cardiovascular health by regulating plasma lipids, lipoproteins, low-density lipoprotein (LDL) oxidation susceptibility, and insulin sensitivity when compared to saturated fatty acids (SFAs). The following qualities distinguish canola oil: low levels (~7 percent) SFAs and significant values of MUFAs and polyunsaturated fatty acids (PUFAs), comprising 61 percent oleic acid, 21 percent linoleic acid, 11 percent alpha-linolenic acid (ALA), ~0.97 percent plant sterols, and ~1200 ppm tocopherols.

The US FDA approved the health claim for canola oil in 2006: "Limited and not conclusive scientific evidence suggests that eating about 1.5 tablespoons (19 grammes) of canola oil daily may reduce the risk of coronary heart disease due to canola oil's unsaturated fat content" (Lin et al., 2013). Canola oil will be able to replace the same quantity of saturated fat while not increasing the overall number of calories consumed per day to reach this putative advantage. The validity of total cholesterol (TC) and low-density lipoprotein (LDL-C) as biomarkers for coronary heart disease were used to support this argument. Recent investigations suggest that canola oil consumption has a positive impact on biomarkers other than blood lipids (Petrie et al., 2020).

11.7 Trade in the global market

Many countries are presently cultivating transgenic crops on a commercial scale. In 2001, 13 million hectares of land were dedicated to transgenic pest-resistant cultivars around the world. These cultivars have a lot of advantages, but they also have a lot of drawbacks. Before they are permitted for commercial use, they must undergo environmental assessments, as mandated by regulatory biosafety standards (Abbas, 2018).

Canola is one of numerous oilseeds that compete for market share on the international stage. Canola products, such as oil and meal, compete with other oils and meals, like those derived from soybeans, cottonseed, peanuts, and sunflower seed – all of which are significant competitors. Soybeans are a substantial source of meal since they are low in oil, and soybean meal has proven to be particularly effective in specific animal feedstuff regimens. Globally, palm oil is another key market rival to be aware of. Canola oil competes with this in some applications. Palm oil is distinct from yearly oil-rich seeds in that it is influenced by the financial side of enormous oil palm plantations, primarily in Malaysia and Indonesia, that are going to provide year-round supply. Canola as well as canola oil and meal are sold as undifferentiated commodities in large quantities on the global market.

The marketing of HT canola in 1998 and HT alfalfa and sugar beets in 2005 was made possible due to the success of these GE crops. Although legal and regulatory difficulties slowed the spread of GE sugar beets and GE alfalfa in the first decade of the twenty-first century, approvals have risen dramatically in recent years. In 2013, GE seeds carrying HT canola characteristics were sown on 95 percent acres and over 99 percent acres of sugar beet harvested in the United States. As mentioned previously, prior to 1980 rapeseed oil was mostly utilised as an industrial lubricant. It was unfit for human or animal consumption due to undesirable by-products such as erucic acid. Both of these issues were solved by Canadian scientists utilising modern (non-GE) plant breeding techniques, with Canada designating the modified, edible rapeseed plants as "canola" in 1979. The US FDA declared rapeseed and canola to be independent crops in 1985, formalising the separation. Despite the fact that it is only known as canola in North America, edible rapeseed varieties have been commercialised in Australia, China, Europe, and India.

Europe is the world's leading importer of canola. Another net importer is China, which is also a significant producer, and Australia has sold effectively into both markets. While there is a widespread shortage of canola and canola oil in India, domestic markets are safeguarded by tariffs imposed on canola and canola oil meant for import. Canada is the world's third-largest canola producer. The majority of canola produced in Canada (about 80 percent) is genetically modified. In 2005–06, around 1.4 million tonnes of canola were produced by Australia. Drought was anticipated to limit production to around 0.8 million tonnes (or less) in 2006–07. Prior to the drought, production for 2006–07 was expected to be at 1.25 million tonnes.

Traditionally, the Australian consumption was roughly 500,000 tonnes of canola domestically and the rest of the crop was exported. To date, Australia's most major and steady export markets have been Japan and Pakistan. Australia was frequently distributing over 1 million tonnes of canola before to the 2002–03 drought. However, exports fell to 880,000 tonnes in 2005–06. Given the recent streak of dry seasons and predicted increases in demand, particularly for meal, Australia is actually projected to import canola seed and meal in the future. Around 550,000 tonnes of oils and fats are consumed in Australia each year. Soft oils (canola, sunflower, cottonseed, and soybean) account for about half of this, of which Canola accounts for more than half. Palm oil and tallow being imported are credited to account for approximately 200,000 tonnes of the total amount of oils and fats. The trade segment continues to be significant, with the commercial and food service sections intended for the majority of usage (ACIL Tasmin, 2007).

Australia's oil consumption can be divided into three categories:

1 **Commercial** – Commercial items like biscuits, salad dressings, snacks, and frozen foods account for around 40 percent of yearly fats and oil consumption in Australia. Some of these products contain canola oil.
2 **Retail** – accounts for 35 percent of domestic demand and fats or oils, including those used in margarines and cooking oils. Although

margarine accounts for half of retail usage, its percentage has been rapidly dropping. In this industry, canola oil is commonly used.

3 **Food services** – This industry is dominated by oils which are hard like tallow fat and palm oil which are best suitable for deep frying and are mostly influenced by price and flavour. Canola oil isn't as well adapted to frying as other oils so it is not widely utilised in Australia. However, the oil can be hydrogenated or inter-esterified.

The equivalency of canola oil produced by GM and non-GM cultivars appears to have benefitted the latest spike in investments in canola processing equipment in Canada as well as the European Union's approval of both types of oils. For example, Cargill announced on July 17, 2006 that it will increase the capacity of their processing plant of canola in Clavet, Saskatchewan to 3000 tonnes per day from 2200 tonnes per day. As a result, this indicates that Canada has not experienced any big commercial crunch as a result of its adoption of GM canola, at least in international markets (Unger, 2011).

In the face of weather fluctuations, the speedy progression of phytopathogens, unwanted plants, and a growing global population, the canola crop is seen as a cost-effective "multipurpose" product that can be used in a variety of markets, sectors, and sustainable development programs. Local legislation, research funding, and improvement goals influence the technique used to generate and improve canola types, such as GM canola. Although existing GM canola varieties benefit from increased product value and lower labour costs (HT canola), the transgenes may have an impact on the performance of other genomic areas. As a result, removing non-essential transgenes from GM canola, conducting an eco-friendly risk assessment study as well as procuring jurisdictive sanctions for marketable release of these varieties are all prerequisites for producing a GM variation; and each of these procedures lengthens the time it takes for a GM canola to reach the market. The EU, the world's top canola importer, has established strict controls on GM organisms, limiting technological options for canola improvement (Puchta, 2017).

11.8 Health and ethical issues/risks

Canola oil is now ranked the third-largest vegetable oil by volume globally, trailing only palm oil and soybean oil, and in the USA it is one of the most extensively used oils for human consumption, second only to soybean oil. While numerous studies have looked at the link between olive oil with disease frequency, mortality, and biomarkers for diseases, canola oil studies have primarily looked into biomarkers. Although little and inconclusive scientific evidence suggests that canola oil consumption has some benefits, outcomes from studies using canola oil-containing diets in experimental animal models have supplied satisfactory data.

Canola oil is used for direct human consumption as well as in meals for human consumption (cooking oil, margarine, etc). Refined canola oil is virtually completely odourless and colourless thanks to modern vegetable oil refining and bleaching procedures. The oil also contains a low content of saturated fatty acids – a property which has helped it in gaining popularity in recent years. As a result, with the introduction of 'double-zero' variations, consumers in OECD countries have steadily expanded their acceptance of refined canola oil as a cooking oil.

The meal from canola is used in animal feed, resulting in food items such as meat, eggs, milk, etc. which are then consumed by humans. This chain commences at the level of seed production and farming and once the seed is crushed, the resulting meal is transferred to feed formulators. The formulated feed comes back to the farm gate, and the finished result then moves from the farm gate to abattoirs, to food processors, and finally to human consumers in the form of dairy, meat, and eggs or processed foods containing these items in the 'second iteration'. Canola has varied applications where the supply chain connects agriculturalists to industrial processors that change the seed, oil, or meal using chemical or other methods (for example by interesterification, fractionation, or extrusion, etc). Canola can be used in a variety of applications once it has been converted in this way: cosmetic items, food additives, ointments, greases, slipping agents, plastics, etc.

Rapeseed oil (commonly referred to as canola oil) has been acknowledged to be a viable olive oil alternative due to its similar monounsaturated fatty acid concentration and overall beneficial fatty acid profile. As a result, owing to its less cost as compared to olive oil and similar health friendly aspects, the use of canola oil by non-Mediterranean countries has expanded significantly in recent years. However, compared with olive oil, scientific evidence is limited and inconclusive. Another cause for caution is that canola oil may contain trans-fats, which can lead to major health issues. In reality, however, all deodorised oils, including canola oil, have comparatively low trans-fat levels. Deodorization is the penultimate stage in the purification of *all* vegetable oils. Customers prefer a bland flavour, which is achieved with this strategy (Bawa & Anilakumar, 2013).

Fat derived from cattle and sheep contains approximately 2–5 percent trans-fat in natural form as a percentage of total amount of fat present. Canola oil is deodorised at temperatures above 200°C (235°C) under vacuum for varying durations of time in order to wipe out volatiles such as free fatty acids and phospholipids. When exposed at high temperatures, the essential ω-6-linoleic and ω-3-linolenic acids get transformed into trans-fatty acid isomers. Trans-isomers of linoleic acid constitute 0.2–1.0 percent of total fatty acids in canola and soybean oils, whereas trans-isomers of linolenic acid may account for up to 3 percent. Linolenic acid isomerizes 12–15 times faster than linoleic acid when heated.

Diets high in canola oil have been proven helpful in diminishing plasma cholesterol levels in comparison to those consuming high saturated fats.

Canola oil also affects biological activities which are associated with other disease risk biomarkers. The health benefits of specific constituents of canola oil have been the subject of previous studies. Canola oil has immediate ramifications for consumers, nutritionists, and anyone who is determining which oil to eat or recommend. In comparison to other food sources of fats, research studies suggest there is an adequate reduction in both total cholesterol and low-density lipoprotein cholesterol along with increased levels of tocopherol as well as insulin sensitivity. In conclusion, from its action on lipid reduction, rising scientific evidence has supported the utilization of canola oil as a primary part of a healthy diet (see Figure 11.1) (Lin et al., 2013).

Plant breeding strategies have been used to create canola oil from rapeseed. These processes, which were launched in the early 1970s, ensured the safe manufacture of canola oil for human use by lowering erucic acid levels to less than 5 percent from an original value of 50 percent – an amount considered potentially hazardous to health. Glucosinolate levels were also lowered as a result of these breeding practices.

Canola oil is also an excellent storehouse of phytosterols and vitamin E, contributing to its nutritional profile. High consumption of canola oil has led to a considerable increase in dietary fatty acid intake in accordance with dietary recommendations. Several findings have also reported lipid-lowering effects due to increased consumption of canola oil compared with oils rich in saturated fatty acids. Additionally, there is indirect evidence of its nutritional benefits related to its high monounsaturated fatty acid and ALA

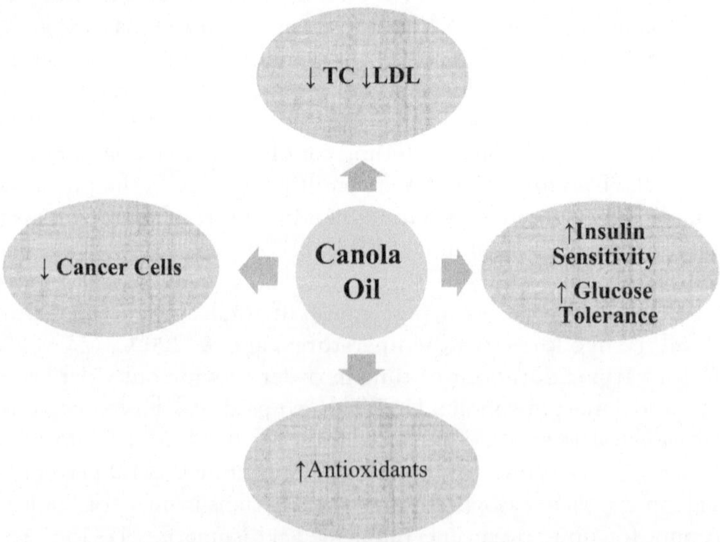

Figure 11.1 Canola oil's effect on health factors.
Source: Authors.

content. Yet, further research is needed to explore its role in improving other aspects of health and to determine the role of this oil in the future.

11.9 Current status and future prospects

Owing to a higher inclination towards the Mediterranean diet, epidemiological and clinical research has shown there is a lower chance of acquiring cognitive impairment and Alzheimer's disease. Out of the various nutritive ingredients which characterize the Mediterranean diet, much emphasis has been placed on the consumption of olive oil on daily basis as the primary source of additional fat; consumption that is considered to be very high in Mediterranean people compared with those living in other geographical locations across the globe. Although limited and inconclusive, scientific evidence suggests that canola oil consumption has some health benefits; yet outcomes from studies using foods containing canola oil in experimental animal models have produced conflicting results.

During the 1990s, customised profiles of canola fatty acids were established using traditional breeding and mutagenesis for certain industries and industrial applications. For example, to meet the demands of the growing cooking oil industry, canola cultivars must have higher levels of oleic acid (C18:1 or n-9) and lower levels of linolenic acid (C18: 3 or n-3) and SFAs, with particular criteria differing by nation. In Australia, for example, the goal is for oleic acid to be up to 67–75 percent and linolenic acid to be less than 3 percent. Expanding the area of high oleic and specialty canola types (to 33 percent of canola acreage) is also part of Canada's canola crop plan for the period 2015–2025, as is lowering SFAs to below 6.8 percent, particularly palmitic and stearic levels to below 4 percent. Researchers also suggest that use of canola oil should be maximised in biofuel and food applications (Salisbury et al., 2016).

11.10 Conclusions

GM technologies have a bright future in terms of enhancing our efforts to satisfy future global food, feed, and fibre needs in a consistent and ethical manner. Several breeding methods have been used traditionally to both produce and exploit genes to isolate effective variants of genes leading to increases in yield, disease resistance, and pest resistance. In more modern times, a major role is now being played by genome level technologies (Nikzad et al., 2020). Agriculture is a diverse endeavour, and we must accept that diversity if we are to succeed. Despite the fact that rapeseed is an important oilseed crop, its production is limited by biotic and abiotic factors such as disease, pests, heat, drought, and cold stress. Plants can cope with stress by slowing their growth and development, lowering their yield, and modifying their morphological, physiological, biochemical, and molecular features; something that, if accepted, genetic modification could address.

References

Abbas, M. S. T. (2018). Genetically engineered (modified) crops (*Bacillus thuringiensis* crops) and the world controversy on their safety. *Egyptian Journal of Biological Pest Control*, 28, 52.

ACIL Tasman (2007, August). *GM Canola: An Information Package Prepared for the Department of Agriculture, Fisheries and Forestry with the Assistance of Innovation Dynamics. Australian Government, Department of Agriculture, Fisheries, and Forestry.* https://www.agriculture.gov.au/sites/default/files/sitecollectiondocuments/ag-food/biotech/gm-canola-info-package.pdf

Anderson, J. A., Ellsworth, P. C., Faria, J. C., Head, G. P., Owen, M. D. K., Pilcher, C. D., … Meissle, M. (2019). Genetically engineered crops: Importance of diversified integrated pest management for agricultural sustainability. *Frontiers in Bioengineering and Biotechnology*, 7, 24.

Bawa, A. S., & Anilakumar, K. R. (2013). Genetically modified foods: Safety, risks and public concerns – a review. *Journal of Food Science and Technology*, 50(6), 1035–1046.

Bebber, D., Ramotowski, M., & Gurr, S. (2013). Crop pests and pathogens move poleward in a warming world. *Nature Climate Change*, 3. doi:10.1038/nclimate1990.

Belide, S., Zhou, X.-R., Kennedy, Y., Lester, G., Shrestha, P., Petrie, J. R., & Singh, S. P. (2013). Rapid expression and validation of seed-specific constructs in transgenic LEC2 induced somatic embryos of Brassica napus. *Plant Cell, Tissue and Organ Culture*, 113, 543–553.

European Commission (2017). *Oilseeds and Protein Crops Market Situation*. Committee for the Common Organisation of Agricultural Markets.

FAO (2012). *Undernourishment Around the World in 2012. The State of Food Insecurity in the World 2012*. https://www.fao.org/3/i3027e/i3027e02.pdf

Fernandez-Cornejo, J., Wechsler, S., Livingston, M., & Mitchell, L. (2014). *Genetically Engineered Crops in the United States*. US Department of Agriculture, Economic Research Service, ERR-162, February.

Fisher, J., & Tozer, P. (2009). *Evaluation of the Environmental and Economic Impact of Roundup Ready® Canola in the Western Australia Crop Production System* (Northam, WA: Curtin University of Technology, Muresk School of Agriculture and Environment).

Fontes, E. M. G., Pires, C. S. S., Sujii, E. R., & Panizzi, A. R. (2002). The environmental effects of genetically modified crops resistant to insects. *Neotropical Entomology*, 31(4), 497–513.

Foreign Agricultural Service/USDA (2022). Global market analysis. https://apps.fas.usda.gov/psdonline/circulars/production.pdf

Hasan, J., Shaikh, R., Megha, S., Herrmann, D. T., Kebede, B., & Rahman, H. (2021) Mapping of flowering time, seed quality and clubroot resistance in rutabaga × spring canola populations and their association. *Euphytica*, 217, 160. doi:10.1007/s10681-021-02889-7.

Hu, Q., Hua, W., Yin, Y., Zhang, X., Liu, L., Shi, J., … Wang, H. (2017). Rapeseed research and production in China. *The Crop Journal*, 5(2), 127–135.

James, C. (2011). *Global Status of Commercialized Biotech/GM Crops*. ISAAA Brief No. 43 (Ithaca, NY: ISAAA).

Lin, L., Allemekinders, H., Dansby, A., Campbell, L., Durance-Tod, S., Berger, A., & Jones, P. J. (2013). Evidence of health benefits of canola oil. *Nutrition Reviews*, 71(6), 370–385.

Liu, J., Hao, W., Liu, J., Fan, S., Zhao, W., Deng, L., … Wang, H. (2019). A novel chimeric mitochondrial gene confers cytoplasmic effects on seed oil content in polyploid rapeseed (Brassica napus). *Molecular Plant*, 12(4), 582–596.

Lu, K., Wei, L., Li, X., Wang, Y., Liu, M., Zhang, C., … Li, J. (2019). Whole-genome resequencing reveals *Brassica napus* origin and genetic loci involved in its improvement. *Nature Communications*, 10, 1154. doi:10.1038/s41467-019-09134-9.

Nikzad, A., Kebede, B., Pinzon, J., Bhavikkumar, J., Wang, X., Yang, R.-C., & Rahman, H. (2020). Potential of the c genome of the different variants of brassica oleracea for heterosis in spring b. *Napus* canola. *Frontiers in Plant Science*, 10, 1691. doi:10.3389/fpls.2019.01691.

Oliver, M. J. (2014). Why we need GMO crops in agriculture. *Missouri Medicine*, 111 (6), 492–507.

Petrie, J. R., Zhou, X.-R., Leonforte, A., McAllister, J., Shrestha, P.. Kennedy, Y., … Singh, S. P. (2020). Development of a Brassica napus (Canola) crop containing fish oil-like levels of DHA in the seed oil. *Frontiers in Plant Science*, 11, 727. doi:10.3389/fpls.2020.00727.

Puchta, H. (2017). Applying CRISPR/Cas for genome engineering in plants: The best is yet to come. *Current Opinion in Plant Biology*, 36, 1–8.

Salisbury, P. A., Cowling, W. A., & Potter, T. D. (2016). Continuing innovation in Australian canola breeding. *Crop Pasture Science*, 67, 266–272.

Scarth, R., & McVetty, P. (1999). Designer oil canola – A review of new food-grade brassica oils with focus on high oleic, low linolenic types. In *Proceedings of the 10th International GCIRC Congress, Canberra, Australia*, p. 10.

Shahbandeh, M. (2022). Rapeseed production volume worldwide 2020/2021, by country. https://www.statista.com/statistics/263930/worldwide-production-of-rapeseed-by-country/

Smyth, S. J. (2014). The state of genetically modified crop regulation in Canada. *GM Crops Food*, 5(3), 195–203.

Tan, H., Yang, X., Zhang, F., Zheng, X., Qu, C., Mu, J., …, Zuo, J. (2011). Enhanced seed oil production in canola by conditional expression of *Brassica napus* LEAFY COTYLEDON1 and LEC1-LIKE in developing seeds. *Plant Physiology*, 156(3), 1577–1588.

Ton, L. B., Neik, T. X., & Batley, J. (2020). The use of genetic and gene technologies in shaping modern rapeseed cultivars (Brassica napus L.). *Genes*, 11(10), 1161. doi:10.3390/genes11101161.

Unger, E. H. (2011). Processing. In *Canola*, edited by J. Daun, N. A. M. Eskin, & D. Hickling (London: Elsevier), 163–188.

US Department of Agriculture (USDA), Animal and Plant Health Inspection Service (APHIS) (2010). *Glyphosate-Tolerant Alfalfa Events J101 and J163: Request for Non-regulated Status. Final Environmental Impact Statement – December 2010*. Available at: http://www.aphis.usda.gov/biotechnology/downloads/alfalfa/gt_alfalfa%20_feis.pdf (accessed May 27, 2015).

US Department of Agriculture (USDA), Animal and Plant Health Inspection Service (APHIS) (2015). *Petitions for Determination of Nonregulated Status*. Available at: http://www.aphis.usda.gov/biotechnology/petitions_table_pending.shtml (accessed May 20, 2015).

Villanueva-Mejia, D., & Alvarez, J. C. (2017). Genetic improvement of oilseed crops using modern biotechnology. In *Advances in Seed Biology*, edited by J. C. Jimenez-Lopez (London: IntechOpen).

Zhang, C., Wohlhueter, R., & Zhang, H. (2016). Genetically modified foods: A critical review of their promise and problems. *Food Science and Human Wellness*, 5(3), 116–123.

12 Genetically modified sugar beet

Making to marketing

Nancy Bhura and Jeena Gupta

12.1 Introduction

The growth of plants and their development can be adversely affected by environmental factors (e.g., drought, cold, alkalinity, and salinity) (Luković et al., 2009). Among these, the major abiotic factor is salt stress which limits crop production (Munns & Tester, 2008). Recent studies have reported that plants respond to these abiotic stresses by metabolising polyamines (Romero et al., 2018). For example, overexpression of sugar beet BvM14-SAMS2 gene in *Arabidopsis* elevates H_2O_2 (hydrogen peroxide) and salt tolerance of the plants (Ma et al., 2017).

Sugar beet (*Beta vulgaris* ssp. *vulgaris*) is a dicot plant commonly cultivated in the temperate zone, having major production regions as Europe and the USA (Jaggard & Qi, 2006). This crop comprises a carbohydrate-enriched taproot and is also a source of biofuel in addition to its high sugar concentration (Duraisam, Salelgn, & Berekete, 2017; H. Liu et al., 2008; W. Kong, Yang, Wang, Bendahmane, & Fu, 2017; De Laporte & Ripplinger, 2018). Other uses include cattle feed (Evans & Messerschmidt, 2017) and pharmaceuticals such as blood substitutes (Leiva-Eriksson et al., 2014). Sugar beet is a biennial crop where carbon is translocated from the roots to the leaves during the generative stage and vice versa during the vegetative phase (Fondy, Geiger, & Servaites, 1989). It is a root crop and, by having a long growing season, it is thus vulnerable to pathogen attacks which include soil microbes that get attracted to its carbohydrate-rich root system (Holmquist et al., 2021). Sugar beet genes have already been reported against salt resistance (Kanhonou, Serrano, & Palau, 2001) and drought (Mulet, Alemany, Ros, Calvete, & Serrano, 2004). Currently, around 20 percent of the global sugar production comes from sugar beet (*Beta vulgaris* L. subsp. *vulgaris*); the rest (80 percent) is from sugar cane (*Saccharum* hybrids L.) (Abaszadeh, Sheibani, & Mola, 2020).

To overcome the drawbacks associated with farming of sugar beet, some biotechnological applications are required and since its introduction, acceptance of pesticide-resistant GM sugar beets has increased significantly in the United States. Farmers have found they can reduce the previously frequent spraying and subsequently save more time and money by controlling weeds

DOI: 10.4324/9781003278566-12

that grow in areas where insecticide-resistant sugar beets are grown. Because of its cost-effectiveness, farmers have adopted different types of biotechnology in a short space of time (Dillen, Demont, Tillie, & Cerezo, 2013). This chapter gives an overview of genetically modified (GM) sugar beet and the development of resistant and/or tolerant varieties. It also highlights the trade value of the GM crops in the global market.

12.2 Why are genetic modifications required in sugar beet?

We need genetic modifications to overcome the environmental drawbacks such as temperature, microorganisms, and heavy metals. The following subsections throws a light on this aspect.

12.2.1 Cold-resistant sugar beet

The majority of losses in sugar beet productivity are due to environmental stress. Sugar beet is cultivated in temperate areas which can be susceptible to frost or low temperatures. The optimal temperature for plant germination is 25°C; however, they can also germinate in the temperature range of 3–25°C (Villarias Moradillo, 1999). Low-temperature stress is a significant cause of crop quality and yield reduction; after a period of low temperatures there is a decrease in the accumulation of sugar and in the growth of plant (Moliterni et al., 2015).

A lot of research on crop improvement is currently dedicated to marker-associated breeding or genomic selection where the important resource is the sugar beet genome (Dohm et al., 2014; Funk, Galewski, & McGrath, 2018). In the case of *B. vulgaris*, a comprehensive description of cold-induced transcriptomic changes is available. For example, there is a transcription factor for the homologue of the Arabidopsis CBF3 which is up-regulated upon cold stress (Moliterni et al., 2015), but no gene has been found in sugar beet that has been demonstrated to confer tolerance against cold by overexpression in any organism. The only commercially available GMO sugar beet variety that has been transformed with a bacterial gene shows tolerance to the herbicide glyphosate (Dewar et al., 2003).

12.2.2 Fungus-resistant sugar beet

The *R. solani* AG2–2IIIB fungus is a anastomosis pathogen causing major problems in production of sugar beet and its required inoculation in soil is predicted to escalate in regions where sugar beet and maize overlap in crop rotation schemes. This is because maize can act as a host for the fungus which can result in pathogen propagation (Buddemeyer, Pfähler, Petersen, & Märländer, 2004; Schulze, Koch, Märländer, & Varrelmann, 2016). Moreover, this fungal pathogen produces sexual spores only occasionally and never any asexual spores (Cubeta & Vilgalys, 1997). This absence of spore formation hinders screening work against resistance as the inoculum amount cannot be accurately controlled, whether in field trials or in indoor experiments. To sum up, altogether these issues make *R. solani* disease control rather complex and, thus, a lot of work is required on crop improvement. On top of this, European regulation is very strict on the use of agrochemicals and forbids their application on the canopy or the soil to reduce *R. solani*-induced damage.

The last and the only way left is to control the disease is by using various cultivation practices, out of which the most important one is the availability of resistant varieties (Buhre, Kluth, Bürcky, Märländer, & Varrelmann, 2009).

12.2.3 Phytoremediation and heavy metal pollution

In one such study, sugar beet was transformed with StGCS-GS (γ-glutamylcysteine synthetase-glutathione synthetase) to generate three transgenic lines which are stable in expression, and which were then subjected to Cd, Zn, and Cu at specified concentrations, individually or simultaneously. The results obtained showed a critical role for StGCS-GS in the bioremediation of single as well as multiple toxic metal ions in transgenic sugar beet. Thus, overexpression of StGCS-GS induced transgenic sugar beets seem make them perfect candidate plants for phytoremediation use against heavy metal pollution (D. Liu, An, Mao, Ma, & Lu, 2015).

12.3 Genes involved in agronomic trait development

Based on their transcriptome analysis, Holmquist et al. (2021) found three proteins known as major latex proteins (MLP) encoding genes – BvMLP1, BvMLP2, and BvMLP3 – which showed that genotypes of sugar beet possess increased transcriptional activity and were partly resistant to *Rhizoctonia solani* even 5 days after inoculation. Two genes (BvMLP1 and BvMLP3) confirmed by quantitative RT-PCR have increased expression in infected sugar beets. Improved resistance against *R. solani* infection was also seen when BvMLP1 and BvMLP3 were cloned and overexpressed in *Arabidopsis thaliana* (Holmquist et al., 2021).

Another significant pest of sugar beet is the beet cyst nematode (BCN) (*Heterodera schachtii* Schm.) which causes major reductions in yield. Unlike most cyst nematodes, the BCN has a wide host range and can infect more than 218 plant species (Lilley, Atkinson, & Urwin, 2005). Hs1pro-1 gene is the first BCN resistance gene cloned from the sugar beet translocation line (A906001) by a map-based cloning strategy. The translocation line carries the locus from *B. procumbens* chromosome 1 that confers resistance to *H. schachtii* in sugar beet (*Beta vulgaris* L.) (Cai et al., 1997). Hunger et al. (2003) cloned 47 resistance gene analogs (RGAs) from the genomic DNA of sugar beet. Most cloned resistance genes belong to the nucleotide-binding site leucine-rich repeat (NBS-LRR) gene family (Leister et al., 1998; Timmerman-Vaughan, Frew, & Weeden, 2000).

With the help of genetic engineering many other genes have been revealed. The gene BvCOLD1 is a gene from sugar beet which can provide resistance to biotic stress tolerance (cold, drought, and salt) and boron homeostasis (Porcel, Bustamante, Ros, Serrano, & Mulet Salort, 2018). DRR206 is a gene that provides resistance against bacterial and fungal infections in pea and *Brassica napus* and it may contribute to the defense system of sugar beet (Daniels, Fristensky, Wagoner, & Hadwiger, 1987; Wang & Fristensky, 2001; Holmquist et al., 2021). Similarly, the sugar beet gene BvM14-SAMS2 provides tolerance in plants against salt and H_2O_2 (Ma et al., 2017). Genes that are responsible for conferring resistance to sugar beet are summarized in Table 12.1.

Table 12.1 Genes involved in conferring resistance to sugar beet along with their functions

Short gene name	*Full gene name*	*Function*
BvMLP1, BvMLP2, BvMLP3	Major latex proteins	Partially resistant to *Rhizoctonia solani*
Hs1pro-1	Hs1pro-1 protein C-terminus (Hs stands for *Heterodera schachtii*)	Provides resistance to *H. schachtii*
BvCOLD1	*Beta vulgaris* COLD1	Provides resistance biotic stress tolerance
DRR206	Disease resistance response gene 206	Provides resistance against bacterial and fungal infections
BvM14-SAMS2	*Beta vulgaris* M14 line S-adenosylmethionine synthase2	Increases salt and H_2O_2 tolerance in plants
RF-OMA1	Restorer-of-fertility-Overlapping with the m-AAA protease 1 homolog	Increases pollen fertility
BvPGIP	*Beta vulgaris* polygalacturonase-inhibiting protein	Overexpression leads to fungal resistance in Nicotiana plants, resistance against *F. oxysporum* in sugar beet
BvBTB	*Beta vulgaris* bric-a-brac, tramtrack, broad complex	Involved in the response of sugar beet against *C. beticola* infection
BvCIPK	*Beta vulgaris* CBL (Calcineurin B-like protein)-interacting protein kinases	Involved in ion homeostasis
ZxNHX	*Zygophyllum xanthoxylum* sodium hydrogen exchanger	Enhances salt tolerance
AtGRF5	*Arabidopsis thaliana* Growth-Regulating Factors	Increases transformation efficiency, accelerated shoot organogenesis, in sugar beet

Source: Authors.

Transformation to sugar beet hairy roots was also made, giving antibody-mediated resistance to rhizomania disease by means of *Agrobacterium rhizogenes* expressing single-chain variable fragment (scFv). This was done to evaluate the efficiency of four antibody-based constructs for interfering with the Beet necrotic yellow vein virus infection which is mainly conferred by dominant Rz1 and Rz2 genes (Jafarzade et al., 2019). Overexpression of BvPGIPs (*Beta vulgaris* polygalacturonase-inhibiting protein) in transgenic sugar beet inhibited the soil-borne fungus *Fusarium oxysporum* and improved Fusarium yellows' resistance in sugar beet (Li & Smigocki, 2019). Differently expressed BvBTB genes in leaves and roots of sugar beet in resistant and susceptible plants provides the theoretical basis for developing new resistant varieties against C. *beticola* infection (Q. Yang et al., 2022). BvCIPKs are anticipated to be associated with K^+ and Na^+ transporters such as NHX8, AKT1, and HAK5, and possibly be involved in the homeostasis of ions (Wu, Xie, Wang, Wang, & Li, 2022). Expression of ZxNHX gene along with ZxVP1–1 enhances the salt tolerating ability of transformed sugar beet (Wu et al., 2015). Overexpression of AtGRF5 in sugarcane fragile callus significantly promoted and accelerated shoot development, reducing throughput time and increasing the efficiency of transgenic plant production (J. Kong et al., 2020). The high transgenic shoot formation rates suggest that sugar beet callus cells overexpressing AtGRF5 might have acquired a regenerative advantage over the large population of noncompetent cells, which is in harmony with the function of GRF genes in regulating cell division and the pluripotent competence of cells and tissues in model plant species (Horiguchi, Ferjani, Fujikura, & Tsukaya, 2006; Lee et al., 2018).

There are several reports showing that when the cry gene is inserted into the plant genome, it can efficiently control the insects but very few show the usage of multiple genes for pest and herbicide resistances in a single cassette simultaneously (Kimoto & Shimamoto, 2001; Sedighi, Rezapanah, & Aghdam, 2011). However, one such study has used this method for generating transgenic sugar beet (Moazami-Goodarzi, Mortazavi, Heidari, & Norouzi, 2020).

12.4 Mechanism of genetic modifications leading to desired trait

Transgenic sugar beets have been produced through Agrobacterium-mediated transformation, microprojectile-mediated DNA delivery, and DNA uptake by protoplasts using polyethylene glycol or electroporation (Biancardi, 2005).

12.4.1 Cell culture mediated modification

Sugar beet monosomic addition line M14 was obtained through hybridization between sugar beet (*Beta vulgaris* L.) and wild sugar beet (*Beta corolliflora* Zoss.) (Guo, Kang, Liu, & Li, 1999). This M14 line demonstrates several remarkable characteristics such as salt and cold tolerance, as well as apomixis. These traits may be attributed to the retention of the 9th wild sugar beet chromosome in the M14 (L. Yang, Ma, Wang, Chen, & Li, 2012).

12.4.2 Agrobacterium tumefaciens mediated modification

Sugar beets tolerant against heavy metals were made using *Agrobacterium tumefaciens* (D. Liu et al., 2015). The general procedure regarding the preparation of transgenic crops is shown in Figure 12.1. Theoretically, Agrobacterium cells can endure in a selection medium consisting of antibiotics and may linger for up to two years between the plant cells in the roots, stems, and leaves of transgenic plants (Palla & Pijut, 2015; A. Yang, Duan, Gu, Gao, & Zhang, 2005). Evaluation, identification, and expression of these transgenes is carried out via RT-PCR but this is only feasible when the lack of Agrobacterium cells in transformed plants are confirmed (Kalbande & Patil, 2016). This process is known as *in planta* transformation, where

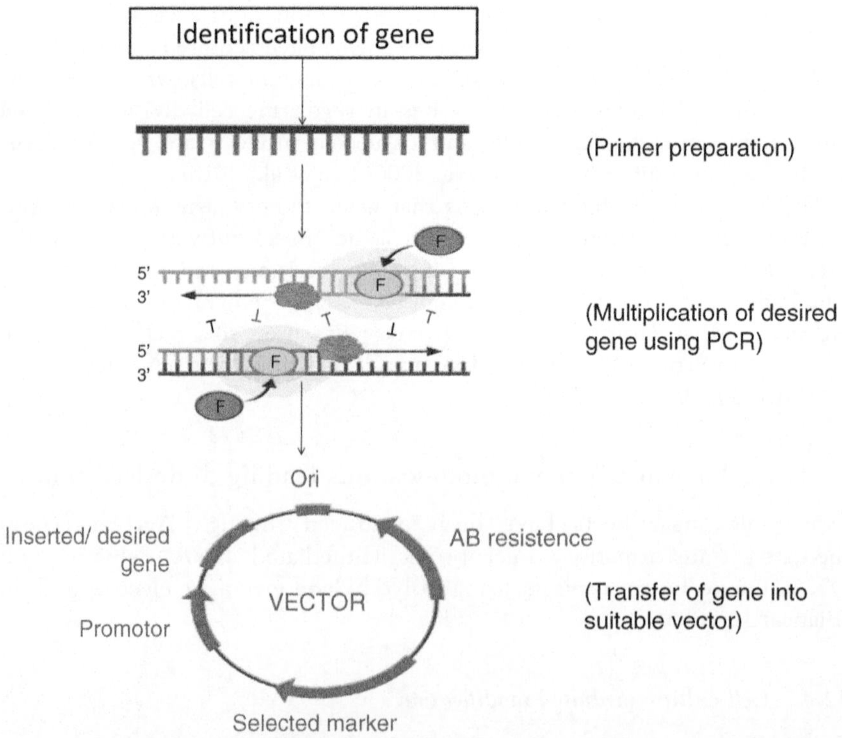

Figure 12.1 Preparation of vector.
Source: Authors.

Agrobacterium, despite infecting the explant, does not get involved in in-vitro culture and regeneration of plant cells or tissues."

Using genetic engineering to produce stress-resistant varieties is an easy, cost effective, and sustainable way to control pests and their diseases (Nyaboga, Njiru, & Tripathi, 2015). When transforming higher dicot plants, Agrobacterium-mediated methods are generally preferred over direct gene transfer methods like particle bombardment, PEG mediated transformation, electroporation, or liposome mediated transformation (Dai et al., 2001; Gelvin, 2003). This is due to its ease of handling and because it is inexpensive. Moreover, with this technique DNA rearrangements are very low, the single copy insertion rate is very high, and the target transgenes are reliably expressed in following generations (Lei et al., 2012). However, despite, *Agrobacterium tumefaciens*' high sensitivity towards infection, regeneration of transgenic sugar beet is mainly dependent on the genotype of the plant and the strain of bacteria (Krens, Zijlstra, vd Molen, Jamar, & Huizing, 1988).

As shown in Figure 12.1 and Figure 12.2, initially a vector is prepared to contain our desired gene. Most of the time we use *Agrobacterium tumefaciens*, which is a Gram-negative, soil bacterium. This bacterium possesses a Ti plasmid which is used for creating transgenic plants and is also known as a natural genetic engineer. Using a selection marker, we select the vector possessing our gene. This vector containing our gene of interest is then allowed to infect our sugar beet cells, followed by co-cultivation, which allows the integration of the vector in the plant cells, followed by their cultivation in a greenhouse to study the traits. After confirming the desired trait and its efficacy on environment, it is then released into the fields. One research group optimized the genetic transformation procedure for sugar beet against glyphosate and insect pest resistance and their associated genes, which was

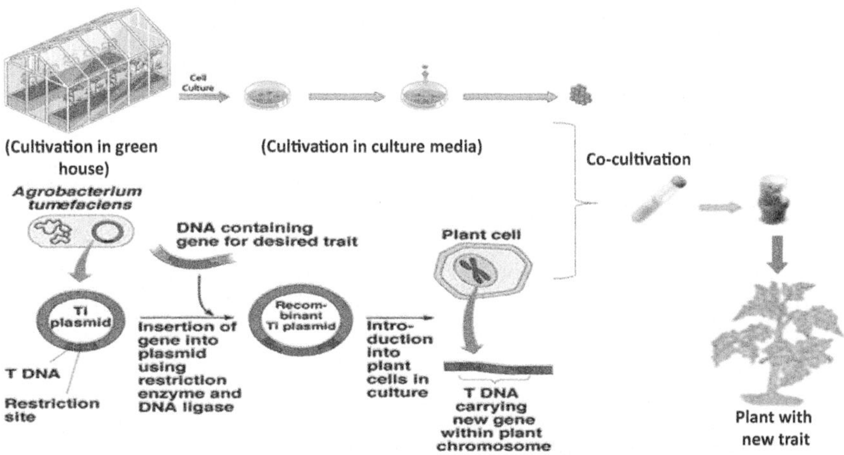

Figure 12.2 Preparation of transgenic sugar beet.
Source: Authors.

reproducible in nature and could be applied on wide range of genotypes (Moazami-Goodarzi et al., 2020). Three important factors are crucial for successful a transformation system: a selectable marker, efficient gene delivery, and a responsive tissue culture (Che et al., 2018).

12.5 Impact on modern agriculture – trade of GM crops

The cultivation of sugar beet is vital in 52 agricultural countries from an agricultural point of view, blooming in temperate climate zones like the USA and central and southern Europe. The price of producing cane sugar is lower than that of sugar beet owing to its perennial nature and also because the bagasse provides free energy for its processing. (For sugar beet, on the other hand, a minimum of 15 to 20 Mw/h capacities, totally produced from non-renewable energy sources, is required for a sugar beet plant processing plant 10,000 tons per day.) Its perennial habitat additionally reduces losses through nitrogen percolation, increases nutrient and water uptake efficiency as the result of a deeper root system supporting soil microbial activity, reduces soil surface loss to water and wind erosion thus lower nitrogen leaching, and also decreases herbicide consumption. However, this is not the case with sugar beet which is a biennial crop, though it does use five times less water than sugar cane (Glover, Cox, & Reganold, 2007; Stevanato et al., 2019). Although the cost of production is high, and it is not preferred in many countries, in the nineteenth century there was upsurge in sugar beet cultivation (Biancardi, 2005) as it was suitable for crop rotations (to reduce soil-borne diseases) of cereals. However, the above-mentioned points became a hurdle, and so to overcome these limitations of sugar beet, various techniques were incorporated like mechanization, high-yielding varieties, and chemical inputs, and gradually by genetic engineering to lessen the fertilizer usage and to make disease-resistant sugar beet.

12.5.1 Global demand for sugar beet

Due to the growing demand for sugar and the need to increase the effectiveness of the sugar beet sector, a sustainable and more intensive crop cultivation system is required to expand its productivity (sugar yield per unit area). Current average sugar yields in the most productive countries are only half the estimated potential of about 24 tonnes per hectare, so there is still much to do (Loel, Kenter, Märländer, & Hoffmann, 2014). Many alternatives can be incorporated in the cultivation of sugar beet for increasing sugar production and decreasing energy consumption (Martindale, 2013).

12.5.2 Agricultural practice

To meet all sustainability requirements, agriculture needs to find the right compromise between present and future production levels so as not to endanger long-term potential. To do this, efficient use of fertilizers or nourishment is required to prevent their release into groundwater and the atmosphere. There is also a need to search for effective and environmentally friendly pesticides and herbicides and to increase the use of available new strains for resistance or tolerance to biotic and abiotic stress. Advances in molecular genetics and biotechnology are anticipated to significantly contribute to this purpose (Stevanato et al., 2019). Improvement in in-vitro cell culture and genetic engineering aids in breeding programs to combat insects and thus increasing pest resistance (Gurel, Gurel, & Lemaux, 2008).

12.6 Benefits to the society

Sugar beet farmers face many challenges when it comes to weed control. Farmers have welcomed the commercialization of sugar beet and are very pleased with the cheap and easy-to-cultivate biotechnology products that provide better weed control than traditional sugar beet. Farmers have recouped many advantages from using RR® (Roundup Ready) sugar beet over traditional varieties: it saves half of the crops needs (less money, weeds, chemicals), 30 percent less fuel, and significantly lowers labor costs and time, including that needed for additional weeding. It also offers an incentive and facilitates the adoption of minimum or no-till cultivation methods, leading to less soil compaction; a reduction in the number of herbicide applications as well as the convenience of being reliant on fewer types of herbicides; and generally is more profitable and convenient to cultivate than conventional sugar beet. Even as far back as 2008, farmers understood the value of RR® beets and were supporting the development of other important properties, such as disease, nematode and insect resistance, and tolerance against cold and drought (James, 2011).

12.7 Trade in the global market

Crops can be infected with nematodes and these cause considerable economic damage around the world. Estimated global losses from plant nematodes amount to approximately US$125 billion annually (Chitwood, 2003). An assimilated strategy, including trap crop planting, cultivation of resistant sugar beet varieties, and nematicide application, might control BCN infestation in sugar beets (Hauer, Koch, Krüssel, Mittler, & Märländer, 2016).

12.7.1 Roundup Ready sugar beet and its trade

The herbicide-tolerant Roundup Ready (RR®) sugar beet was first cultivated in 2006 to test new technologies and to market the sugar, pulp, and molasses. Another small acre was planted in 2007, but due to the very limited availability of biotechnology seeds, only one sugar beet company was able to switch to Roundup Ready (RR®). With increased seed production, in 2008 an estimated 59 percent of the 437,246 hectares cultivated in the United States, or 257,975 hectares, was with RR® sugar beet (James, 2011). Herbicide-resistant RR® sugar beet was quickly and widely adopted by US and Canadian producers in 2009. For the first time in 2009, herbicide-resistant RR® sugar beet seeds were finally provided to farmers in sufficient quantities. An estimated 95 percent of sugar beet grown in the United States in 2009, or about 485,000 hectares, was of these improved varieties developed using biotechnology. In the United States, the sugar beet planted area in 2010 and 2011 was approximately 485,000 hectares, 95 percent of which were produced using biotechnology. Canadian producers planted approximately 15,000 hectares of biotechnology varieties in 2009. It accounted for almost 96 percent of the country's sugar beet harvest, and in 2011 biotechnology adoption was almost at 100 percent with an area of 18,000 hectares. 2011 was the third year of commercial planting in Eastern Canada and the second year of commercial production in Western Canada (James, 2011).

12.8 Health and ethical issues/risks

The sustainability of GM sugar beet is very important for food security but risk assessment is still a major concern due to the unpredictable insertion of the gene introduced which may alter the metabolism or change in metabolic concentrations (Gong & Wang, 2013). Many presume that any crop that has been genetically modified, possesses cells (or their contents) which would not be present naturally (Kamle & Ali, 2013). Thus, a comparison of the molecular properties between the GM crop and the non-GM parental crop becomes very crucial for the identification of potential hazards and for maintaining their biosafety (Baker et al., 2006). Obviously, the mere variation between them is not a reason for transgenic crops to be considered unsafe. However, the changes must be in the range of natural alterations and, in particular, new toxic or allergenic compounds should not be produced (Herman & Price, 2013).

12.8.1 Ethical approval in the international market

Based on the report of Genetically Modified Organisms of the European Food Safety Authority (GMO Panel), sugar products derived from GM sugar beet (which is herbicide-tolerant) have been authorized for sale in the EU (EFSA Panel on Genetically Modified Organisms, 2017). According to independent scientific analysis, the sugar obtained from RR® sugar beet is identical to sugar from other non-GM sugar beets at the molecular level. Note that because the sugar in RR® sugar beet does not contain the DNA from the biotech transformation process, and is thus similar to conventional sugar beet sugars, it is not subject to labeling in the US or overseas markets like Japan. Since the United States is one of the largest sugar importers in the world, most of the sugar and by-products from sugar beet production are consumed domestically. However, the sugar, pulp, and molasses from RR® sugar beet are approved in all important export markets including Japan, Canada, Mexico, the European Union, South Korea, Australia, New Zealand, Colombia, Russia, China, Singapore, and the Philippines (James, 2011). Hejri et al. (2022) compared the metabolic content of transgenic sugar beet resistant to rhizomania as a part of a risk assessment study and concluded that the transgenic crop was safe when produced using the RNA silencing mechanism. Similar studies were also conducted previously where they used gene silencing and RNA silencing mechanisms to confer the resistance against rhizomania in greenhouse experiments and field trials (Zare et al., 2015; Safar et al., 2021; Lennefors et al., 2006; Pavli, Panopoulos, Goldbach, & Skaracis, 2010), and was quite successful. Currently, the herbicide-tolerance gene has been used as an eco-friendly selectable marker in sugar beet transformation. Genetic engineering procedures have been developed for several sugar beet varieties, but existing methods are based on the application of the antibiotics kanamycin and hygromycin (Ivic-Haymes & Smigocki, 2005; Jafari et al., 2009; Lytvyn et al., 2014). Because of the widespread usage of antibiotic-resistance genes, public safety remains a matter of concern, and because of this, application of a safe selection system for crop transformation and a thorough risk assessment continue to be crucial areas of research in plant and agricultural biotechnology.

12.9 Current status and future prospects

GM sugar beet varieties which are resistant against the non-selective herbicide glyphosate are considered a novel approach for weed control and were released in the market 20 years ago. Although it has been permitted in the USA for several years, for the largest sugar beet producer – the European Union – this GM crop is forbidden and this is still the case at the time of writing (DeFrancesco, 2013). Recently, a fungus–apple interaction RNAseq study was conducted, where one MLP gene was found to impact a handful of defense-related genes including transcription factors (He et al., 2020). There is a possibility that MLP genes are responsible for the defense system of many crops including sugar beet; however, its mechanism needs to be explored. Regeneration of whole transgenic sugar beet plants using *A. tumefaciens* mediated transformation is a very lengthy and time-consuming process and therefore cultivation of resistant cultivars should be our main focus as an alternative method. BvM14-SAMDC provides salt tolerance, but the mechanism by which expression happens is still under research (Ji et al., 2019).

12.10 Conclusion

Many genes confer resistance against different pathogens or improve sugar content in sugar beet, however, their mode of action is yet to be established.

Most GM crops are produced using *Agrobacterium tumefaciens* which is a time consuming process. Instead, it is recommended to try cultivating resistant cultivars or try reproducing the existing GM sugar beet instead. Although there have been very few health hazards associated with GM sugar beet crop, a thorough risk assessment still needs to be carried out every time a new GM sugar beet is introduced. In conclusion, GM sugar beet is the future for the places where very low sugar is produced.

References

Abaszadeh, M., Sheibani, H., & Mola, A. (2020). Theoretical and experimental investigations into the structural, electronic, and molecular properties of 1,5-dihydropyrano 2,3-c chromene derivatives. *Biointerface Research in Applied Chemistry*, 10 (3), 5586–5591.

Baker, J. M., Hawkins, N. D., Ward, J. L., Lovegrove, A., Napier, J. A., Shewry, P. R., & Beale, M. H. (2006). A metabolomic study of substantial equivalence of field-grown genetically modified wheat. *Plant Biotechnology Journal*, 4(4), 381–392.

Biancardi, E. (2005). *Genetics and Breeding of Sugar Beet* (Boca Raton, FL: CRC Press).

Buddemeyer, J., Pfähler, B., Petersen, J., & Märländer, B. (2004). Genetic variation in susceptibility of maize to *Rhizoctonia solani* (AG 2–2IIIB) – symptoms and damage under field conditions in Germany/Genetische Variation in der Anfälligkeit von Mais gegenüber *Rhizoctonia solani* (AG 2–2IIIB)–Symptome und Schäden im Feld in Deutschland. *Zeitschrift für Pflanzenkrankheiten und Pflanzenschutz/Journal of Plant Diseases and Protection*, 111(6), 521–533.

Buhre, C., Kluth, C., Bürcky, K., Märländer, B., & Varrelmann, M. (2009). Integrated control of root and crown rot in sugar beet: Combined effects of cultivar, crop rotation, and soil tillage. *Plant Disease*, 93(2), 155–161.

Cai, D., Kleine, M., Kifle, S., Harloff, H.-J., Sandal, N. N., Marcker, K. A., ... Stiekema, W. J. (1997). Positional cloning of a gene for nematode resistance in sugar beet. *Science*, 275(5301), 832–834.

Che, P., Anand, A., Wu, E., Sander, J. D., Simon, M. K., Zhu, W., ... Liu, D. (2018). Developing a flexible, high-efficiency Agrobacterium-mediated sorghum transformation system with broad application. *Plant Biotechnology Journal*, 16(7), 1388–1395.

Chitwood, D. J. (2003). Research on plant-parasitic nematode biology conducted by the United States Department of Agriculture – Agricultural Research Service. *Pest Management Science: Formerly Pesticide Science*, 59(6–7), 748–753.

Cubeta, M., &Vilgalys, R. (1997). Population biology of the *Rhizoctonia solani* complex. *Phytopathology*, 87(4), 480–484.

Dai, S., Zheng, P., Marmey, P., Zhang, S., Tian, W., Chen, S., ... Fauquet, C. (2001). Comparative analysis of transgenic rice plants obtained by Agrobacterium-mediated transformation and particle bombardment. *Molecular Breeding*, 7(1), 25–33.

Daniels, C., Fristensky, B., Wagoner, W., & Hadwiger, L. (1987). Pea genes associated with non-host disease resistance to Fusarium are also active in race-specific disease resistance to Pseudomonas. *Plant Molecular Biology*, 8(4), 309–316.

De Laporte, A. V., & Ripplinger, D. G. (2018). Economic viability of energy beets (*Beta vulgaris*) as advanced biofuel feedstocks. *Industrial Crops and Products*, 111, 254–260.

DeFrancesco, L. (2013). US Congress moves to 'protect' GM crop plantings. *Nature biotechnology*, 31(6), 479–479.

Dewar, A. M., May, M. J., Woiwod, I. P., Haylock, L. A., Champion, G. T., Garner, B. H., ... Pidgeon, J. D. (2003). A novel approach to the use of genetically modified herbicide tolerant crops for environmental benefit. *Proceedings of the Royal Society of London, Series B: Biological Sciences*, 270(1513), 335–340.

Dillen, K., Demont, M., Tillie, P., & Cerezo, E. R. (2013). Bred for Europe but grown in America: The case of GM sugar beet. *New Biotechnology*, 30(2), 131–135.

Dohm, J. C., Minoche, A. E., Holtgräwe, D., Capella-Gutiérrez, S., Zakrzewski, F., Tafer, H., ... Reinhardt, R. (2014). The genome of the recently domesticated crop plant sugar beet (*Beta vulgaris*). *Nature*, 505(7484), 546–549.

Duraisam, R., Salelgn, K., & Berekete, A. K. (2017). Production of beet sugar and bio-ethanol from sugar beet and it bagasse: A review. *International Journal of Engineering Trends and Technology*, 43(4), 222–233.

EFSA Panel on Genetically Modified Organisms (2017). Assessment of genetically modified sugar beet H7-1 for renewal of authorisation under Regulation (EC) No 1829/2003 (application EFSA-GMO-RX-006). *EFSA Journal*, 15(11), e05065.

Evans, E., & Messerschmidt, U. (2017). Sugar beets as a substitute for grain for lactating dairy cattle. *Journal of Animal Science and Biotechnology*, 8(1), 1–10.

Fondy, B. R., Geiger, D. R., & Servaites, J. C. (1989). Photosynthesis, carbohydrate metabolism, and export in *Beta vulgaris* L. and *Phaseolus vulgaris* L. during square and sinusoidal light regimes. *Plant Physiology*, 89(2), 396–402.

Funk, A., Galewski, P., & McGrath, J. M. (2018). Nucleotide-binding resistance gene signatures in sugar beet: Insights from a new reference genome. *The Plant Journal*, 95(4), 659–671.

Gelvin, S. B. (2003). Agrobacterium-mediated plant transformation: The biology behind the "gene-jockeying" tool. *Microbiology and Molecular Biology Reviews*, 67(1), 16–37.

Glover, J. D., Cox, C. M., & Reganold, J. P. (2007). Future farming: A return to roots? *Scientific American*, 297(2), 82–89.

Gong, C. Y., & Wang, T. (2013). Proteomic evaluation of genetically modified crops: Current status and challenges. *Frontiers in Plant Science*, 4, 41.

Guo, D., Kang, C., Liu, L., & Li, Y. (1999). Study of apomixis in the allotriploid beet (VVC). *Agric. Sci. China*, 32, 1–5.

Gurel, E., Gurel, S., & Lemaux, P. G. (2008). Biotechnology applications for sugar beet. *Critical Reviews in Plant Sciences*, 27(2), 108–140.

Hauer, M., Koch, H.-J., Krüssel, S., Mittler, S., & Märländer, B. (2016). Integrated control of *Heterodera schachtii* Schmidt in Central Europe by trap crop cultivation, sugar beet variety choice and nematicide application. *Applied Soil Ecology*, 99, 62–69.

He, S., Yuan, G., Bian, S., Han, X., Liu, K., Cong, P., & Zhang, C. (2020). Major latex protein MdMLP423 negatively regulates defense against fungal infections in apple. *International Journal of Molecular Sciences*, 21(5), 1879.

Hejri, S., Salimi, A., Malboobi, M. A., Fatehi, F., & Yousefiara, M. (2022). Investigation of possible changes induced by RNA silencing in some leaf metabolites of transgenic sugar beet events. *Food Chemistry: Molecular Sciences*, 4, 100073.

Herman, R. A., & Price, W. D. (2013). Unintended compositional changes in genetically modified (GM) crops: 20 years of research. *Journal of Agricultural and Food Chemistry*, 61(48), 11695–11701.

Holmquist, L., Dölfors, F., Fogelqvist, J., Cohn, J., Kraft, T., & Dixelius, C. (2021). Major latex protein-like encoding genes contribute to *Rhizoctonia solani* defense responses in sugar beet. *Molecular Genetics and Genomics*, 296(1), 155–164.

Horiguchi, G., Ferjani, A., Fujikura, U., & Tsukaya, H. (2006). Coordination of cell proliferation and cell expansion in the control of leaf size in *Arabidopsis thaliana*. *Journal of Plant Research*, 119(1), 37–42.

Hunger, S., Gaspero, G. D., Möhring, S., Bellin, D., Schäfer-Pregl, R., Borchardt, D. C., … Salamini, F. (2003). Isolation and linkage analysis of expressed disease-resistance gene analogues of sugar beet (*Beta vulgaris* L.). *Genome*, 46(1), 70–82.

Ivic-Haymes, S. D., & Smigocki, A. C. (2005). Biolistic transformation of highly regenerative sugar beet (*Beta vulgaris* L.) leaves. *Plant Cell Reports*, 23(10), 699–704.

Jafari, M., Norouzi, P., Malboobi, M. A., Ghareyazie, B., Valizadeh, M., Mohammadi, S. A., & Mousavi, M. (2009). Enhanced resistance to a lepidopteran pest in transgenic sugar beet plants expressing synthetic cry1Ab gene. *Euphytica*, 165(2), 333–344.

Jafarzade, M., Ramezani, M., Hedayati, F., Mokhtarzade, Z., Zare, B., Sabet, M., … Malboobi, M. (2019). Antibody-mediated resistance to rhizomania disease in sugar beet hairy roots. *The Plant Pathology Journal*, 35(6), 692.

Jaggard, K., & Qi, A. (2006). Agronomy. In *Sugar Beet*, edited by A. Draycott (Oxford: Blackwell), Chapter 7.

James, C. (2011). *Global Status of Commercialized Biotech/GM Crops, 2011*. ISAAA Brief No. 44 (Ithaca, NY: ISAAA).

Ji, M., Wang, K., Wang, L., Chen, S., Li, H., Ma, C., & Wang, Y. (2019). Overexpression of a S-adenosylmethionine decarboxylase from sugar beet M14 increased *Araidopsis* salt tolerance. *International Journal of Molecular Sciences*, 20(8), 1990.

Kalbande, B. B., & Patil, A. S. (2016). Plant tissue culture independent *Agrobacterium tumefaciens* mediated in-planta transformation strategy for upland cotton (*Gossypium hirsutum*). *Journal of Genetic Engineering and Biotechnology*, 14(1), 9–18.

Kamle, S., & Ali, S. (2013). Genetically modified crops: Detection strategies and biosafety issues. *Gene*, 522(2), 123–132.

Kanhonou, R., Serrano, R., & Palau, R. R. (2001). A catalytic subunit of the sugar beet protein kinase CK2 is induced by salt stress and increases NaCl tolerance in *Saccharomyces cerevisiae*. *Plant Molecular Biology*, 47(5), 571–579.

Kimoto, Y., & Shimamoto, Y. (2001). Difference in toxicity to larvae of cabbage armyworm between transgenic sugar beet lines with Cry1Ab and Cry1C. *Proceedings of the Japanese Society of Sugar Beet Technologists*, 43, 20–23.

Kong, J., Martin-Ortigosa, S., Finer, J., Orchard, N., Gunadi, A., Batts, L. A., … Stuiver, M. (2020). Overexpression of the transcription factor GROWTH-REGULATING FACTOR5 improves transformation of dicot and monocot species. *Frontiers in Plant Science*, 11, 1389.

Kong, W., Yang, S., Wang, Y., Bendahmane, M., & Fu, X. (2017). Genome-wide identification and characterization of aquaporin gene family in *Beta vulgaris*. *PeerJ*, 5, e3747.

Krens, F., Zijlstra, C., vd Molen, W., Jamar, D., & Huizing, H. (1988). Transformation and regeneration in sugar beet (*Beta vulgaris* L.) induced by 'shooter' mutants of *Agrobacterium tumefaciens*. *Euphytica*, 39(3), 185–194.

Lee, S.-J., Lee, B. H., Jung, J.-H., Park, S. K., Song, J. T., & Kim, J. H. (2018). GROWTH-REGULATING FACTOR and GRF-INTERACTING FACTOR specify meristematic cells of gynoecia and anthers. *Plant Physiology*, 176(1), 717–729.

Lei, J., Li, X., Wang, D., Shao, L., Wei, X., & Huang, L. (2012). Agrobacterium-mediated transformation of cotton shoot apex with SNC1 gene and resistance to cotton fusarium wilt in T 1 generation. *Cotton Genomics and Genetics*, 3.

Leister, D., Kurth, J., Laurie, D. A., Yano, M., Sasaki, T., Devos, K., ... Schulze-Lefert, P. (1998). Rapid reorganization of resistance gene homologues in cereal genomes. *Proceedings of the National Academy of Sciences*, 95(1), 370–375.

Leiva-Eriksson, N., Pin, P. A., Kraft, T., Dohm, J. C., Minoche, A. E., Himmelbauer, H., & Bülow, L. (2014). Differential expression patterns of non-symbiotic hemoglobins in sugar beet (*Beta vulgaris* ssp. vulgaris). *Plant and Cell Physiology*, 55(4), 834–844.

Lennefors, B.-L., Savenkov, E. I., Bensefelt, J., Wremerth-Weich, E., van Roggen, P., Tuvesson, S., ... Gielen, J. (2006). dsRNA-mediated resistance to *Beet Necrotic Yellow Vein Virus* infections in sugar beet (*Beta vulgaris* L. ssp. vulgaris). *Molecular Breeding*, 18(4), 313–325.

Li, H., & Smigocki, A. C. (2019). Suppression of *Fusarium oxysporum* with recombinant polygalacturonase inhibiting proteins (BvPGIPs) extracted from sugar beet roots. *Plant Cell, Tissue and Organ Culture (PCTOC)*, 136(1), 197–203.

Lilley, C. J., Atkinson, H. J., & Urwin, P. E. (2005). Molecular aspects of cyst nematodes. *Molecular Plant Pathology*, 6(6), 577–588.

Liu, D., An, Z., Mao, Z., Ma, L., & Lu, Z. (2015). Enhanced heavy metal tolerance and accumulation by transgenic sugar beets expressing *Streptococcus thermophilus* StGCS-GS in the presence of Cd, Zn and Cu alone or in combination. *PLoS One*, 10(6), e0128824.

Liu, H., Wang, Q., Yu, M., Zhang, Y., Wu, Y., & Zhang, H. (2008). Transgenic salt-tolerant sugar beet (*Beta vulgaris* L.) constitutively expressing an *Arabidopsis thaliana* vacuolar Na+/H+ antiporter gene, AtNHX3, accumulates more soluble sugar but less salt in storage roots. *Plant, Cell & Environment*, 31(9), 1325–1334.

Loel, J., Kenter, C., Märländer, B., & Hoffmann, C. M. (2014). Assessment of breeding progress in sugar beet by testing old and new varieties under greenhouse and field conditions. *European Journal of Agronomy*, 52, 146–156.

Luković, J., Maksimović, I., Zorić, L., Nagl, N., Perčić, M., Polić, D., & Putnik-Delić, M. (2009). Histological characteristics of sugar beet leaves potentially linked to drought tolerance. *Industrial Crops and Products*, 30(2), 281–286.

Lytvyn, D., Syvura, V., Kurylo, V., Olenieva, V., Yemets, A., & Blume, Y. B. (2014). Creation of transgenic sugar beet lines expressing insect pest resistance genes cry1C and cry2A. *Cytology and Genetics*, 48(2), 69–75.

Ma, C., Wang, Y., Gu, D., Nan, J., Chen, S., & Li, H. (2017). Overexpression of S-adenosyl-L-methionine synthetase 2 from sugar beet M14 increased Arabidopsis tolerance to salt and oxidative stress. *International Journal of Molecular Sciences*, 18(4), 847.

Martindale, W. (2013). The sustainability of the sugar beet crop – the potential to add value. *British Sugar Beet Review*, 81, 49–52.

Moazami-Goodarzi, K., Mortazavi, S. E., Heidari, B., & Norouzi, P. (2020). Optimization of agrobacterium-mediated transformation of sugar beet: Glyphosate and insect pests resistance associated genes. *Agronomy Journal*, 112(6), 4558–4567.

Moliterni, V. M. C., Paris, R., Onofri, C., Orrù, L., Cattivelli, L., Pacifico, D., ... Mandolino, G. (2015). Early transcriptional changes in *Beta vulgaris* in response to low temperature. *Planta*, 242(1), 187–201.

Mulet, J. M., Alemany, B., Ros, R., Calvete, J. J., & Serrano, R. (2004). Expression of a plant serine O-acetyltransferase in *Saccharomyces cerevisiae* confers osmotic tolerance and creates an alternative pathway for cysteine biosynthesis. *Yeast*, 21(4), 303–312.

Munns, R., & Tester, M. (2008). Mechanisms of salinity tolerance. *Annual Review of Plant Biology*, 59, 651–681.

Nyaboga, E. N., Njiru, J. M., & Tripathi, L. (2015). Factors influencing somatic embryogenesis, regeneration, and Agrobacterium-mediated transformation of cassava (*Manihot esculenta* Crantz) cultivar TME14. *Frontiers in Plant Science*, 6, 411.

Palla, K. J., & Pijut, P. M. (2015). Agrobacterium-mediated genetic transformation of *Fraxinus americana* hypocotyls. *Plant Cell, Tissue and Organ Culture (PCTOC)*, 120 (2), 631–641.

Pavli, O. I., Panopoulos, N. J., Goldbach, R., & Skaracis, G. N. (2010). BNYVV-derived dsRNA confers resistance to rhizomania disease of sugar beet as evidenced by a novel transgenic hairy root approach. *Transgenic Research*, 19(5), 915–922.

Porcel, R., Bustamante, A., Ros, R., Serrano, R., & Mulet Salort, J. M. (2018). BvCOLD1: A novel aquaporin from sugar beet (*Beta vulgaris* L.) involved in boron homeostasis and abiotic stress. *Plant, Cell & Environment*, 41(12), 2844–2857.

Romero, F. M., Maiale, S. J., Rossi, F. R., Marina, M., Ruíz, O. A., & Garriz, A. (2018). Polyamine metabolism responses to biotic and abiotic stress. In *Polyamines*, Methods in Molecular Biology, edited by R. Alcázar & A. F. Tiburcio (New York: Springer), 37–49.

Safar, S., Bazrafshan, M., Khoshnami, M., Behrooz, A. A., Hedayati, F., Maleki, M., … Malboobi, M. A. (2021). Field evaluation for rhizomania resistance of transgenic sugar beet events based on gene silencing. *Canadian Journal of Plant Pathology*, 43(1), 179–188.

Schulze, S., Koch, H.-J., Märländer, B., & Varrelmann, M. (2016). Effect of sugar beet variety and nonhost plant on *Rhizoctonia solani* AG2–2IIIB soil inoculum potential measured in soil DNA extracts. *Phytopathology*, 106(9), 1047–1054.

Sedighi, L., Rezapanah, M., & Aghdam, H. R. (2011). Efficacy of Bt transgenic sugar beet lines expressing cry1Ab gene against *Spodoptera littoralis* Boisd. (Lepidoptera: Noctuidae). *Journal of the Entomological Research Society*, 13(1), 61–61.

Stevanato, P., Chiodi, C., Broccanello, C., Concheri, G., Biancardi, E., Pavli, O., & Skaracis, G. (2019). Sustainability of the sugar beet crop. *Sugar Tech*, 21(5), 703–716.

Timmerman-Vaughan, G., Frew, T., & Weeden, N. (2000). Characterization and linkage mapping of R-gene analogous DNA sequences in pea (*Pisum sativum* L.). *Theoretical and Applied Genetics*, 101(1–2),241–247.

Villarias Moradillo, J. L. (1999). *Compendio práctico del cultivo de la remolacha azucarera* (Madrid: Ediciones Agrotécnicas).

Wang, Y., & Fristensky, B. (2001). Transgenic canola lines expressing pea defense gene DRR206 have resistance to aggressive blackleg isolates and to *Rhizoctonia solani*. *Molecular Breeding*, 8(3), 263–271.

Wu, G.-Q., Feng, R.-J., Wang, S.-M., Wang, C.-M., Bao, A.-K., Wei, L., & Yuan, H.-J. (2015). Co-expression of xerophyte *Zygophyllum xanthoxylum ZxNHX* and *ZxVP1–1* confers enhanced salinity tolerance in chimeric sugar beet (*Beta vulgaris* L.). *Frontiers in Plant Science*, 6, 581.

Wu, G.-Q., Xie, L.-L., Wang, J.-L., Wang, B.-C., & Li, Z.-Q. (2022). Genome-wide identification of CIPK genes in sugar beet (*Beta vulgaris*) and their expression under NaCl stress. *Journal of Plant Growth Regulation*, January, 1–15.

Yang, A., Duan, X., Gu, X., Gao, F., & Zhang, J. (2005). Efficient transformation of beet (*Beta vulgaris*) and production of plants with improved salt-tolerance. *Plant Cell, Tissue and Organ Culture*, 83(3), 259–270.

Yang, L., Ma, C., Wang, L., Chen, S., & Li, H. (2012). Salt stress induced proteome and transcriptome changes in sugar beet monosomic addition line M14. *Journal of Plant Physiology*, 169(9), 839–850.

Yang, Q., Liu, Y., Zhao, C., Wang, X., Ding, G., Li, Y., & Chen, L. (2022). Genome-wide identification and expression analysis of the BTB domain-containing protein gene family in sugar beet. *Agronomy*, 12(2), 253.

Zare, B., Niazi, A., Sattari, R., Aghelpasand, H., Zamani, K., Sabet, M., ... Norouzi, P. (2015). Resistance against rhizomania disease via RNA silencing in sugar beet. *Plant Pathology*, 64(1), 35–42.

13 Genetic modification for 4G biofuel production

Sayantan Chakraborty and Himjyoti Dutta

13.1 Introduction

Among various types of biofuels, the petroleum-replica molecules such as alcohols, hydrocarbons, fatty acid derived molecules, and isoprenoids currently hold the most promise to replace existing fossil fuels. The third generation of biofuels produced from algae partially alleviated the drawbacks of the first two generations resulting in reduced competition with crops for land use as well as absence of any costly pre-treatments being involved. However, commercial production of third generations biofuels was not feasible due to the high cost of algae cultivation, difficult downstream processing, and the only meager output of microbial oils from the algal cells. This prompted the extension of research into fourth generation (4G) biofuels that utilize genetically modified organisms (GMOs) such as cyanobacteria, microalgae, bacteria, and yeast to improve the process efficiency and product yield.

An overview of the four generations of biofuel are illustrated in Figure 13.1. Microbial cell factories using GMOs have been already utilized to produce pharmaceuticals and enzymes and researchers consider that these techniques can be applied to the production of advanced 4G biofuels (Mehmood et al., 2021).

However, achieving a solution through the production of 4G biofuels is not simple. There exists a plethora of issues that have to be addressed before it attains commercial viability. The major areas of focus to improve biofuel production from GMOs have been on optimizing and regulating the cellular pathways responsible for product synthesis, improving stress response mechanism in the cells, reducing cellular toxicity of the biofuel compounds, and redirecting the ATP and carbon flux from undesirable to desirable compounds. Additionally, there has to be considered debate on the environmental risks regarding the use of GMOs for biofuel production (Liao et al., 2016). The objective of the current chapter is to shed light on the processes that are employed for 4G biofuel production with an emphasis on strategies used to improve their production processes, their benefits to society, and their potential with reference to environmental and socioeconomic issues.

DOI: 10.4324/9781003278566-13

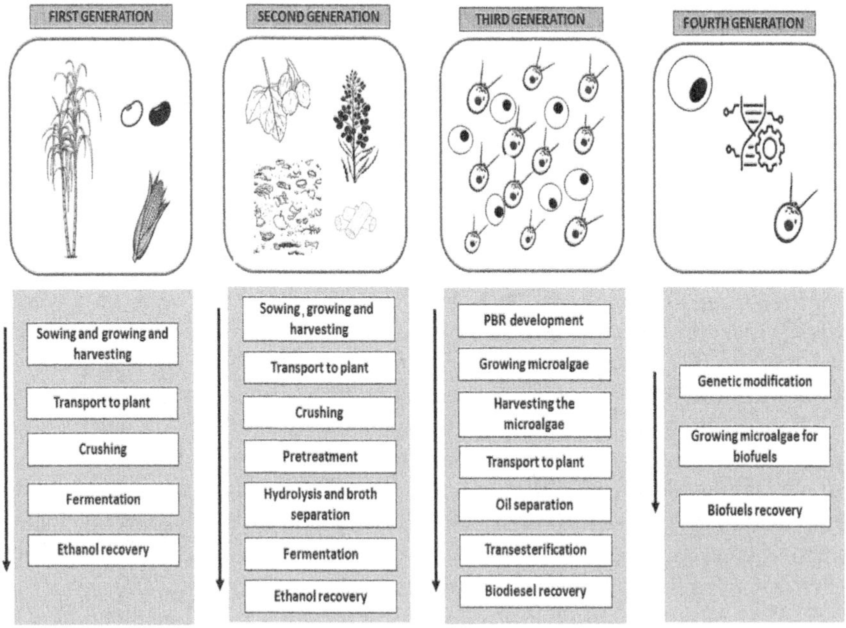

Figure 13.1 An overview of the production process of four generations of biofuel. Source: Authors.

13.2 Genetic modifications in the fuel sector

Biofuels such as alcohols, alkanes, isoprenoids, and derived biomolecules are produced by a variety of microorganisms using different substrates as the precursor molecules. There are several species of microorganisms, namely *Clostridium sp.*, *Pseudomonas putida*, and *Bacillus subtilis*, which have been observed to naturally produce biofuels like butanol. Several oleaginous microbes such as bacterial species of *Mycobacterium*, *Streptomyces*, *Rhodococcus* and eukaryotes such as yeast and algae produce high amounts of triacylglycerols (TAG) that can be used as a precursor for the production of biodiesel (Shields-Menard et al., 2018). Even so, these natural biofuel producers are unable to produce fuel in appreciable quantities due to their slower growth rates and low production titers and yield. Utilization of such natural biofuel producers for genetic engineering is also hindered due to unknown endogenous pathways for gene regulation and fuel production, incompetent genetic elements, loss of metabolic activity due to evolution, presence of inhibitors, a lower redox potential, and an intolerance to cellular toxicity caused by increasing concentration of the biofuels (Phulara et al., 2018).

Yet, improvement in synthetic biology has enabled us to redesign native microbial pathways to improve the yield of biofuels in microorganisms. Some of these include utilization of new pathways, improvement of the existing fermentation processes, optimization of the microbial metabolism, overexpression of the desired genes, heterologous gene expression and modification of the target enzymes, heterologous synthesis of desired products, blocking competitive pathways, and synthesis of desired products using orthogonal pathways. Furthermore, utilization of GMOs as platforms for biofuel production is an environment-friendly method when compared with chemical synthesis.

Microalgae possess a substantial surface area to mass ratio compared to other terrestrial plants. They are highly efficient photosynthesizers and thus produce huge amounts of starch and lipids (up to 70–90 percent dry cell weight) that can be converted into various types of biofuels. Cultivating algae may seem to be a simple process due to their ease of access. However, different species of algae have varying conditions for their optimal growth. Similar to microalgae, photosynthetic prokaryotic cyanobacteria are also one of the leading platforms for genetic manipulation for the production of biofuel. Cyanobacteria are highly efficient photo convertors and CO_2 sequestrates and can be fabricated to produce the desired biochemical products. Biohydrogen, ethanol, butanol, isoprene, and fatty alcohols are some of the products that have been efficiently produced using GM cyanobacteria (Moravvej et al., 2019).

Additionally, yeasts such as *Saccharomyces cerevisiae* have been identified to be effective GMOs for biofuel production as the tools for their genetic conversion are easily available. Yeasts also possess genes which can be modified easily compared to other oleaginous microbes. Numerous bacterial strains have also been investigated for the production of biofuels. *Escherichia coli*, *Corynebacterium glutamicum*, *Zymomonas mobilis*, *Lactococcus lactis*, and *Bacillus subtilis* are among the few strains that have been assessed for their potential to produce biofuels. The ease of using *E. coli* is due to the fact that it is a very user-friendly model organism and its genetic regulation has been extensively studied, which makes editing its metabolic and regulatory pathways simple (Malik et al., 2018).

Genetic modification of organisms gives us the means to produce highly selective 4G biofuel. Even so, microbial cell factories have not as yet been able to meet commercial demands. The reason behind this 'inefficiency' is due to the myriad of problems that need solving before genetic modification can be thought of as a viable method of biofuel production. The cytotoxic nature of biofuels – especially at higher concentrations, the low viability of the engineered cells, complexity in formulating the fermentation mixture, multiple physicochemical impurities in the fermentation media, and the non-selective bioconversion of the products are some of the bottlenecks that are arresting its commercial viability (Mercer & Armenta, 2011; Dunlop et al., 2011).

13.3 Genes involved in agronomic trait development

Genetic modification offers diverse options for the mass production of biofuels. This is achieved with the assistance of genetically engineered industrially competent GMOs having faster growth rates, increased CO_2 sequestration, and the ability to utilize a broad range of substrates. They have the potential to produce high titers of the target compound with easy downstream processing.

13.3.1 Butanol

Butanol is a commercially important solvent commonly utilized as a gasoline substitute and as a solvent in paints, perfumes, and plastics. The *Clostridium* species of bacteria is known to produce butanol naturally through the Acetone–Butanol–Ethanol (ABE) pathway (Nawab et al., 2020). Synthesis of butanol through fermentation is a carbon-neutral process and has the potential to be utilized commercially. The major hindrance in using *Clostridium* sp. for butanol production is on account of its slower growth rate, complexity in substrate formulation, impurities in the product, difficulties in isolating and purifying the product, its complex life cycle as well as difficulties in its genetic manipulation. Therefore, genetic model organisms such as *E. coli, S. cerevisiae* and *Lactobacillus brevis* have been genetically engineered to introduce and express the butanol producing gene and produce commercially viable titers of butanol (Nawab et al., 2020). The natural 1-butanol synthesis pathway from *Clostridium* was identified and introduced in *E. coli* which resulted in a titer of 30 g/L of 1-butanol. Similarly, the 2-keto acid pathway was constructed in *E. coli* for the purpose of isobutanol production. The 2-keto acid pathway produced a titer of 22 g/L of isobutanol in biofermentation systems (Liu et al., 2021). An increased concentration of butanol (19.12 g/L) was obtained by double overexpression of the 6-phosphofructokinase and pyruvate kinase enzymes in *Clostridium acetobutylicum*. In another study, 12.15 g/L of butanol titer was obtained on overexpressing the heat shock protein in *Clostridium tyrobutyricum* cells (Mehmood et al., 2021).

13.3.2 Hydrocarbons

Addition of acyl carrier protein (ACP)/CoA reductase and fatty aldehyde deformylating oxygenase from cyanobacteria to *E. coli* is carried out to heterologously synthesize alkanes. Expression of these enzymes resulted in the formation of fatty acid intermediates to alkanes and alkenes of the range C_{13} to C_{17} with a yield of 0.025 g/L in *E. coli* (Schirmer et al., 2010). In *E. coli*, an artificially constructed molecular pathway was inserted to express a variety of fatty acid reductase and fatty aldehyde deformylating oxygenase (FADO) enzymes responsible for the synthesis of hydrocarbons. An artificial molecular pathway was successful in achieving 0.58 g/L titer of hydrocarbons of the range C_{13} to C_{17} in *E. coli* (Howard et al., 2013). In microalgae, the CAR gene was replaced with UndB of *Pseudomonas* and the enzyme aldehyde deformylating oxygenase was replaced with fatty acid photodecarboxylase from *Chlorella*. This resulted in a one-step conversion of CO_2 to hydrocarbons. An 8-fold increase (0.008 g/L) in alkene production in *Chlamydomonas reinhardtii* and a 19-fold increase in alkane production (0.077 g/L) in *Synechocystic* sp. was observed when genetically engineered by the aforementioned method (Yunus et al., 2018).

13.3.3 Isoprenoids

Isoprenoids are excellent sources of jet fuels, fuel additives, and biodiesel. Isoprenoids are naturally synthesized by plants and bacteria through the 2-C-methyl-D-erythritol 4-phosphate (MEP) pathway and mevalonate (MVP) pathway (Figure 13.2). To increase its production, common hosts such as *S. cerevisiae*, *E. coli*, and cyanobacteria have been utilized (Walls & Rios-Solis, 2020; Rodrigues & Lindberg, 2021). Biosynthesis of squalene in *S. cerevisiae* increased by a factor of 71 (634 mg/L) as a result of overexpressing the INO2 gene which increases the size of the endoplasmic reticulum in the cell. Synthesis of farnesol, geranylgeraniol, and nerolidol was also observed to increase in *S. cerevisiae* due to the overexpression of the gene encoding β-hydroxy β-methylglutaryl-CoA reductase of mevalonate and prenyl diphosphate pathways. On insertion of geraniol synthase from *Ocimum basilicum* in *S. cerevisiae*, the yeast cells produced geraniol. Interestingly, it was also observed that expression of geraniol synthase in *S. cerevisiae* disrupted farnesyl diphosphate and increased geraniol synthesis (Fischer et al., 2011).

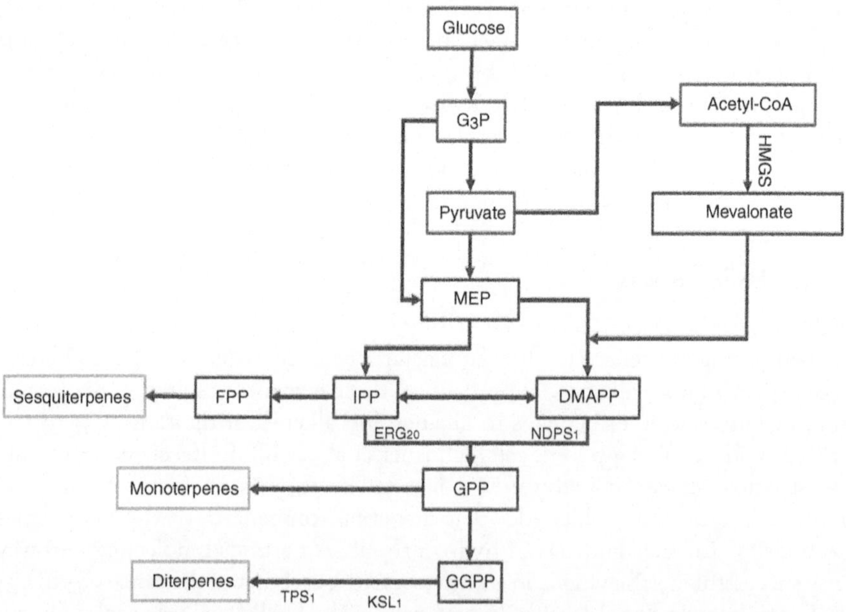

Figure 13.2 MEP/MVP pathway for isoprenoid production.
Source: Authors.

13.4 Mechanism of genetic modification leading to desired trait

13.4.1 Mechanism involved in butanol production

The clostridial CoA-dependent 1-butanol pathway (Figure 13.3) has been inserted in many model organisms such as *S. cerevisiae*, *E. coli*, and *S. elongatus*. The 1-butanol pathway starts with the condensation of two molecules of acetyl-CoA catalyzed by thiolase to form acetoacetyl-CoA. Acetoacetyl-CoA then converts to butyryl-CoA by undergoing a succession of dehydration and reduction catalyzed by the enzymes 3-hydroxybutyryl-CoA dehydrogenase, crotonase and butyryl-CoA dehydrogenase-electron transport complex. Butyryl-CoA then finally converts to 1-butanol, catalyzed by the enzyme bifunctional aldehyde/alcohol dehydrogenase, which is encoded by either aldehyde/alcohol dehydrogenase (adhE1) gene or butyraldehyde-butanol dehydrogenase (adhE2) gene (Nawab et al., 2020). *Clostridium* species such as *C. tyrobutyricum*, *C. cellulovorans*, and *C. cellulolyticum* can produce butanol without any genetic modification albeit producing lower titers; however, on overexpressing the adhE2 gene, it has been observed that the concentration of butanol produced by *Clostridium* species increases many fold compared to genetically modified model microbes such as *E. coli*

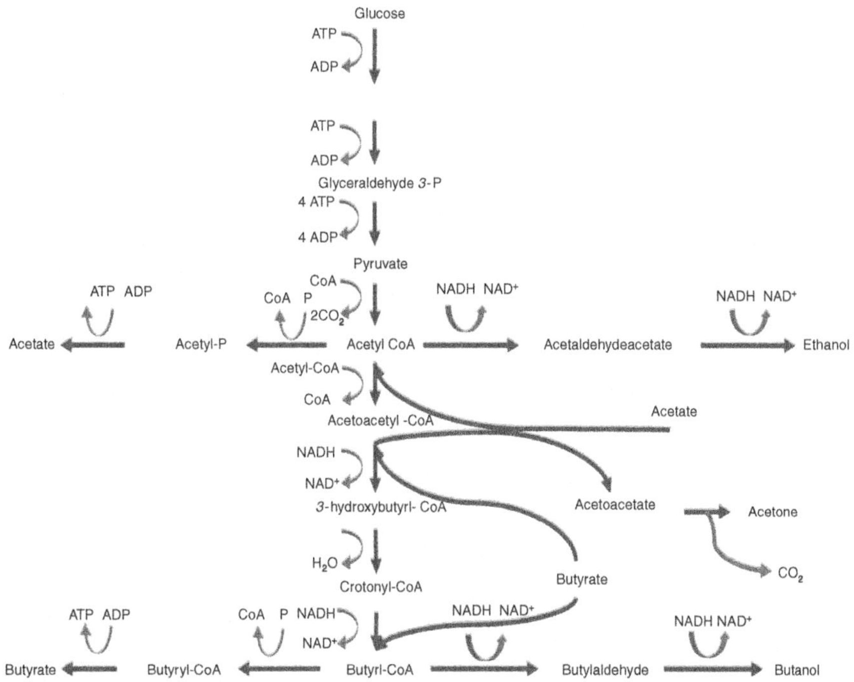

Figure 13.3 Acetone–Butanol–Ethanol pathway for butanol synthesis.
Source: Authors.

13.4.2 Mechanism involved in propane production

E. coli strain BW25113 can be utilized to produce propane using a novel valine synthesis pathway as illustrated in Figure 13.4. Acetolactate synthase gene from *B. subtilis* was expressed in *E. coli* to convert pyruvate into 2-acetolactate. 2-acetolactate was further converted to 2-ketoisovalerate by the overexpression of two native *E. coli* genes – ketol-acid reductoisomerase and dihydroxy-acid dehydratase. Alpha-ketoisovalerate decarboxylase gene from *L. lactis* and aldehyde-deformylating oxygenase from *P. marinus* MIT9313 introduced in *E. coli*, converts 2-ketoisovalerate into isobutyraldehyde and then to propane. For the efficient conversion of isobutyraldehyde to propane, 13 different endogenous genes in *E. coli*, namely yqhD, adhE, eutG, adhP, yjgB, yiaY, yahK, fucO, idhA, frdABCD, pfiB, fnr, and dkgA, which encode for the aldehyde reductase, are knocked out, as aldehyde reductase stops propane synthesis by shifting isobutyraldehyde conversion from propane to iso-butanol (Rodriguez & Atsumi, 2014).

13.5 Benefits to society

13.5.1 Environmental

Biofuels provide a host of benefits to society. One of the primary benefits of biofuel usage is the reduction of greenhouse gas emissions. Algae is one of the most efficient photo-synthesizers responsible for around 50 percent of the global carbon fixation and is an efficient means of converting sunlight to energy and transforming CO_2 to biomass. Similarly, combustion of microalgae-derived biofuels has reduced greenhouse gas emissions in combination with high CO_2 sequestration. High lipid content is achieved in microalgae at high CO_2 concentrations. Biofuel produced from such genetically engineered microbes has the benefit of being carbon negative. For example, in *Chlamydomonas reinhardtii*, mutating LHC translation repressor NAB1 resulted in

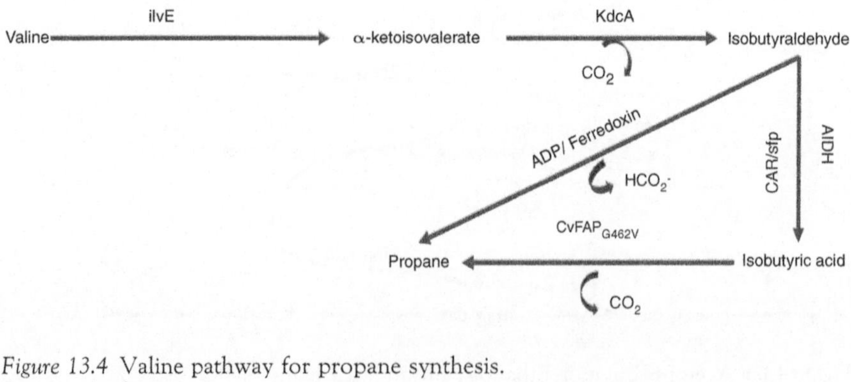

Figure 13.4 Valine pathway for propane synthesis.
Source: Authors.

the reduction of the PSII antennae size, leading to an increased photosynthesis efficiency (by 50 percent). Genetically engineered *Synechococcus elongatus* can generate and secrete carbonic anhydrase in the growth medium. Carbonic anhydrase converts CO_2 to HCO_3, which is further absorbed by the cyanobacteria and converted to biomass. This can increase the carbon intake of the cyanobacteria cells by 41 percent (Chen et al., 2012; Khan et al., 2021). Another benefit from cultivating algae is its ability to utilize wastewater as a growth medium, making algae an efficient and economically viable option for treating organic and inorganic contaminants in waste water. Organic compounds in waste water, namely carbon containing compounds and inorganic compounds like nitrates, phosphates, and various heavy metals, are utilized by algae for their growth, thus extracting them from the waste water. One example is the use of *Ascophyllum nodosum*, a type of brown seaweed that is used to extract indium from waste water. Indium is an important metal in modern electronics, utilized particularly in making screens. This alleviates the problem of combusting circuit boards for removing this rare metal (Khan et al., 2021).

Yet another major reason for using biofuels is the reduction in the tail pipe emission of greenhouse gases. The USEPA and the California Air Resource Board found that tail pipe emission of greenhouse gases was reduced by 50 percent when biofuel was used as a replacement for fossil fuels. Blending 80 percent low-sulphur diesel with 20 percent biodiesel achieved a 50–85 percent reduction in greenhouse gas emissions and could therefore successfully remove 18 Mt of CO_2 (Ambaye et al., 2021).

13.5.2 Water

One of the drawbacks associated with first and second generation biofuels was the need for a large water supply. Since fresh water is a limited resource, there is a need to use seawater or brackish water for the purpose of biofuel generation in order to make it an environmentally viable option. Water footprint refers to the total volume of water required for a process. The water footprint of microalgae was found to be one quarter of the water footprint of terrestrial plants. For example, the water footprint of biodiesel obtained from microalgae was 14–87 m^3 ton^{-1}, whereas the water footprint of bioethanol produced from sugar cane was 139 m^3 ton^{-1} and from rape seed was 165 m^3 ton^{-1} (Khan et al., 2021). Production of biodiesel from microalgae has a water footprint of 3726 kg water/kg of biodiesel. However, 84 percent of this water is recycled, which takes the actual water footprint down to 591 kg water/kg of biodiesel. Another advantage of using algae is that they can utilize the nutrients in waste water and sea water which reduces the need for fresh water in biodiesel production. Therefore, around 90 percent of the water discharged can be restored back to the manufacturing process (Feng et al., 2016).

13.5.3 Land

The major benefit of 4G biofuel production is land conservation. In the current scenario of increasing world population, converting land use for biofuel production in place of food crops is highly impractical. Along with reducing food security, growth of biofuel producing crops also deteriorates soil quality and harms biodiversity. Plus, it has been also observed that, globally, increased biofuel production coincides with increased prices of food and vegetable oil. This was one of the major drawbacks of producing first and second generations biofuels. Microscopic algae have a higher growth rate and efficiency as compared to crop production, agroecosystems, and other aquatic plants, and require significantly less surface area compared to other biofuel feedstocks (Chisti, 2007). For instance, sugarcane roughly produces 10 kg dry weight $m^{-2}yr^{-1}$ whereas brown seaweed produces 13.1 kg dry weight $m^{-2}yr^{-1}$. Additionally, the high growth rate of microalgae and its cultivation on non-arable land, eliminates the competition between biofuel production and food cultivation. Another advantage of microalgae cultivation and their limited land dependency is that it leads to less habitat destruction and improved biodiversity conservation (Rajkumar et al., 2013). For example, coastal areas and underutilized rice lands can be used for algae cultivation. Marine microalgae can be also cultivated in lands rendered unproductive due to the inflow of salt water (Abomohra et al., 2017). The US Department of Energy has estimated that for microalgae biofuel to replace petroleum fuel, it would require only 15,000 square miles of cultivation land, which is one-seventh the land required for corn cultivation. Plus, oil yields per hectare from microalgae remarkably surpass the yield from oil plants such as rapeseed, sunflower, and palm (Khan et al., 2021).

13.5.4 Socio-economic

Development of sustainable and economically viable biofuels is dependent on their ability to generate jobs and revenue in local communities. For example, in 2018 around 2,063,000 jobs related to the field of biofuel were created worldwide; with maximum employment in Brazil with 832,000 new jobs, followed by the USA (311,000) and the European Union (208,000). In India, around 35,000 new jobs were created in the biofuel sector in that year (Ambaye et al., 2021). Hence, fourth generation biofuels have the advantage of providing food security along with an opportunity for generating capital amongst local communities. Harvesting wild seaweed is a vital cultural and economic practice in many nations such as Japan, the Philippines, and Brazil. The majority of the harvesters and cultivators in these cases are mostly unskilled or semiskilled laborers, of which a higher proportion are women. As cultivation practices are easy, little support is needed from the government to train a sufficient workforce to be employed in 4G biofuel production from microalgae. In addition, the by-products of microalgae cultivation such as biofeed and biofertilizers can generate extra income for the harvesters (Alam et al., 2012). Therefore, 4G biofuel production offers a good entrepreneurship opportunity. Many companies have decided to invest in this sector, which has been also referred to as 'green crude' owing to the fact that biofuels can be serious options for fossil fuel replacement. 4G biofuel production can address the rising fuel prices, serve as a cleaner option for biofuel production, and generate about 50 times more fuel per unit area as compared to first and second generation biofuels. Focused research on small scale biodiesel production will popularize small scale biodiesel entrepreneurship and will help reduce dependency on imported fossil fuels and thus increase energy security (Singhal & Prashant, 2020).

13.6 Trade in the global market

Global biofuel production is continuously expanding, but in terms of usage their current share is only 2.3–2.5 percent of the global oil use. According to the RFA database, the major global producer of ethanol is the USA (62 percent) followed by Brazil (27 percent), the European Union (6 percent), and China (3 percent). In the case of biodiesel production, the major producers are the European Union with a market share of 34 percent, followed by the USA with 19 percent, Brazil with 13 percent, Indonesia with 11 percent, and then Argentina and Thailand at 8 percent and 5 percent, respectively (Mizik, 2020). Yet, overall, international trade in biofuel is relatively low in volume. Exports of ethanol and biodiesel were reported to be less than 9 percent and 14 percent of the global ethanol and biodiesel production respectively over the period 2017–2019. The USA is a major biofuel consumer and is the only country with high domestic use along with being a net biofuel exporter. The other major biofuel producers, namely Brazil, Argentina, and Indonesia, are net exporters of biofuel because domestically their biofuel usage has been limited. However, international trade in biofuel is expected to reduce before it improves due to recent regulations for the reduction of first and second generation biofuel usage.

In the future, the majority of the key biofuel exporters such as China, Indonesia, and Brazil are aiming to domestically increase biofuel consumption. With the current leaps made in the electric vehicle sector and the demand for highly efficient fuels, focus is on the development of advanced biofuels from microbial sources. The greatest advancements in biofuel technology are expected to come from the USA and the European Union. The continuous demand for biofuel in these countries could pave the way for African and Asian countries like India and South Africa to take the lead in biofuel production and export. Advanced 4G biofuels could generate national energy security and shift global focus from fossil fuel- to biofuel-based power generation. This will result in improving the current uncertainties on fuel prices, increase employment opportunities, and meet the ever-increasing energy demand (Khan et al., 2021; Singhal & Prashant, 2020).

But, the major hurdle for 4G biofuel trade is the negative social perception towards GMOs and the strict policies surrounding them relating to public concerns. Additionally, along with sanitary and phyto-sanitary concerns, there are also certain technical trade barriers such as high import and export tariffs, promotion of domestically produced biofuel over exported biofuel, and logistical issues with biofuel transportation. For trade in GMOs and their products, there has to be several socio-economic considerations to be taken into account before any decision can be made to make the trade easier between countries. Therefore, a strong international consensus is required to solve the issues regarding the safety of GMOs and their use for biofuel production and trade (Shokravi et al., 2021).

13.7 Health and ethical risks

Biofuel governance and policies should be formulated by considering the three pillars of sustainability: the economic, the environmental, and the social. For any kind of product to be successful, maintaining the balance between these three aspects of sustainability is itself a challenge and accommodating them comes with various issues. Several policies regarding the use of GMOs for biofuel production have failed. This is mainly because policies have been made focusing just on economically favorable technical solutions, overlooking the social and environmental aspects of biofuel production.

13.7.1 Environmental risks

One of the major issues involving GMOs is their potential effect on the environment. Microalgae are geographically distributed over a broad range of habitats. Marine macroalgae are less distributed compared to microalgae and phytoplanktons. Different species of micro- and macroalgae require different temperatures, light, substrate, and nutrients for their optimum growth. Therefore, native species of algae can easily be rendered extinct if GM algae escape into the environment. This is complicated by the fact that GM algae have characteristics such as a rapid growth mechanism, vegetative propagation, and pest resistance, which also contribute to them replacing local species and disturbing local and regional ecosystems. For example, GM algal strains can produce toxins which mimic algal blooms and prevent other microbial competitors from growing. GM algae also have enhanced nutrient uptake mechanisms, which ensure rapid growth and thereby lead them to outgrow potential competitors. Anti-apoptosis genes in algae prevent their deaths even when the density of certain toxic cellular by-products increase in the growth media due to utilization of all the nutrients in the media (Bertani & Lu, 2021; Plöhn et al., 2021; Reynolds et al., 2021). The risk of diffusion of genetically engineered chromosomal and plasmid DNA and potentially causing horizontal gene transfer in microorganisms is very high as chromosomal and plasmid DNA remain active even after death of the host organism. Therefore, escape of such GMOs could potentially create antibiotic resistant, rapidly expanding super bugs, which can potentially devastate a healthy ecosystem. These risks stem not only from algal and cyanobacterial species but from a variety of species which are genetically modified. However, the risks associated with algal species are high due to their way of culturing. For the purpose of biofuel production, algae are most commonly grown on open culture systems such as open ponds. Therefore, if proper care is not maintained, there is a high possibility of GM algal cells escaping to the environment during the feedstock preparation or the harvesting and drying stages (Shokravi et al., 2021).

13.7.2 Health risks

Similarly, health risks associated with GMOs are still not properly understood. There are two schools of thought in this matter. One considers that there are no health risks associated with GMOs and that it is a significant resource which, if properly employed and managed, has the answer to many of our current problems associated with food security, greenhouse gas emissions, replacement of fossil fuels, and curing certain diseases. The other considers GMOs as presenting a grave threat to our ecosystem, biodiversity, and health. It can be opined that such confusion stems from insufficient studies being conducted on the safety of these materials. Most of the studies are short term and carried out on rats and therefore fail to portray the effect of consuming artificially designed genetic material on the human body. For the time being, there are no significant studies on the potential of horizontal gene transfer from microbes engineered to produce biofuels, or their effect on the environment and humans in general, yet some studies regarding consumption of GM crops have been reported. One study found that feeding of NK603 maize to rats had no negative effect on the animals' health (Steinberg et al., 2019), whereas another study found that feeding rats the same variety of maize resulted in organ failure, increased occurrence of tumors, and premature death (Séralini et al., 2014). Therefore, many of the studies in this regard are inconclusive and contradictory and a multidisciplinary approach should be taken to fully understand the effect of designed genetic material on the human body and the ecosystem as a whole.

13.7.3 Socio-economic risks

One socio-economic issue regards the profiteers of the said technology and governments' intention behind GMO regulations. A good example of this is the difference in the regulations regarding GMOs between the USA and the EU. The USA, which is the major exporter of genetically modified products, has adopted a coordinated framework for the commercial production of GM products. It does not identify between biotechnologically designed organisms and their natural counterparts. Whereas in the EU and several other countries, including India, there are stringent laws against cultivating GMOs and using GM products. The USA has repeatedly argued against such policies and for the relaxation of such stringent policies. Hence, a precautionary principle should be maintained while enacting such regulations for making practical decisions to avoid health and environmental related disasters as scientific evidence related to the effects of GMOs is still lacking.

13.8 Current status and future prospects

The annual global production of biofuels in 2019 increased by 5 percent, with an increase of 13 percent in biodiesel and 2 percent in ethanol production. Annual production of biofuel is projected to increase by 3 percent each year for the next five years according to the International Energy Agency (IEA). Worldwide employment in biofuels was estimated at 2.5 million in 2019 (IRENA, 2022). The liquid biofuel market, which is one of the most cost-effective means of producing high efficiency fuels such as aviation fuel, was heavily affected by Covid-19. Biofuel production for transportation was reported to be 144 billion liters in 2020, which was estimated to be 11.6 percent less than in 2019 (IRENA, 2020; OECD/FAO, 2020).

Energy security is seen as a crucial developmental index in India. India processes around 650 million liters of biodiesel every year. It currently has six plants for producing biodiesel, with capacity ranging from 11 to 280 million liters. India aims to shift towards biofuel for the purpose of electricity production, setting a target share of 40 percent to non-fossil-based fuel by the end of 2030. Through coordinated national programs like the Ethanol Blended Petrol Program, the National Biodiesel Project, and the Biodiesel Blending Program, the government is preparing a strategic road map to reduce fossil fuel import. Other strategies adopted by the government to promote the role of biofuel include policies and incentives for the production of biofuel, adoption of biofuels, promoting research towards increasing fuel efficiency, improving the refinery cycle in biofuel production, and slow integration of biofuels in the market as a substitute for fossil-based fuels. Biofuel production in India also goes hand in hand with Make in India and Swachh Bharat Abhiyan. It also has the potential of increasing farmers' incomes, increasing biofuel exports, generating employment, and creating capital. India aims to increase the concentration of ethanol in diesel from 2 percent to 20 percent and the concentration of biodiesel in diesel from 0.1 percent to 5 percent by 2030. This goal can be realized only by increasing the domestic production of biofuels, innovation in new types of feed stock for biofuel production, introduction of state-of-the-art biofuel conversion technology, development of infrastructure for promotion, and efficient production of biofuels (Singhal & Prashant, 2020).

Despite the reported advantages of microbial biofuel, to be commercially viable several hurdles must be addressed. Some of the areas of improvement include: improving culture parameters to allow faster growth, high biomass generation, and the ability to convert nutrients to higher content of lipid in the cells. Additionally, easy biomass harvesting and product isolation procedures have to be identified along with recycling of water, nutrients, and spent biomass (Faried et al., 2017). Algal biofuel production by The National Renewable Energy Laboratory (NREL) route, which consists of fermentation, lipid extraction, distillation, and hydrodeoxygenation steps, had the lowest selling price (LSP) of \$0.96, whereas the UA route consisting of the traditional cultivation, harvesting, lipid extraction, transesterification, and product purification steps for biodiesel production had an LSP of \$2.32. The difference in LSP between the two methods is due to the production of additional by-products in the NREL route that generate additional revenue (Dutta et al., 2016). The cost of algal biodiesel in more recent studies has been estimated to be €0.33/L and for biomass production to be €2.01/kg in a 15.247-hectare plant. At a scaled up biodiesel production plant of 100 hectares, the total investment cost was reported to be \$48/m^2 for open raceways and \$66/m^2 for tubular photobioreactors (Branco-Vieira et al., 2020). Therefore, the reported outcomes of these studies indicate that the costs of biodiesel from algal sources are currently higher compared to fossil fuels. Combining microalgae cultivation with waste water treatment could reduce production costs by reducing the money spent on utilities such as harvesting equipment and centrifuges, cutting labor costs, circumventing fertilizers, increasing CO_2 sequestration and the construction of efficient photobioreactors and open raceway systems. Additionally, efficient recovery of by-products from the algal biomass – protein, carbohydrates, rare metals and therapeutic compounds – can reduce production costs, thereby reducing the cost of algal biodiesel.

13.9 Conclusions

The 4G biofuels produced from GMOs such as *E. coli*, *S. cerevisiae*, *P. putida*, and *S. elongatus* have the potential to address many of the problems that have hampered the commercialization of the previous generations. They are synthesized by introducing the common pathways for biofuel synthesis like the ABE pathway, valine pathway, and the MEP/MVP pathway in the microbial hosts. However, production of biofuels from metabolically engineered microbes currently suffers from low product titers, cellular toxicity, reduced cellular expression of the modified genes, and complexity in product isolation and purification. Therefore, the focus of research has been on optimizing the production process. The course of action for rectifying the said problems has been on up-regulating the pathways and the enzymes responsible for synthesizing the desired product or its precursors, deleting pathways responsible for competing and or product inhibition, redirecting metabolic flux and modifying membrane proteins for secretion of the product to reduce cellular toxicity and simplify downstream processing.

Advanced 4G biofuels can assist in improving people's socio-economic status by generating employment and by generating capital in local and rural communities. Being carbon negative, they can also tackle the problems associated with greenhouse gas emissions and, as they are microbial in nature, production of 4G biofuels can go hand in hand with food crop cultivation. But due to the environmental risks associated with GMOs and purported health risks, stringent government policies have been formulated regarding their cultivation and use. Obviously, this has impeded the international trade in 4G biofuels. In spite of that, domestic production of biofuels in developing countries like India, Brazil, and Bangladesh is expected to grow. This is because of government regulations which promote biofuel production, the relaxation of tariffs, mandatory blending with fossil fuels, and funding of research and development. 4G biofuels are expected to take over the market of fossil fuels in the future. However, significant development in 4G biofuel processing still required for its commercial viability across the world.

References

Abomohra, A. E.-F., El-Sheekh, M., & Hanelt, D. (2017). Screening of marine microalgae isolated from the hypersaline Bardawil lagoon for biodiesel feedstock. *Renewable Energy*, 101, 1266–1272.

Alam, F., Date, A., Rasjidin, R., Mobin, S., Moria, H., & Baqui, A. (2012). Biofuel from algae – is it a viable alternative? *Procedia Engineering*, 49, 221–227.

Ambaye, T. G., Vaccari, M., Bonilla-Petriciolet, A., Prasad, S., van Hullebusch, E. D., & Rtimi, S. (2021). Emerging technologies for biofuel production: A critical review on recent progress, challenges and perspectives. *Journal of Environmental Management*, 290, 112627. doi:10.1016/j.jenvman.2021.112627.

Bertani, P., & Lu, W. (2021). Cyanobacterial toxin biosensors for environmental monitoring and protection. *Medicine in Novel Technology and Devices*, 10, 100059. doi:10.1016/j.medntd.2021.100059.

Branco-Vieira, M., Mata, T. M., Martins, A. A., Freitas, M. A. V., & Caetano, N. S. (2020). Economic analysis of microalgae biodiesel production in a small-scale facility. *Energy Reports*, 6, 325–332.

Chen, P.-H., Liu, H.-L., Chen, Y.-J., Cheng, Y.-H., Lin, W.-L., Yeh, C.-H., & Chang, C.-H. (2012). Enhancing CO_2 bio-mitigation by genetic engineering of cyanobacteria. *Energy & Environmental Science*, 5(8), 8318. doi:10.1039/c2ee21124f.

Chisti, Y. (2007). Biodiesel from microalgae. *Biotechnology Advances*, 25(3), 294–306.

Dunlop, M. J., Dossani, Z. Y., Szmidt, H. L., Chu, H. C., Lee, T. S., Keasling, J. D., … Mukhopadhyay, A. (2011). Engineering microbial biofuel tolerance and export using efflux pumps. *Molecular Systems Biology*, 7(1), 487. doi:10.1038/msb.2011.21.

Dutta, S., Neto, F., & Coelho, M. C. (2016). Microalgae biofuels: A comparative study on techno-economic analysis & life-cycle assessment. *Algal Research*, 20, 44–52.

Faried, M., Samer, M., Abdelsalam, E., Yousef, R. S., Attia, Y. A., & Ali, A. S. (2017). Biodiesel production from microalgae: Processes, technologies and recent advancements. *Renewable and Sustainable Energy Reviews*, 79, 893–913.

Feng, P.-Z., Zhu, L.-D., Qin, X.-X., & Li, Z.-H. (2016). Water footprint of biodiesel production from microalgae cultivated in photobioreactors. *Journal of Environmental Engineering (New York, NY)*, 142(12), 04016067. doi:10.1061/(asce)ee.1943–7870.0001150.

Fischer, M. J. C., Meyer, S., Claudel, P., Bergdoll, M., & Karst, F. (2011). Metabolic engineering of monoterpene synthesis in yeast. *Biotechnology and Bioengineering*, 108 (8), 1883–1892.

Howard, T. P., Middelhaufe, S., Moore, K., Edner, C., Kolak, D. M., Taylor, G. N., … Love, J. (2013). Synthesis of customized petroleum-replica fuel molecules by targeted modification of free fatty acid pools in *Escherichia coli*. *Proceedings of the National Academy of Sciences of the United States of America*, 110(19), 7636–7641.

IRENA (2020). *Renewable Energy and Jobs – Annual Review 2020*. https://www.irena.org/publications/2020/Sep/Renewable-Energy-and-Jobs-Annual-Review-2020

Khan, N., Sudhakar, K., & Mamat, R. (2021). Role of biofuels in energy transition, green economy and carbon neutrality. *Sustainability*, 13(22), 12374. doi:10.3390/su132212374.

Liao, J. C., Mi, L., Pontrelli, S., & Luo, S. (2016). Fuelling the future: Microbial engineering for the production of sustainable biofuels. *Nature Reviews – Microbiology*, 14(5), 288–304.

Liu, X., Xie, H., Roussou, S., & Lindblad, P. (2021). Current advances in engineering cyanobacteria and their applications for photosynthetic butanol production. *Current Opinion in Biotechnology*, 73, 143–150. doi:10.1016/j.copbio.2021.07.014.

Malik, S., Afzal, I., Mehmood, M. A., Al Doghaither, H., Rahimuddin, S. A., Gull, M., & Nahid, N. (2018). Heterologous synthesis and recovery of advanced biofuels from bacterial cell factories. *Protein and Peptide Letters*, 25(2), 120–128.

Mehmood, M. A., Shahid, A., Malik, S., Wang, N., Rizwan Javed, M., Haider, M. N., … Boopathy, R. (2021). Advances in developing metabolically engineered microbial platforms to produce fourth-generation biofuels and high-value biochemicals. *Bioresource Technology*, 337, 125510. doi:10.1016/j.biortech.2021.125510.

Mercer, P., & Armenta, R. E. (2011). Developments in oil extraction from microalgae. *European Journal of Lipid Science and Technology: EJLST*, 113(5), 539–547.

Mizik, T. (2020). Impacts of international commodity trade on conventional biofuels production. *Sustainability*, 12(7), 2626. doi:10.3390/su12072626.

Moravvej, Z., Makarem, M. A., & Rahimpour, M. R. (2019). The fourth generation of biofuel. In *Second and Third Generation of Feedstocks*, edited by A. Basile & F. Dalena (Oxford: Elsevier), 557–597.

Nawab, S., Wang, N., Ma, X., & Huo, Y.-X. (2020). Genetic engineering of non-native hosts for 1-butanol production and its challenges: A review. *Microbial Cell Factories*, 19(1), 79. doi:10.1186/s12934-12020-01337-w.

OECD, & Food and Agriculture Organization of the United Nations (2020). *OECD–FAO Agricultural Outlook 2020–2029* (Paris: OECD).

Phulara, S. C., Chaurasia, D., Diwan, B., Chaturvedi, P., & Gupta, P. (2018). In-situ isopentenol production from *Bacillus subtilis* through genetic and culture condition modulation. *Process Biochemistry*, 72, 47–54.

Plöhn, M., Spain, O., Sirin, S., Silva, M., Escudero-Oñate, C., Ferrando-Climent, L., … Funk, C. (2021). Wastewater treatment by microalgae. *Physiologia Plantarum*, 173(2), 568–578.

Rajkumar, R., Yaakob, Z., & Takriff, M. S. (2013). Potential of micro and macro algae for biofuel production: A brief review. *Bioresources*, 9(1),1606–1633.

Reynolds, D., Huesemann, M., Edmundson, S., Sims, A., Hurst, B., Cady, S., … Gao, S. (2021). Viral inhibitors derived from macroalgae, microalgae, and cyanobacteria: A review of antiviral potential throughout pathogenesis. *Algal Research*, 57, 102331. doi:10.1016/j.algal.2021.102331.

Rodrigues, J. S., & Lindberg, P. (2021). Metabolic engineering of *Synechocystis* sp. PCC 6803 for improved bisabolene production. *Metabolic Engineering Communications*, 12, e00159. doi:10.1016/j.mec.2020.e00159.

Rodriguez, G. M., & Atsumi, S. (2014). Toward aldehyde and alkane production by removing aldehyde reductase activity in *Escherichia coli*. *Metabolic Engineering*, 25, 227–237.

Schirmer, A., Rude, M. A., Li, X., Popova, E., & del Cardayre, S. B. (2010). Microbial biosynthesis of alkanes. *Science (New York, NY)*, 329(5991), 559–562.

Séralini, G. E., Mesnage, R., Defarge, N., & Spirouxde Vendômois, J. (2014). Conclusiveness of toxicity data and double standards. *Food and Chemical Toxicology: An International Journal Published for the British Industrial Biological Research Association*, 69, 357–359.

Shields-Menard, S. A., Amirsadeghi, M., French, W. T., & Boopathy, R. (2018). A review on microbial lipids as a potential biofuel. *Bioresource Technology*, 259, 451–460.

Shokravi, H., Shokravi, Z., Heidarrezaei, M., Ong, H. C., Rahimian Koloor, S. S., Petrů, M., ... Ismail, A. F. (2021). Fourth generation biofuel from genetically modified algal biomass: Challenges and future directions. *Chemosphere*, 285, 131535. doi:10.1016/j.chemosphere.2021.131535.

Singhal, A., & Prashant (2020). Biofuels: Perspective for sustainable development and climate change mitigation. In *Biotechnology for Biofuels: A Sustainable Green Energy Solution*, edited by N. Kumar (Singapore: Springer), 1–22.

Steinberg, P., van der Voet, H., Goedhart, P. W., Kleter, G., Kok, E. J., Pla, M., ... Wilhelm, R. (2019). Lack of adverse effects in subchronic and chronic toxicity/carcinogenicity studies on the glyphosate-resistant genetically modified maize NK603 in Wistar Han RCC rats. *Archives of Toxicology*, 93(4), 1095–1139.

Walls, L. E., & Rios-Solis, L. (2020). Sustainable production of microbial isoprenoid derived advanced biojet fuels using different generation feedstocks: A review. *Frontiers in Bioengineering and Biotechnology*, 8, 599560. doi:10.3389/fbioe.2020.599560.

Yunus, I. S., Wichmann, J., Wördenweber, R., Lauersen, K. J., Kruse, O., & Jones, P. R. (2018). Synthetic metabolic pathways for photobiological conversion of CO_2 into hydrocarbon fuel. *Metabolic Engineering*, 49, 201–211.

14 Detection methods of GM crops

Manish Bakshi, Prerna Gupta and Ranjeet Singh Jamwal

14.1 Genetically modified plants: detection of GMOs

Genetically modified organisms (GMOs; plants or animals) have received special attention as a lot of research has been done regarding raising a progeny of plants with a modified genetic makeup suitable for specific purposes. Up until now, the practical application of GMO plants is for herbicide tolerance, insect virus resistance, and other beneficial traits such as delayed ripening in the case of fruits and vegetables or the quality content of crops (Kjellsson & Strandberg, 2001). There are 29 different crops and fruit trees in 42 countries which have been successfully modified for various traits like herbicide tolerance, insect/pest resistance, disease resistance, and quality improvement (Nazir et al., 2019). For example, increasing host plant resistance has been used in cotton for combating insect attacks (Trapero et al., 2016).

However, in keeping with food safety and ethical norms, some GMO crops have been recommended for cultivation and some are still awaiting scientific approval. Therefore, standard mechanisms to detect genetic modifications in plants are necessary. Several different methods have been put to use for such identification: the detection of genetic insertions in the DNA; analysis of the mRNA of the newly introduced gene; and protein, metabolite, or phenotypic analysis. For detection of the inserted DNA, the PCR (polymerase chain reaction) method is usually employed.

The criteria for selecting a particular detection method are discussed below.

14.1.1 Ideal characteristics of analytical methods

An ideal GMO testing method should have the following characteristics:

- Universality – i.e. the ability to detect all GMOs.
- Quantifiable – it must quantify the results as to what amount of GMO content is present in the sample being analyzed.
- Interpretable – results delivered should be easy to interpret for a variety of agricultural products.

DOI: 10.4324/9781003278566-14

- Reliable and reproducible – it must be reliable and reproducible and must not give faux results.

Fundamentally, assessment of samples for the presence of GMOs is broken down into three stages:

1) *Detection*: The principal aim of this process is to assess if a sample has any GMOs in it. Any method that can screen a food product and yield a positive/ negative statement can be utilized to achieve this goal. PCR is commonly used in screening approaches (Dong et al., 2008).

2) *Identification*: If GMOs are detected, more investigation is then necessary to determine the type of GMO and its authorization. PCR-based approaches have again been most efficient in identifying the GMO varieties.

3) *Quantification*: After detection and identification, the next step is to determine the quantity of the GMO insertion and its compliance with the approved threshold level. Semi-quantitative PCR or real-time PCR is commonly used for this stage.

14.2 Sampling

14.2.1 Sampling plan

Selection of the right sampling procedure is critical to the outcome, irrespective of the analytical method employed. An error in accuracy of GMO detection mainly results from: 1) the sampling strategy, 2) the sample treatment, and/or 3) the accuracy and precision of the analytical method. GMO contamination in a seed batch or a field may not be uniformly distributed. Sampling can be accurate only if a requisite sample size is considered. While sampling a lot of seed or crops in a field, a discrete population (i.e. one seed in 1,000) should be sampled. To identify minute contamination levels, some experts propose seed samples weighing between 1.500 and 3.000 grams, whereas others prescribe a quantification based on mass equivalent (e.g. 10,000 seeds). Sample size will be determined by the choice of the detection method and the GMO in question.

Two criteria determine the authenticity of sample:

1 Normally, the sample should be a statistical representation of the whole batch and must be large enough to allow sensitivity analysis.
2 The sample to be analyzed should be typical of the field sample submitted for analysis.

14.2.2 Principles for sampling live plants

Sampling of plants should be carried out at random across the crop rather than being restricted to one region of the area to be sampled. Also, the trial plan should indicate the sample location and the plants should be marked in a way so that the respective position of the plants may be determined later on, if required. Fresh leaves from the apex of the plant should be harnessed and kept in a tidy container, using clean gloves – no separate bags are required for multiple samples from actively growing plants.

14.2.3 Principles for sampling seeds for GMO

The ISTA (International Seed Testing Association) has laid down procedures for sampling of food items to be checked for GMOs. Sample size to be used for detection should be in accordance with the established standards for testing procedures. A portion of the supplied sample – i.e. a working sample produced using ISTA procedures – should be used. The minimum size of the sample should not be less than 3000 seeds. Under the auspices of the Standing Committee on Seed, working groups have been tasked with harmonizing standards on seed lot sampling techniques for the development of working samples for GMO analysis and the selection of appropriate analysis methods. They have created a number of documents relating to GMO identification in traditional seed lots, including testing plans, sequential tests, and statistical elements. In order to save time and reduce the costs of detection, performing the scheduled tests stepwise is recommended.

14.3 GMO testing methods

It's critical to recognize that GMO detection approaches differ on a variety of levels. Currently, only PCR allows for the broad screening for an inserted gene and the identification of specific "events". Protein analysis and phenotypic characterization detect specific features that may be present in a variety of GMOs.

Testing methods, both conventional and advanced will be discussed below. Herbicide bioassays and immunoassays are considered "conventional methods" since they can be performed in most laboratories, but the PCR technique and microarrays are considered "advanced methods" because they are technologically superior and require more equipment and skilled personnel.

14.3.1 Conventional methods

14.3.1.1 Herbicide bioassays (phenotypic characterization)

Different analytical approaches are used depending on the efficacy level required. Seed firms utilize herbicide bioassays for quality assurance since they are said to be accurate, inexpensive, and simple to use. The specific trait can be detected through phenotypic characterization. So far, only testing for herbicide resistance and tolerance are available. Herbicide bioassays are tests that can be performed to check for the herbicide-resistant GMO cultivars.

They involve comparative germination tests, with non-GMO and GMO seeds displaying distinct characteristics, on a germination substrate in the presence of herbicide.

Seed germination determines the detection level, and germination techniques should guarantee the germination of all healthy seeds in the sample. Seeds once marked positive should be tested again for viability. As a preventive test for seed enterprises, herbicide bioassays are considered to be precise, cost effective, and reliant. Companies utilize herbicide bioassays to validate specific deliveries as part of the quality assurance programme. For every analytical sample, negative and positive phenotype seeds should be used.

Usually, 400 seeds are tested but this may vary as per the seed lot size or germination rate of seeds and can be adjusted to 1000 to 2000 seeds/seed lot. The number of seeds tested determines the detection level and quantification. Currently, herbicide bioassays for soybean, maize, cotton, oilseed rape, and several other crops are commercially available (Ladics et al., 2015). Bioassays for resistance to insects or other types of GMOs may be developed in the future.

14.3.1.2 Protein identification methods

For identifying and estimating unique (foreign) proteins inserted into plants by genetic modification, immunoassay is usually used. An immunoassay relies heavily on the specificity of an antibody for a specific target (antigen). These are often very specific and sample preparation needs to be carried out very shortly prior to use. Furthermore, immunoassay application is not restricted to concentration range. Immunoassays, like herbicide bioassays, necessitate independent tests for each attribute of interest.

The availability of specific proteins for the production of antibodies, which is at the core of the immunoassay, is crucial for the valid identification of foreign proteins in GMOs. Proteins developed by the firm that created the plant variety may be proprietary and hence not widely available. Furthermore, certain features of the antigen utilized for antibody formation, such as size, hydrophobicity, and tertiary structure, influence the likelihood of developing a successful immunoassay. Antibodies can be either polyclonal or monoclonal, meaning they were either raised in animals or in cell cultures. Rabbits, goats, and sheep are frequently used to create commercially available polyclonal antiserum. Monoclonal antibodies have a number of advantages over polyclonal antibodies, including persistent adherence and specificity for a single motif or antigenic component, as well as the ability to be produced in vast quantities.

To improve sensitivity, antibodies may have to go through additional purification processes. The antibodies' specificity must be critically examined to rule out any cross-reactivity with related compounds, which could lead to false positive results. Immunoassays rely on an antibody's specific binding to the antigen. As a result, availability of antibodies with the required affinity and specificity is the most significant component in setting up test systems.

A second antibody, especially one that interacts with another epitope on the antigen, can be used to detect the antigen–antibody response. The second antibody is either labelled or does not have a detectable label. Immunoassays can be utilized in the field as well as in the lab.

14.3.1.3 ELISA

On a solid phase, the antigen–antibody reaction is created (microtiter plates). Antigen and antibody react and produce a stable complex, which can be visualized by addition of a second antibody linked to an enzyme. Addition of a substrate for that enzyme results in a colour formation, which can be measured photometrically. Some ELISA kits include calibrators, which contain target analyte concentrations in solution, as well as a negative control for interpreting the data.

At different quantities, these standards (calibrators and controllers) display different colours of blue. The concentration range of the sample can be visibly determined by comparing the colour of the sample with the standards. This is a semi-quantitative interpretation.

For recording the exact optical density of the samples, a microplate reader can be used. The user then uses the reader's software to compute the sample concentration from the standard-curve. With a precision efficiency of less than 0.1 percent, ELISA test kits deliver quantitative findings in hours. Some businesses, however, use significantly higher quantification levels, such as 0.3 percent.

14.3.1.4 Lateral flow sticks

GMOs can be detected in plant samples using dipstick devices (lateral flow sticks). The capture antibody is supported by paper strips or plastic paddles, which ultimately serve as the site of reaction. The paper strip/paddle is dipped in vials filled with various liquids. After each dip, there is a rinse phase. Reaction is judged by a change in colour of the vial. The lateral flow technique is augmented by the recent dipstick format in which capillary force moves the reactants through the channels. The experiment can be completed in a single step, with reagent performance controls incorporated. The lateral flow strip contains antibodies that are specific to the foreign protein, as well as a colour reagent. The linked antibody binds to the protein as soon as the lateral flow strip is inserted in a small volume of an extract from plant tissue harboring an alien protein. Some of the antibody bound to the colour reagent is used to construct a sandwich, but not all of it. Two capture zones are found on the membrane: one for the bound foreign protein and the other for the colour reagent. When the sandwich and/or nonreacted coloured reagent are secured in the designated zones on the membrane, these capture zones become crimson in colour. A single line on the membrane indicates a negative sample, while a positive sample is indicated by two lines. Qualitative or semi-quantitative lateral flow approaches are used. By following proper sampling techniques, an accuracy of 99 percent at a confidence level of less than 0.15 percent GMO can be achieved.

14.3.2 Advanced technological methods

14.3.2.1 PCR

The most widely used method for identifying GMOs is PCR. PCR includes DNA extraction and purification, PCR amplification of the introduced DNA, and confirmation of the amplified PCR result.

In theory, a single target molecule may be detected by PCR in a complex DNA mixture. Nevertheless, a distinction must be made between the PCR approach's potential sensitivity and the actual sensitivity of PCR-based GMO detection techniques. Other steps in PCR-based assays have a significant impact on the method's overall reliability and sensitivity. Each step of the PCR has an impact on the assay's reliability and sensitivity and its optimization is essential in creating a high-end GMO analytical approach. The most difficult part in PCR analysis is extracting high-yield DNA from the sample, ideally 100 percent, carefully avoiding degradation of DNA and removing contaminants that can interfere with amplification process.

Moreover, appropriate controls and standards in a PCR are a must in assay verification, ensuring that the process is functioning smoothly and that the data obtained are reliable. This includes having access to the necessary reference material as well as material testing procedures. In PCR, a set of positive control primers for a gene found naturally in all types of the applicable crop is included. Any of the inserted genetic material can be detected using PCR tests: promoter, structural gene, stops signal, or marker gene. Any test's specific design is determined by the requirements. Screening of GMOs using PCR can be performed using primers that recognize common DNA, such as the promoter genes 35S for Cauliflower Mosaic Virus (CaMV) or the nos promoter of *Agrobacterium tumefaciens*, or the nos terminator. Utilizing PCR to make a valid and unique identification requires knowledge of the insertion sequences. Plant–construct junction sequences as primer targets are the only unambiguous technique. In an indirect manner, this reflects the entire inserted sequence, including the promoter, functional gene, and enhancer/terminator.

14.3.2.2 Qualitative analysis using PCR

As previously stated, the choice of target for the primers is critical for GMO detection by PCR. In general, there are three approaches to selecting a suitable target:

1 The nos promoter, the nos terminator, and the kanamycin resistance gene nptII are all found in the 35S or nos promoter and are prevalent genetic components found in GMOs. This is known as a general GMO screening, and it indicates the possibility of GMO presence in the event of a positive answer. The 35S promoter of CaMV, for example, has been detected in 32 GMOs.
2 To identify a specific gene construct, the connector sequences between two contiguous DNA segments can be used as a destination for specific genetic construct recognition. However, these two joint components can be transplanted into various organisms, producing several GMOs with the same genetic design. As a result, identifying a single occurrence of GMO contamination is difficult.
3 The junction sequences in the integration site can be used to identify a specific event, such as a transformation event (plant–construct junction fragment). When a GMO is the result of non-homologous recombination, the integration site is unique. If two GMOs with the same gene construct are developed, this will be the only way to tell them apart. Nonetheless, the development of such techniques necessitates knowledge of the GMO's sequence as well as access to appropriate reference material.

The application of a screening PCR based on fragment amplification shared by several GMOs relies on the intended outcome, and it cannot be used as a common technique in all agricultural goods because certain GMOs lack the universal element. As a result, the use of a broad screening PCR must be carefully defined. The presence of a GMO can be detected using a broad screening PCR; after initial screening, further identification of the GMO is required.

Broad screening PCR approaches, as previously indicated, take advantage of genetic components like promoters and terminators commonly found in GMOs. Nevertheless, because the same sequences can be found in other GMOs, PCR methods based on internal insertion sequences should only be used as an initial screening tool. A positive or negative list is created as a result of the PCR screening.

Moreover, screening 35S promoter or nos-terminator, which is extensively utilized, will not distinguish between components naturally occurring in infected plants and those found in GMO genetic linkages. Because rape seed and other *Brassica* members can be infected by the cauliflower mosaic virus in nature, a positive result from a 35S promoter screening could be a false positive. False positives induced by virus-infected plants can be avoided by employing a CaMV-specific PCR based on genes that aren't found in GMOs.

An unambiguous, unequivocal signature should be used to identify a specific GMO variety. Because insertion of genes into plant genomes is currently still a haphazard process, border fragments (the junction between the plant's genome and the inserted gene) are unique to each GMO and are the only way to unequivocally identify and quantify each transformation event. In the current state of GMO approvals, certain sequences within the insert (construct-specific) can be used to identify and quantify approved varieties (MacCormick et al., 1998). The problem is much more complicated when it comes to detecting non-EU authorized GMOs.

Screening samples for species-specific DNA is a common standard sequential test technique for GMO detection. If DNA is found, samples are tested using general genetic elements for a variety of GMO-DNA types. Positive screening results are followed up with assays to look for specific genes or constructions identified in GMO crops. The assays used rely on the material (for example, Cry genes, EPSPS genes, pat genes) or plant–construct junction fragments. Thanks to this approach, any published results have been reconfirmed through a variety of testing systems.

Yet, in the event of the presence of GMO content, it is critical to evaluate the amount in order to ensure compliance with the threshold level, and this is considered in the following sections.

14.3.2.3 *Quantitative estimation through PCR*

The most frequent technique of quantification in GMO analytical laboratories is to use broad-spectrum primer sets to detect transgenic elements like the CaMV 35S promoter, the nos terminator, or inserted genes. However, because various transgenic events contain variable numbers of these similar sequence components, a precise assessment of the percentage of GMO cannot be made using these common sequence elements alone. The 35S promoter can be found in 1 to 4 copies in different maize events; hence results can exaggerate the percentage of GMO in the sample. As a result, quantification using event-specific primers yields more exact results not just in terms of the type but also in terms of quantity.

14.3.2.4 *Traditional PCR quantification*

Traditional PCR measures the PCR reaction products at a single point. As a result, "end-point" quantitative PCR can be used to accomplish quantification (Peccoud & Jacob, 1998). However, the precision of conventional PCR quantification is reduced because the connection between the concentration of DNA and PCR signal is not linear and varies from one analytical run to the next. As a result, traditional methods of PCR are considered semi-quantitative, whereas real-time PCR is considered purely quantitative.

Competitive PCR is another option for traditional PCR quantification. A qualitative PCR reaction produces an amplified result from a single primer pair. A second DNA fragment with a known concentration is added to the reaction mixture in a competitive PCR reaction. The competitor fragment has the same primer pair binding sites as the first, but it is smaller. By doing a series of tests with various amounts of synthetic DNA, it is possible to identify the initial amount of template DNA. The original amount of template DNA matches the initial amount of generated DNA if the PCR products have the same intensity as determined by gel electrophoresis.

This method is used twice in a double-competitive PCR (DC-PCR): once for a generic housekeeping gene and its competitor, and once for a GMO-specific gene and its competitor. The GMO percentage can be computed using the results of both PCR reactions, taking into account the entire amount of the organism used to make the GMO.

The competitive and double-competitive systems have the distinct benefit that laboratories do not need to purchase new equipment if they are already carrying out qualitative PCR procedures because they are based on the same equipment. Because they require a standard to compare to, the competitive and DC-PCR procedures are considered semi-quantitative. In this situation, the standard is the known amount of synthetic DNA added. As a result, the findings can only show a value that is less than, equal to, or more than a predetermined standard concentration. For this reason, real-time PCR is projected to increasingly replace these competitive PCR techniques for quantification.

14.3.2.5 Real-time PCR quantification

Real-time PCR quantification is a technology that monitors PCR products in real time. A fluorometric measurement of an inside probe is used to do this during the reaction. This probe is made up of a fluorophor and a quencher in a short DNA segment. The quencher segment suppresses all of the fluorescence of the irradiation fluorophor due to its proximity. The probe's DNA sequence is engineered to probe precisely in the area to be amplified. The DNA is first denatured in each PCR cycle to separate the two strands. The primers and internal probe are annealed in the second phase, and the DNA polymerase enzyme replicates the DNA in the last step. The polymerases used in this technique have a high 5'-3' exonuclease activity that cause the internal probe to break down into nucleotides. As a result of the quencher and fluorophor no longer being in close proximity, the fluorescence signal increases.

14.3.2.6 Microarray technology

This technology also known as DNA chip technology has been developed in recent years for the rapid screening of huge number of samples with respect to gene expression and sequence variation. The classic DNA hybridization approach is used in microarray technology, with the exception that many of specialized probes adhere to a solid surface. Macroarrays, high-density oligonucleotide arrays (gene chips or DNA chips), and micro-electronic arrays are all examples of different forms of DNA chip technology (Freeman et al., 2000). Here, short-chain oligonucleotides are synthesized on a solid surface in DNA chips, whereas the PCR products relating to genomic DNA sequences are settled onto solid glass slide (microarray) DNA arrays (macroarray). A thin layer of agarose is coated on sets of electrodes (capable of generating current) and connected to an affinity moiety. These strategies are fast evolving and offer numerous benefits, though for the time being they also have significant drawbacks. For example, because the procedures are highly sensitive and still in development, they are only available in specialized facilities. With the new DNA chip technology, GMOs can be screened and identified in raw materials, food, and animal feed.

In a single test, the GMO chip kit integrates previously developed screening and identification methods. The chip offers findings from a total of 14 different analyses, including additional testing for plant species. All reagents are included in the GMO chip kit, as well as analysis software (Signalyse®). For the detection of permitted and non-licensed plant types, the current GMO chip kit is able to identify the species-specific DNA of plants and viruses commonly used as genetic insertion elements and especially genetic alterations.

The GMO chip kit also allows for the detection of all GMOs with the CaMV 35S promoter, nos terminator, bar gene, and pat gene. However, with the use of microarrays, adequate quantification is now a challenge. It has become more difficult to quantify and compare data within and across microarray platforms as the microarray technology has advanced. The lack of uniform references and the range of data analysis methodologies in use are the key causes for this. Novel data normalization and statistical interpretation methodologies will improve microarray data standardization and validation, allowing for precise and exact data extraction and quantitative comparison of microarray data across studies.

14.3.3 Methods of advanced detection

14.3.3.1 Walking the DNA

The method of discovering new nucleotide sequences next to known DNA sections in a genome using particular primers for a previously known sequence and further sequencing of the final product using Sanger sequencing equipment is known as DNA walking (Spalinskas et al., 2013).

The three types of DNA walking are:

1 The restriction approach entails digesting genomic DNA with a restriction enzyme of choice, which targets regions near the sequence under consideration, such as the union between unknown and known sequences (Fraiture et al., 2015, 2015b). Inverted PCR is the process of self-circularizing or ligating the produced fragments to a DNA cassette.

2 Amplification of a sequence-specific primer using a sequence-specific primer amplification technique (Spalinskas et al., 2013; Fraiture et al., 2015, 2015a).
3 Primer-based approaches combine with PCR procedures to target specific primers (Spalinskas et al., 2013).

The DNA walking approach is critical for detecting GMOs in both processed as well as unprocessed foods and feed. This method uses two semi-nested PCRs to boost GMO target yield and specificity, which is critical when a low level of GMO is present (Spalinskas et al., 2013). This method has been successfully adopted and developed for routine GM crop analysis in accordance with regulatory criteria for a variety of crops, and it is both time-saving and precise.

As a result, regulatory components like the promoter (P35S) and terminator (nos) are frequently targeted (Raymond et al., 2010; Yang et al., 2005; Wang et al., 2011; Fraiture et al., 2015a, 2015b; Ruttink et al., 2010; Kok et al., 2014). This method, however, is not recommended for detecting unknown components in GM crops.

14.4 Conclusions and future aspects

Inspection authorities face a significant barrier in detecting, identifying, and quantifying GMO presence in seed samples and food or feed products. The current analytical methods for GMO testing provide inspectors with a variety of options and compromises. The approach chosen is determined by what information is required. In general, the higher the demand from customers, the higher the cost of delivering information. Furthermore, as the technique becomes more precise, even more cost and time become attached to it. Also, it is evident that worldwide coordination is required to ensure fulfillment of claims by various companies regarding GMO testing. This can be accomplished through:

- Additional benchmarking of the applicability of DNA and protein-based techniques;
- Method to method compatibility;
- Appropriate validation and proficiency testing techniques; and
- By providing easy access to the necessary reference materials.

There are currently no methods that can authenticate the absence of non-approved GMOs in seed samples – even at the 0.1 percent level. If US corporations who manufacture GMOs define the constructions and make them available to national governing organizations, the chances of identifying non-approved GMOs would be greatly increased by inspection authorities.

Furthermore, there are basic flaws in the current screening methods. The most generally used approach is PCR, and the well-known 35S and nos promoters are frequently used as targets in the initial GMO screening. However, the presence of GMOs with other promoters will not be revealed by this test. Furthermore, because spontaneous infection with the CaMV is frequent in *Brassicas*, the 35S test does not perform dependably, causing the results to be muddled. All types of promoters are expected to be used in the future.

As the requirement is to develop GM plants with promoters that turn on specific traits only when they are required, a difficult issue in the standardization of future analytical identification methods will be to establish methods that enable the easy screening of all the promoters used around the world. In addition, as the number of different GMOs grows, it will be necessary to change and upgrade the analysis procedures. Under the initial progeny of GM plants, only a few favourable genes producing herbicide tolerance and insect/pest resistance were used. In the second progeny of GM plants, the genes and traits are more varied.

The various new qualities to be included are:

- High tolerance for stress;
- Modified growth features and phonological attributes;
- Desirable chemical composition;
- Pharmaceutical manufacturing;
- Industrial compound production.
- Bioremediation – the process of removing pollutants from the environment (e.g. the decontamination of toxic soils) – and phytoremediation (Martínez et al., 2006).

The number of different features introduced into the same plant has a tendency to rise and, in fact, many of the internationally authorized cultivars already have multiple traits. With this increasing complexity, it is likely that GMO detection will grow more difficult in the near future. In addition, though currently only a few GM varieties have been allowed for sale, many more are about to get approval or are currently on the market. As a result, more powerful screening and detection technologies will need to be developed or implemented soon in order to meet European regulatory requirements.

Among the many techniques, a particular advantage of DNA microarray technology is that more types of GMOs can be introduced to the screening system by simply inserting the required sequences into the array. Because this means that sequence availability is crucial, there have been suggestions that GMO-producing companies should exchange sequences in conjunction with their bids for new GM variety approval.

References

Dong, W., Yang, L., Shen, K., Kim, B., Kleter, G. A., Marvin, H. J. P. ... Zhang, D. (2008). GMDD: A database of GMO detection methods. *BMC Bioinformatics*, 9, 260. https://doi.org/10.1186/1471-2105-9-260

Fraiture, M. A., Herman, P., & Taverniers, I. (2015). Validation of a sensitive DNA walking strategy to characterize unauthorized GMOs using model food matrices mimicking common rice products. *Food Chemistry*, 173, 1259–1265.

Fraiture, M. A., Herman, P., Lefevre, L., Taverniers, I., De Loose, M., Deforce, D., & Roosens, N. H. (2015a). Integrated DNA walking system to characterize a broad spectrum of GMOs in food/feed matrices. *BMC Biotech*, 15, 76. doi:10.1186/s12896-015-0191-3.

Fraiture, M. A., Herman, P., Taverniers, I., De Loose, M., Deforce, D., & Roosens, N. H. (2015b). Current and new approaches in GMO detection: Challenges and solutions. *Biomed Research International*, 392872. doi:10.1155/2015/392872.

Freeman, W. M., Robertson, D. J., & Vrana, K. E. (2000). Fundamentals of DNA hybridization arrays for gene expression analysis. *BioTechniques*, 29, 1042–1055.

Kjellsson, G., & Strandberg, M. (2001). *Monitoring and Surveillance of Genetically Modified Higher Plants. Guidelines for Procedures and Analysis of Environmental Effects* (Basel: Birkhäuser).

Kok, E. J., Pedersen, J., Onori, R., Sowa, S., Schauzu, M., & De, S. A. (2014). Plants with stacked genetically modified events: To assess or not to assess? *Trends in Biotechnology*, 32, 70–73.

Ladics, G. S., Bartholomaeus, A., Bregitzer, P., Doerrer, N. G., Gray, A., & Holzhauser, T. (2015). Genetic basis and detection of unintended effects in genetically modified crop plants. *Transgenic Research*, 24, 587–603. doi:10.1007/s11248-015-9867-7.

MacCormick, C. A., Griffin, H. G., Underwood, H. M., & Mason, M. J. (1998). Common DNA sequences with potential for detection of genetically manipulated organisms in food. *Journal of Applied Microbiology*, 84, 969–980.

Martínez, M., Bernal, P., Almela, C., Vélez, D., García-Agustín, P., Serrano, R., & Navarro-Aviñó, J. (2006). An engineered plant that accumulates higher levels of heavy metals than *Thlaspi caerulescens*, with yields of 100 times more biomass in mine soils. *Chemosphere*, 64(3), 478–485.

Nazir, S., Iqbal, M. Z., & Rahman, S. (2019). Molecular identification of genetically modified crops for biosafety and legitimacy of transgenes. In *Gene Editing – Technologies and Applications*, edited by Y.-C. Chen & S.-J. Chen (London: IntechOpen).

Peccoud, J., & Jacob, C. (1998). Statistical estimations of PCR amplification rates. In *Gene Quantification*, edited by F. Ferré (New York: Birkhauser).

Raymond, P., Gendron, L., & Khalf, M. (2010). Detection and identification of multiple genetically modified events using DNA insert fingerprinting. *Analytical and Bioanalytical Chemistry*, 396(6), 2091–2102.

Ruttink, T., Demeyer, R., & Van Gulck, E. (2010). Molecular tool box for the identification of unknown genetically modified organisms. *Analytical and Bioanalytical Chemistry*, 396(6), 2073–2089.

Spalinskas, R. M., van den Bulcke, M., van den Eede, G., & Milcamps, A. (2013). LT-RADE: An efficient user-friendly genome walking method applied to the molecular characterization of the insertion site of genetically modified maize MON810 and rice LLRICE62. *Food Analytical Methods*, 6(2), 705–713.

Trapero, C., Wilson, I. W., Stiller, W. N., & Wilson, L. J. (2016). Wilson enhancing integrated pest management in GM cotton systems using host plant resistance. *Frontiers in Plant Science*, 7, 500.

Wang, W. X., Zhu, T. H., Lai, F. X., & Fu, Q. (2011). Event-specific qualitative and quantitative detection of transgenic rice Kefeng-6 by characterization of the transgene flanking sequence. *European Food Research and Technology*, 232(2), 297–305.

Yang, L., Xu, S., Pan, A., Zhang, K., Wang, Z., Zhou, Z., & Zhang, D. (2005). Event specific qualitative and quantitative polymerase chain reaction detection of genetically modified MON863 maize based on the 5′-transgene integration sequence. *Journal of Agricultural and Food Chemistry*, 53(24), 9312–9318.

15 Integrated Pest Management through genetically modified crops

Nitika Sharma and Akanksha Bakshi

15.1 Introduction

Agriculture is an inseparable component of our existence and the growing global population has accelerated the demand for superior and resistant crop varieties. In addition to providing a raw material for food, agriculture also gives employment possibilities to a fairly high percentage of the population. In countries with different level of economic development, food security is an important issue and this issue can be resolved with improved food availability (Pawlak & Kolodziejczak, 2020). By 2050, the world's population is predicted to reach 9 billion people (Kennedy, 2014). Given that agricultural land and water supplies are limited commodities, there is little possibility of sustaining this population surge without agricultural research. Despite this, the agriculture sector has been neglected for decades. Agricultural research must be prioritized if we are to combat new insect strains that damage crops, create new crop varieties with higher nutritional value, increase yields, develop disease and drought resistance, and provide environmentally friendly production practices (Kennedy, 2014).

A pest is an insect or organism that harms humans, their livestock, crops, or property (Hill, 1997). Organisms frequently become pests because they escape normal control by natural regulating agents. Due to the lack of population growth measures, an organism can swiftly reach a population level that cause damage, becoming pests. In order to reduce the number of pests to an acceptable level, where they do not cause any harm, pest management methods are employed (see "What is pest management?", napawatersheds.org). Farmers use various pest management tactics to control undesirable plants and pests; i.e., weeds, insect attack, fungal diseases, virus infestation, and bacterial infections of crop plants. Some other measures adopted by farmers are crop rotation, the regular examination of fields, tilling of soil, planting dates, density of plants, and the use of organic as well as synthetic pesticides (USDA ERS – Pest Management).

Integrated pest management (IPM) has been developed to control pests in present day agricultural methods without depending merely on pesticides (Kos et al., 2009). IPM originated more than 50 years ago as a response to

DOI: 10.4324/9781003278566-15

concerns connected with the indiscriminate use of chemical pesticides to control pests, diseases, and weeds (Hokkanen, 2015). During the 1970s, it was figured out that irrational use of chemical pesticides leads to serious environmental damage and health problems in humans, with IPM emerging as a cornerstone for agricultural practice (Anderson et al., 2019). IPM is defined by the Food and Agriculture Organization (FAO) as the

> careful consideration of all available pest control techniques and subsequent integration of appropriate measures that discourage the development of pest populations and keep pesticides and other interventions to levels that are economically justified while reducing or minimising risks to human health and the environment.
>
> (FAO, 2018)

And, for the protection of plants from pests, IPM is a prioritized standard internationally (Anderson et al., 2019).

Despite the fact that effective IPM tactics have been established for a variety of crops around the world, IPM programmes will become even more widely used if they are improved. New cultivars with high levels of insect resistance will be a significant improvement. Traditional breeding for improved host plant resistance takes time and effort and the intended beneficial feature might be related to or connected with negative qualities. For example, in some circumstances, insect resistance reduces the effectiveness of flesh-eating insects, and yet these are an important part of the management of living systems. Conventional breeding systems may also produce varieties which cannot be used as food (Kos et al., 2009). The use of biotechnology in the production of genetically modified (GM) plants which are resistant to insects, as a precious addition to IPM, is discussed in this chapter.

15.2 Why are genetic modifications required?

Pesticide overuse has resulted in the evolution of insects which are resistant to a wide range of insecticides, toxicity to beneficial non-target organisms, residue issues on the marketable crop, and environmental contamination. As a result, multiple complementary tactics are required in the building of IPM programmes, which include keeping a continuous check, aesthetic, tangible and automatic measures, resistance between the host and plant, biotic control, as well as the prudent use of pesticides (Mouden et al., 2017).

The technique of transferring genes (the resistant genes) from unlike or different plants, animals, and microorganisms is revolutionising traditional breeding programmes (for plants) that are being used to improve resistance in plants. These new approaches are quick, refined, and provide an entrance to a much larger number of advantageous genes than conventional techniques. In plants encoding noxious proteins like proteinase inhibitors, lectins, and cytokinins, multiple gene transfers have the potential to create stable resistance (Way & Van Emden, 2000).

Innovative approaches to pest insect prevention and management are constantly sought in pest management programmes. Understanding plant–insect reciprocity, the genetic variation between the insect species and resistant crop cultivars yields important statistics that can lead to new ideas and methods by enhancing our understanding of compound resistance characteristics. While domestication of uncultivated plants increased their taste and yield, it significantly decreased genetic and phenotypic variations, resulting in loss of insect resistance. Therefore, it is the wild ancestors that represent a potential source for breeding pest resistance traits. Furthermore, the significant diversity in resistance to pests between successions, as found in a variety of vegetables and ornamental plants, can be utilized (Mouden et al., 2017). Ultimately, plant properties can be altered using genetic engineering techniques by adding DNA from a different species.

15.3 Genes involved in agronomic trait development

IPM is a holistic method to combat pests (including weeds, herbivores, and pathogens) that uses a combination of curative and preventive measures with the aim of modulating the population of pests to such low levels that they do not cause financial losses. The genotype of plants has a strong influence on the distribution, abundance, and health of herbivores. Furthermore, genetic resistance varies significantly among crop plant species. Functional IPM is built on the foundation of intrinsic crop resistance (Green et al., 2020; Stout & Davis, 2009).

Breeding of plants through genetic engineering has led to the emergence of a number of new varieties with a number of agronomic traits. The genomes of many plant species have already been tampered with, and there are various species-specific transformation procedures available. One of the most frequent methods for introducing heterologous DNA includes genetic transformation facilitated by *Agrobacterium tumefaciens*. Until recently, the genes present in the nucleus were the primary focus. However, given the ability to modify the genes of the chloroplast and the benefits of this technique, both genomes are now modified (Altpeter et al., 2016). Current genetic engineering tools enable the entry and stabilization of a DNA sequence from any creature, such as foreign genes of interest, or promoter and regulatory elements to control endogenous gene expression. Regardless of the manner of transformation, single or multiple copies of foreign DNA are randomly integrated into the genome. Because of the unpredictability of the insertion of the DNA sequences – such as introduction within a critical operator gene – and the presence of numerous copies, unwanted effects can occur. As a result, the genotype of interest must be tailored by using tissue culture procedures as well as genetic engineering.

In recent years, the latest and reliable DNA sequencing tools have changed plant biotechnology research. Strong and productive research teams have acquired large amounts of important information by sequencing complete genomes (mitochondrial, chloroplast, and nuclear), transcriptomes, miRNAs, methylomes, exomes, and small RNAs, in conjunction with bioinformatic techniques and advanced molecular biology methods. Overexpression and silencing of genes linked with the desired phenotypic or agronomic qualities have enabled the development of plants with enhanced attributes.

Overexpression of transcription factors (TFs) (like DREB and AREB to increase tolerance of stress in abiotic conditions), genome editing (by using Cas9/CRISPR), fine-tuning of miRNA, and transcriptional repression or activation (by using dCpf1) are currently employed to improve genes of interest (GOI). Because an orthologous gene present in a species of interest might show noteworthy expression or maybe not show any functional effect similar to the GOI, both cisgene and transgene methods are used in the improvement of plant varieties (Lowder et al., 2015; Altpeter et al., 2016; Castle et al., 2020). Notably, a potential GOI is frequently present in several copies in the genome; hence, it is advised that factors like expression level, structure of the gene, and the existence of domains (conserved) be used to pick amongst them. Furthermore, enhancing a GOI's efficiency for translation in certain crops has necessitated codon use optimization for dicots or monocots. Similarly, the GOI's guanine–cytosine (GC) content can boost its level of expression; both codon use and GC content are factors in protein accumulation and mRNA stability. In the 5⊠-UTR (untranslated region), however, a high GC level can limit ribosome loading and impair translation.

Furthermore, for plant transformation, only the GOI's protein-coding region (often referred to as the coding sequence or CDS) is put into the expression cassette. Because the region of genome encoding proteins usually includes numerous introns in addition to the exon sequence, formation of clones in the entire region is problematic due to its huge size. Additionally, the recipient organism's posttranscriptional processing of the entire transcript may differ, resulting in undesired mRNA and protein; and the fact that the DNA silencing machinery of a cell targets longer GOI sequences more frequently as compared to shorter sequences, can lower transgenic stability. As cis-regulatory elements are necessary for the translation and expression of GOIs, GOIs are sometimes employed in their original form (containing introns, exons, 50- and 30-UTRs) (Basso et al., 2020).

Bacillus thuringiensis (Bt) is a Gram-positive, aerobic, endospore-forming bacillus whose spores and crystal proteins were originally sold as a biopesticide spray in France during the late 1930s. Bt's insecticidal activity is derived from Cry proteins (the protein containing endotoxins) found in crystals generated after sporulation, while "vegetative insecticidal proteins" have also been developed before sporulation. Toxins produced by *B. thuringiensis* are also known as Bt. A vast number of Cry toxins made from one type of genes are specialized for certain insect pest groups – Lepidoptera, which includes butterfly and moth larvae, and Coleoptera, which includes beetle larvae. Bt was initially instigated into plants of tobacco in 1987 after coding genes for crystal proteins were identified by Schnepf and Whiteley (1981). Synthetic genes formed from Bt were designed to be compliant enough with plant expression to create more effective insect-protected plants. Wild type (or modified) Cry1A.105, Cry1Ab, Vip3A, Cry1Ac, Cry1F, and Cry2Ab2 are currently used to control Lepidoptera, whilst wild type or modified Cry3Bb1, Cry35Ab, 1Cry3A, and Cry34Ab are used to control Coleoptera. To kill the target insect, Cry34Ab1 and Cry35Ab1 should be expressed simultaneously as a binary toxin. Bt cotton and maize account for the great bulk of insect-protected cropland, although other Bt products for reducing or more precisely controlling insects in tomato, potato, and, more recently, soybean have also been produced (Duke, 2011).

In horticulture, plant growth agents are also used as active ingredients in commercial growth regulator solutions. Synthetic auxin growth regulators are used on the one hand as a root inducer and yet, on the other, as a herbicide. In pomiculture, blockers of ethylene perceptivity and gibberellin production, as well as chemicals with gibberellin action, are widely used to manage ornamental crop longevity and improve fruit output and quality. However, genetic engineering of the genes responding to plant growth regulators and metabolism in crop plants is uncommon owing to technological challenges encountered during the process of transformation. Monocots are the most exploited plant species by humans, despite the fact that they have long been thought to be resistant to many transformation technologies (Zalabák et al., 2013).

15.4 Mechanism of genetic modification leading to desired trait

Pest control strategies fall into numerous broad categories, with the most common being chemical control, host resistance, biological control, habitat management, and legal control (Irwin, 1999). Wild plants have acquired features that enable them to directly or indirectly withstand pests and pathogens. Secondary compounds, trichomes, and wax coatings operate as direct defences against invaders, making the plant unappealing for eating and thus less prone to pests. In contrast, indirect defences attract and reward pests' natural adversaries, enlisting them as 'bodyguards.' Indirect defences include smells which are often induced by pests to attract natural pest enemies to afflicted plants and nectar that recompenses and helps in their maintenance. In spite of this, with the domestication of crops, many natural resistance qualities have been lost.

There are many reasons for this. For example, during the twentieth century when pest issues were treated with inexpensive organic pesticides, resistance was not taken as a priority. In addition, direct defence features were frequently not favored for the treatment of pests as they restricted the taste and texture of the crop. Indirect defences were also chosen only rarely, but mostly because their relevance had gone unnoticed (Stenberg, 2017).

Herbicide resistance transgenes and the production of Bt toxins are the most commonly used pest management feature. Almost all herbicide-resistant crops include one or more glyphosate-resistant gene. Herbicide resistance is available in four major crops viz. soyabean, cotton, maize, and canola, whereas insect resistance is available in just three – maize, soybean, and cotton. Many countries are now authorizing the cultivation of GM crops and thus the rates of acceptance of these crops have progressively increased; a trend that is increasing day by day (Duke, 2011; Turnbull et al., 2021; Endale et al., 2022).

According to Abrol (2013), screening, categorization, breeding, and implementation are the four main steps traditionally used by IPM for host plant resistance. These are discussed in detail below.

15.4.1 Screening

Evaluating or screening crop germplasm to identify genotypes which express resistance to the insect or a phenotype (may be related to resistance) is the first step for the host-plant resistance management programme. Screening entails subjecting a part of the plant (or whole plant) to the same populations of the pest and then evaluating injury, damage, infestation levels, insect performance, or other appropriate endpoints which are directly linked with resistance. Screening can be performed in a greenhouse or in a laboratory but at some point screening or evaluation needs to be carried out in the field. Pest insect 'invasions' may be natural or artificially brought about (Abrol, 2013).

The process of screening is carried out via one of two methods: the standard seedbox method or the modified seedbox method. A good example is the screening of rice genotype for planthopper resistance. The seeds to be tested (of rice genotype) for resistance and susceptibility checks are planted in a row (60 cm × 40 cm × 20 cm) into a seedbox. In the standard seedbox method, after 7 days (20 days in the modified seedbox method) the plantlets in the seedbox are infested with second instar nymphs. When the plantlets undergoing the susceptibility checks are killed by planthoppers the damage is rated on 0–9-point scale. These methods help in enabling the high-throughput of tests of thousands of rice lines for planthopper resistance (Horgan, 2009).

15.4.2 Categorization

Researchers aim to identify resistant genotypes which are then assigned to one or more of Painter's three categories of resistance viz. antibiosis, anti-xenosis or non-preference, and tolerance. When Painter defined this trichotomous framework of resistance, nothing was known about the incidental foundations of resistance. However, because these categories are only vaguely defined, it's impossible to assign a *specific* resistance phenomenon to them. Some plant resistance mechanisms, such as the herbivore-induced release of volatiles which attract parasitoids, cannot be assigned to just one of these three categories. Yet, in the literature these terminologies are still commonly used in applied plant resistance (Abrol, 2013).

15.4.3 Breeding

When a source of resistance has been discovered/found, the resistant gene/s must be introduced into a genetic background suited to commercial agriculture on a broad scale. When resistance has been found in a variety (or genotype) with suitable agronomic qualities, producers in pest-prone areas can then be advised to plant the resistant variety. Moreover, breeding processes are required to generate a genotype which has both resistance and is agronomically acceptable. The genetic foundation of resistance has a considerable influence on the difficulty of transferring resistance into an appropriate agronomic background; for example, it is easier to transfer monogenic resistance than polygenic resistance. Monogenic resistance to arthropods is unfortunately uncommon, and generally includes tiny insects such as aphids and gall midges. In monogenic resistance, the level of resistance imparted by genes is often very high. For example, Russian wheat aphid and Hessian fly resistance, planthopper resistance in rice, and resistance of lettuce-to-lettuce aphid, *Nasonovia ribisnigri*. Usually, however, plant resistance against arthropods is polygenic, which means that resistance is a result of a combination of genes or quantitative trait loci (Duffey & Stout, 1996). In the case of polygenic plant resistance, high levels of resistance are generally rare in germplasm collections. Because viable resistance levels are rare in the collections of germplasm, a significant number of genotypes must be evaluated. After identifying a source of resistance, traditional breeding procedures can make it difficult to incorporate polygenic resistance into an appropriate background since not all genes implicated in the resistance are transferred while crossing over occurs, which can lead to the dilution of the intended resistance. Because of these reasons, the development of polygenic resistant cultivars may take several years and requires considerable expertise and significant infrastructure and resources (Abrol, 2013).

15.4.4 Implementation

Following the development or identification of an agronomically acceptable resistant variety, it needs to be disseminated. Resistance in plants has traditionally been utilized in one of two ways: as a primary or stand-alone strategy, or as a supplement to other management strategies. When a cultivar's resistance is robust enough to keep pest populations below economically viable levels, the former technique is practicable. However, this level of resistance is uncommon, and when it does occur, it is almost always due to monogenic resistance. And, of course, adopting strong, single-gene resistance as a major approach can carry the danger of the target pest evolving to overcome the resistance. Therefore, because the level of resistance identified in a particular cultivar is not usually high enough, plant resistance is often paired or merged with other control methods, such as pesticide sprays, traditional practices, and biotic control, and not employed as a primary strategy. In fact, the integration of plant resistance with some other control strategies is critical to the overall effectiveness of the management program (Abrol, 2013).

In the last five decades, weed management strategies have remained largely unchanged. Despite the adoption of genetically engineered crops with herbicide-tolerant features, herbicides are still the main way of controlling weeds (Anderson et al., 2019). In IPM, herbicide-resistant transgenes and transgenes for the generation of Bt toxins are the most common pest management characteristics that are now marketed. Due to their higher effectiveness and ease of pest management for farmers, these two features have been quickly embraced wherever they are accessible. Furthermore, with less ploughing and fewer visits to the field to spray pesticides, they have significantly reduced the use of environmentally and toxicologically hazardous pesticides while lowering the carbon footprint generated by pest management.

Glyphosate-resistant crops have been the most effective of these products (Duke, 2011). Despite the fact that bromoxynil-resistant crops (like cotton and canola) were launched in 1995 and were available in the market just years after their launch, the majority of crops planted are those with glyphosate and glufosinate resistance (Duke & Cerdeira, 2010); in fact, bromoxynil-resistant crops were never very popular and were eventually phased out. Glufosinate-resistant crops, which were first launched in 1995, have showed more promising effects, but they still account for a small percentage of the transgenic agricultural market. Since the commercial introduction of glyphosate-resistant crops in 1996, the mixture of these glyphosate-resistant crops and the herbicide glyphosate become the favored weed management technique for farmers. The method has proven to be so effective that farmers have continued to utilize it year after year. But this method also failed, mainly due to two reasons. First, continuing to use and rely on a highly effective insecticide is unsustainable. For example, it produces a vacuum, which is something that nature despises. The glyphosate-treated fields with minimal weeds and glyphosate-resistant crops contributed an ideal environment for weed species with natural glyphosate resistance (tolerance). These species had not been able to survive in the agro-ecological environment inhabited by weeds which were susceptible to glyphosate or were well managed by tillage before glyphosate-resistant crops. Most weed experts refer to this movement in troublesome weed species as "weed shifts" (Owen, 2008; Prince et al., 2012). Second, is the evolution of glyphosate resistance in weeds – a topic that has attracted a lot of attention. In 1996, *Lolium rigidum* became the first documented case of a weed which is resistant to glyphosate – the same year that glyphosate-resistant crops were first commercialized. For some farmers, the glyphosate resistance problem is severe, prompting them to turn to other weed management strategies, such as glufosinate-resistant crops, to achieve satisfactory weed control. Herbicides with new action mechanisms are desperately required to avoid ever-worsening weed resistance (Duke, 2011; Ghanizadeh & Harrington, 2020).

15.5 Impact on modern agriculture – trade of GM crops

Despite an unhurried and often arduous procedure, traditional plant breeding has made significant advancements over a period of time. This strategy is primarily used to introduce a single elementary feature to an otherwise perfect variety. Genetic engineering, on the other hand, represents a useful tool for crop plants, allowing for the horizontal insertion of desirable genes for desired features. The collaboration of genetic engineering techniques with traditional plant breeding efforts has resulted in more precise and efficient crop improvement. To now, only two types of GMO traits have shown to be successful – Bt crops and herbicide-resistant crops –with transgenic techniques being either ineffective or unnecessary for improving some quantitative traits in agricultural plants (Basso et al., 2020).

Transgenes that enhance or replace conventional farm inputs are known as input characteristics. Insect-, herbicide-, and disease-resistance traits are examples of pest-controlling crop protection features that are input traits and that can be used instead of pesticides (Castle et al., 2006). Since their first commercial launch, transgenic crops with input features have been quickly embraced by various significant agricultural markets.

Insects have a remarkable ability to build resistance to a variety of insecticides. Chemical pesticides have resulted in the development of resistance to one or more insecticides in more than 1,000 bug/insecticide resistance combinations, as well as of at least 17 insect species resistant to major insecticide classes. Indiscriminate use of biopesticides also raises concerns about the problem of resistance development. However, despite the widespread use of Bt toxin-based products in agriculture, forestry, and civil life, there have as yet been few recorded incidences of insect resistance to Bt (Kumar et al., 2008; Gassmann, 2021).

15.6 Benefits to society

The purpose of an IPM approach is to support the long-term production of high-quality crops while minimising pest or pest management-related environmental consequences (Anderson et al., 2019). The need to enhance food production has become urgent as a result of global population growth and the high rates of malnutrition. These issues must be addressed by developing comprehensive responses aimed at increasing food supply and incorporating changing socioeconomic, geopolitical, and demographic factors into food supply system design. Transgenic crops have been advocated as a way to increase food production, meet rising food demand, and improve the nutritional and genetic properties of food products, allowing them to effectively tackle harsh environmental circumstances or making them less sensitive to diseases (Barragán-Ocaña et al., 2019).

One of the worries raised by those opposed to transgenic crops is that the transgene could affect the quality and/or safety of the plant's edible parts; something that may be due to the toxicity of the transgene's protein. However, transgenic crops are subjected to significantly more scrutiny than traditional crops when it comes to regulatory approval, with analytical, nutritional, and toxicological tests being used (Duke & Cerdeira, 2010; Shetty et al., 2018). For instance, Bt crops have undergone extensive research, which has shown that they are safe for humans, animals, and the environment. Cry proteins present in currently authorized Bt crops are subjected to mammalian toxicological and digestive fate investigations, which have demonstrated that they are harmless to humans and represent no major allergic risk. Food for humans and feed for animals made from Bt crops is equally nutritious and safe with a proven track record (Duke, 2011; Ku & Ha, 2020).

15.7 Trade in the global market

Biotechnology advancements have improved the production of a number of agricultural commodities in key exporting countries such as the United States, Argentina, Australia, and Canada (James, 1999; Marra et al., 2002). By means of public–private partnerships, GM cotton, maize, cassava, and cowpea have now been approved in some African countries (Nigeria, Kenya, Ethiopia, Malawi, Eswatini, and Sudan) (Endale et al., 2022). Some emerging countries, such as China and Mexico, have adopted this technology in their home markets, although these products are not widely exported. To date, maize, soybean, and cotton have benefited the most, with research ongoing into other crops (Marra, 2001). This new technology, according to proponents, can boost production, minimize pesticide use, and improve food nutritional quality. Opponents, on the other hand, are concerned about the safety of eating GM foods as well as the environmental impact of the biotechnology (Sheldon, 2002). Hence, the introduction of new GM technology has created plenty of problems and challenges for the global trading system. Much of the argument has been very politicized up to this point. Better information is desperately needed, both regarding the underlying science and the economic implications of the new technology. The dispute over GM foods raises a variety of issues that have ramifications for commerce and the global trading system (Diaz-Bonilla & Robinson, 2001; Srivastava, 2022; Turnbull et al., 2021).

Currently, different countries have greatly varying GM product policies. GM foods are not considered significantly different from regular foods by US regulators (e.g., they are substantially *equivalent*). The European Union, however, follows the precautionary principle. Because of the alleged hazards, EU regulators see GM products and conventional products as *distinct*. Despite the fact that in the 20 years since their introduction, there have been numerous studies which have been published on the enormous advantages of GM crops, regulatory and trade hurdles remain. One more point to be mentioned here is that although GM crops have been demonstrated to be safe to use, lots of money is still being spent by the eNGO community on campaigning and battling in opposition to their usage (Byrne, 2015).

The CPB (Cartagena Protocol on Biosafety) is currently the most significant international obstacle facing the breakthrough of agricultural technology, GM crops, and NBTs. Furthermore, as well as forcing developing nations that desire preferential tariff rates to join the CPB, the EU also gives an ultimatum to growing nations that, if they accept GM food/crops, they will no longer have way into the common markets. Once the LLP (Low Level Presence) of accepted and unaccepted GM crops are identified, the EU closes its borders to Canadian and US shipments as they are mixed with GM crops, as witnessed by developing countries. Most developing countries believe the EU and, in particular, EU-based eNGOs have already made the decision for them as to whether GM crops are to be adopted for domestic purposes or not (Ludlow et al., 2014).

If, and only if, two dramatic adjustments are undertaken can global food security be improved. The first adjustment will be to help and assist developing nations in their decision to withdraw from the CPB. Considering that the CPB serves as a vehicle for the EU and its group of eNGOs to speak out and influence security of food crops in growing nations, these countries will carry on enduring famine and malnutrition. Solid reasoning should be involved in calculating the advantages of different forms of GM crop advocacy (including NBTs), as well as the price of non-adoption of GM crops. The second task will be to devise a method of separating Europe from global commodity trading. Countries that have adopted GM crops, such as Canada, the United States, and Australia, must look for new territorial deal agreements with growing nations in order to marketize agricultural commodities. This could be true specifically for items produced through NBTs. Every single tie between the EU and progressive nations in the agriculture sector must be severed. The threat posed by the EU should be put to an end so that evolving countries can decide, without any pressure, whether or not to accept GM crops. The ability of the group of eNGOs to prescribe food options to starving poor countries must come to a stop. Political advocacy is necessary to persuade authorities of the advantages of developing nationwide trade agreements (Symth, 2017).

15.8 Health and ethical issues/risks

In 1999 and 2004, the Nuffield Council on Bioethics (NCOB) in its two reports give an account of the concerns (ethical as well as social) entangled with the use of transgenic crops (Weale, 2010). In both the reports mainly two presumptions were acknowledged. The first was that the strategy in the use of GM/transgenic crops should rely on the results of the performance of an individual modified crop and that no generalized conclusion should be made in favor of or against the crop. In spite of generalization, a careful examination of the benefits and hazards of specific applications in comparison to the available alternatives was also deemed necessary. This notion has a number of ramifications, one of which is managing the efficiency of growing nations. The other major finding was that the GM crops must be used responsibly and that the availability of these crops freely to individuals in developing nations was morally essential. If the profits were accessible, hoarding them to the detriment of the poor would be counter to the values of justice and solidarity (Weale, 2010).

The five main ethical issues about GM crops that have been put forward are: negative effects on human health, possible loss of environment, a deteriorating effect on conventional agricultural methods, unrestricted industrial dominance, and the unnatural behavior of technology. In fact, every plant, no matter how 'well made', might have unfavorable consequences on the environment. Yet, the main concern of users lies in long-term effect of these biotech crops on human beings (Maghari & Ardekani, 2011).

For example, glyphosate-resistant Roundup Ready® soybean, an extensively accessible marketed by-product among GM crops, may show a rise in the level of compounds which hinder the absorption of nutrients by plants (i.e. anti-nutrients) (Padgette et al., 1996). The heat stress resistant anti-nutrients including glucinins, phytoestrogens, and phytic acid have been reported to induce reproductive issues in bovines. Binding of anti-nutrients with inorganic compounds like zinc and phosphorus leads to allergic reactions and also reduces the availability of these compounds to animals (Adams, 1995; Liener, 1994). Since food obtained from GM crops might be ingested uncooked, a rise in anti-nutrient levels must be avoided (Brookes & Barfoot, 2016; Gizaw, 2019).

In the majority of modified crops, the promoter of Cauliflower Mosaic Virus 35S (CaMV35S) is used to turn on the inserted gene. The topic of the efficiency of CaMV35S virus's horizontal spread, its disease-causing capability such as causing cancer, mutagenesis or reactivation of latent viruses and/or generation of new viruses, is well debated (Hodgson, 2000). On the other hand, according to Ho et al. (2000), CaMV present in typical foods is not very contagious and mammals are not able to absorb it. Others, however, claim that despite the fact that humans eat significant doses of CaMV35S and that it neither recombines with human viruses nor causes disease in human beings (Paparini & Romano-Spica, 2004; Tepfer et al., 2004), the transitory production of transgenes generated from the CaMV35S promoter in mammalian cells has been demonstrated, raising the probability that 35S promoter control genes *could* be expressed in mammals. Yet, in their investigations, Paparini and Romano-Spica (2006) were not able to find transcriptional activity of CaMV35S and transfer of DNA in mice using real-time PCR, although they do underline the need for more research.

A body of evidence from trials on animal feeding (though few in number), which investigate effects on health, has started to appear. Ewen and Pusztai (1999) were the first to show that each GM plant product must be thoroughly tested on animal models. Member states should thoroughly evaluate all applications, according to Varzakas et al. (2007), because corporations strive to conceal information about the health effects of genetic modification. Although feeding large quantities of specific "foods" to animals over a lengthy period of time can cause an imbalance in nutritional status, this is the only way the toxicity of a chemical can be exposed (Varzakas et al., 2007). Besides, in compliance with the regulations of drug toxicity testing, animals must be used in large numbers for tests of toxicity. GM crops must be tested more carefully than any drug before making conclusions as a large proportion of human beings might consume these crops after their approval.

It has been claimed that the use of biotechnology for raising the productivity of plants, producing a nutrient-rich staple food crop, will lead to a reduction in food shortages and malnutrition. But the evidence collected so far does not back up these claims (Abbas, 2018).

Apart from ethical problems, human health and the environment are both at risk as a result of the utilization of GM crops in agriculture. Among common people there exists reservation, concern, and worry about genetic engineering techniques. There is no doubt that genetic engineering has noteworthy potential for addressing hunger, food insecurity, and nutritional issues at the global level. On the other hand, some people are of the opinion that genetic engineering is unnatural, and that there is a risk of gene transfer to their children if food obtained from GM crops is consumed by them. In addition, some companies do not even label their products as GM food. In the USA GM food is eaten by the general population, but in Europe it is still not accepted due to ethical concerns.

Before being released onto the market, GMOs should be tested carefully for their use by animals as well as humans. For example, new proteins sometimes behave as allergens. And, also, in producing GMOs wherein genes resistant to antibiotics are commonly used, this raises issues about the transference of such genes via consumption. For these reasons, a major scientific effort is currently underway to fully comprehend and foresee potential effects of GMOs on humans, animals, and the environment (Dona & Arvanitoyannis, 2009; Kumar et al., 2020). To summarize, proof of ill health caused by GM crops would make us more wary of developing and using them, but the evidence presented so far has not stood up to scrutiny.

15.9 Current status and future prospects

Transgenic agricultural products are made using molecular techniques to make it easier to introduce new, valuable features to target crop species

(Mumm & Walters, 2001). In recent years, significant advancements have been made in the modification of a variety of grains. Variation in the plant body tissue, polyploidy, inactivation of modified gene, and hypopigmentation have all been eradicated from transgenic progeny as scientists have enhanced the capacity of transformation and frequency of rejuvenation from the callus of an embryo. However, some varieties of cereals, like rye, or other economically utilized cultivars, have up to now been challenging to change. Another key stumbling block to bringing new GM products to market is public skepticism of recombinant DNA technology. The European Union's legal prohibition on genetically altered food and crops has stymied advancement in scientific development and the environmental evaluation of new modified gene types.

Hence, a significant amount of effort must be put into mass education so that people can make informed decisions about GM crops or food based on scientific certainty rather than rumors and incomprehension. Both their possible risks to the environment and their safety should be continually communicated or supervised. Only then can the public embrace GM crops completely (Zalabák et al., 2013).

15.10 Conclusions

The main role of GM crops in IPM is to reduce the use of synthetic toxicants, extra persistent regulation of pests in environmentally stabilized agricultural production systems, and to minimize hazards to human health and that of the soil and plants. Since the transition from hunter-gatherer to farmer, humans have experimented with GM crops. Through plant breeding, the traits of wild crops (i.e., pest-resistant or herbicide-tolerant traits) are transferred to modern cultivars (which over time have lost many direct and indirect defence traits) in order to boost pest and disease resistance. The number of biotechnology-based pest management alternatives is expected to surge in the next few years. We hope that these products will contribute to pest management diversification that is both sustainable and environmentally friendly.

References

Abbas, M. S. T. (2018). Genetically engineered (modified) crops (Bacillus thuringiensis crops) and the world controversy on their safety. *Egyptian Journal of Biological Pest Control*, 28(1), 1–12.

Abrol, D. P. (ed.). (2013). *Integrated Pest Management: Current Concepts and Ecological Perspective* (San Diego, CA: Academic Press).

Adams, N. R. (1995). Detection of the effects of phytoestrogens on sheep and cattle. *Journal of Animal Science*, 73, 1509–1515.

Altpeter, F., Springer, N. M., Bartley, L. E., Blechl, A. E., Brutnell, T. P., Citovsky, V., ... StewartJr., C. N. (2016). Advancing crop transformation in the era of genome editing. *The Plant Cell*, 28(7), 1510–1520.

Anderson, J., Ellsworth, P. C., Faria, J. C., Head, G. P., Owen, M. D., Pilcher, C. D., … Meissle, M. (2019). Genetically engineered crops: Importance of diversified integrated pest management for agricultural sustainability. *Frontiers in Bioengineering and Biotechnology*, 7, 24. doi:10.3389/fbioe.2019.00024.

Barragán-Ocaña, A., Reyes-Ruiz, G., Olmos-Peña, S., & Gómez-Viquez, H. (2019). Production, commercialization, and intellectual property of transgenic crops in Latin America: A state of the art review. *Journal of Agribusiness in Developing and Emerging Economies*, 9(4), 333–351.

Basso, M. F., Arraes, F. B. M., Grossi-de-Sa, M., Moreira, V. J. V., Alves-Ferreira, M., & Grossi-de-Sa, M. F. (2020). Insights into genetic and molecular elements for transgenic crop development. *Frontiers in Plant Science*, 11, 509. doi:10.3389/fpls.2020.00509.

Brookes, G., & Barfoot, P. (2018). Environmental impacts of genetically modified (GM) crop use 1996–2016: Impacts on pesticide use and carbon emissions. *GM Crops & Food*, 9(3), 109–139.

Byrne, J. (2015). Presentation by President of v-Fluence Interactive at the Agricultural Bioscience International Conference. Melbourne, Australia, September 8.

Castle, L. A., Wu, G., & McElroy, D. (2006). Agricultural input traits: Past, present and future. *Current Opinion in Biotechnology*, 17(2), 105–112.

Diaz-Bonilla, E., & Robinson, S. (2001). *Biotechnology, Trade, and Hunger: IFPRI 2000–2001 Annual Report Essay* (No. 2001Essay1). International Food Policy Research Institute (IFPRI).

Dona, A., & Arvanitoyannis, I. S. (2009). Health risks of genetically modified foods. *Critical Reviews in Food Science and Nutrition*, 49(2), 164–175.

Duffey, S. S., & Stout, M. J. (1996). Antinutritive and toxic components of plant defense against insects. *Archives of Insect Biochemistry and Physiology: Published in Collaboration with the Entomological Society of America*, 32(1), 3–37.

Duke, S. O. (2011). Comparing conventional and biotechnology-based pest management. *Journal of Agricultural and Food Chemistry*, 59(11), 5793–5798.

Duke, S. O., & Cerdeira, A. L. (2010). In Transgenic crops for herbicide resistance. In *Transgenic Crop Plants*, edited by C. Kole, C. H. Michler, A. G. Abbott, & T. C. Hall (Berlin: Springer), 133–166.

Endale, G. K., Karim, M., Joseph, G., & Muffy, K. (2022). Commercialization of genetically modified crops in Africa: Opportunities and challenges. *African Journal of Biotechnology*, 21(5), 188–197.

Ewen, S. W. B., & Pusztai, A. (1999). Effect of diets containing genetically modified potatoes expressing *Galanthus nivalis* lectin on rat small intestine. *Lancet*, 354, 1353–1354.

FAO (2018). Integrated pest management. Briefing note. https://www.fao.org/documents/card/en/c/CB1199EN/

Gassmann, A. J. (2021). Resistance to Bt maize by western corn rootworm: Effects of pest biology, the pest–crop interaction and the agricultural landscape on resistance. *Insects*, 12(2), 136.

Ghanizadeh, H., & Harrington, K. C. (2020). Perspective: Root exudation of herbicides as a novel mode of herbicide resistance in weeds. *Pest Management Science*, 76 (8), 2543–2547.

Gizaw, Z. (2019). Public health risks related to food safety issues in the food market: A systematic literature review. *Environmental Health and Preventive Medicine*, 24(1), 1–21.

Green, K., Stenberg, J. A., & Lankinen, Å. (2020). Making sense of Integrated Pest Management (IPM) in the light of evolution. *Evolutionary Applications*, 13(8), 1791–1805.

Hill, D. S. (1997). Pest definitions. In *The Economic Importance of Insects* (Dordrecht: Springer), 51–63.

Ho, M. W., Ryan, A., & Cummins, J. (2000). Hazards of transgenic plants containing the cauliflower mosaic virus promoter. *Microbial Ecology in Health and Disease*, 12(3), 189–198.

Hodgson, J. (2000). Scientists avert new GMO crisis. *Nature Biotechnology*, 18, 13. doi:10.1038/71838.

Hokkanen, H. M. (2015). Integrated pest management at the crossroads: Science, politics, or business (as usual)? *Arthropod–Plant Interactions*, 9(6), 543–545.

Horgan, F. (2009). Mechanisms of resistance: A major gap in understanding planthopper–rice interactions. In *Planthoppers: New Threats to the Sustainability of Intensive Rice Production Systems in Asia*, edited by K. L. Heong & B. Hardy (Los Baños, The Philippines: International Rice Research Institute), 281–302.

Irwin, M. E. (1999). Implications of movement in developing and deploying integrated pest management strategies. *Agricultural and Forest Meteorology*, 97(4), 235–248.

James, C. (1999). *Global Status of Commercialized Transgenic Crops: 1999*, ISAAA Brief No. 21 (Ithaca, NY: ISAAA).

Kennedy, D. (2014). Building agricultural research. *Science*, 346(6205), 13.

Kos, M., van Loon, J. J., Dicke, M., & Vet, L. E. (2009). Transgenic plants as vital components of integrated pest management. *Trends in Biotechnology*, 27(11), 621–627.

Ku, H. K., & Ha, S. H. (2020). Improving nutritional and functional quality by genome editing of crops: Status and perspectives. *Frontiers in Plant Science*, 11, 577313.

Kumar, S., Chandra, A., & Pandey, K. C. (2008). *Bacillus thuringiensis* (Bt) transgenic crop: An environment friendly insect-pest management strategy. *Journal of Environmental Biology*, 29(5), 641–653.

Kumar, K., Gambhir, G., Dass, A., Tripathi, A. K., Singh, A., Jha, A. K., ... Rakshit, S. (2020). Genetically modified crops: Current status and future prospects. *Planta*, 251(4), 1–27.

Liener, I. E. (1994). Implications of anti-nutritional components in soybean foods. *Critical Reviews in Food Science and Nutrition*, 34, 31–67.

Lowder, L. G., Zhang, D., Baltes, N. J., PaulIII, J. W., Tang, X., Zheng, X., ... Qi, Y. (2015). A CRISPR/Cas9 toolbox for multiplexed plant genome editing and transcriptional regulation. *Plant Physiology*, 169(2), 971–985.

Ludlow, K., Smyth, S. J., & Falck-Zepeda, J. (eds). (2014). *Socio-economic Considerations in Biotechnology Regulations* (New York: Springer).

Maghari, B. M., & Ardekani, A. M. (2011). Genetically modified foods and social concerns. *Avicenna Journal of Medical Biotechnology*, 3(3), 109.

Marra, M. C. (2001). Agricultural biotechnology: A critical review of the impact evidence to date. In *The Future of Food: Biotechnology Markets and Policies in an International Setting*, edited by P. G. Pardey (Washington, DC: International Food Policy Research Institute).

Marra, M. C., Pardey, P. G., & Alston, J. M. (2002). *The Payoffs to Agricultural Biotechnology: An Assessment of the Evidence*, International Food Policy Research Institute (IFPRI), Environment and Production Technology Division (EPTD) Working Paper No. 87, Washington, DC. http://dx.doi.org/10.22004/ag.econ.16093

Mouden, S., Sarmiento, K. F., Klinkhamer, P. G., & Leiss, K. A. (2017). Integrated pest management in western flower thrips: Past, present and future. *Pest Management Science*, 73(5), 813–822.

Mumm, R. H., & Walters, D. S. (2001). Quality control in the development of transgenic crop seed products. *Crop Science*, 41(5), 1381–1389.

Owen, M. D. (2008). Weed species shifts in glyphosate-resistant crops. *Pest Management Science: Formerly Pesticide Science*, 64(4), 377–387.

Padgette, S. R., Taylor, N. B., Nida, D. L., Bailey, M. R., MacDonald, J., Holden, L. R., & Fuchs, R. L. (1996). The composition of glyphosate-tolerant soybean seeds is equivalent to that of conventional soybeans. *The Journal of Nutrition*, 126, 702–716.

Paparini, A., & Romano-Spica, V. (2004). Public health issues related with the consumption of food obtained from genetically modified organisms. *Biotechnology Annual Review*, 10, 85–122.

Paparini, A., & Romano-Spica, V. (2006). Gene transfer and cauliflower mosaic virus promoter 35S activity in mammalian cells. *Journal of Environmental Science and Health B*, 41, 437–449.

Pawlak, K., & Kołodziejczak, M. (2020). The role of agriculture in ensuring food security in developing countries: Considerations in the context of the problem of sustainable food production. *Sustainability*, 12 (13), 5488. doi:10.3390/su12135488.

Prince, J. M., Shaw, D. R., Givens, W. A., Newman, M. E., Owen, M. D., Weller, S. C., … Jordan, D. L. (2012). Benchmark study: II. A 2010 survey to assess grower awareness of and attitudes toward glyphosate resistance. *Weed Technology*, 26(3), 531–535.

Schnepf, H. E., & Whiteley, H. R. (1981). Cloning and expression of the *Bacillus thuringiensis* crystal protein gene in *Escherichia coli*. *Proceedings of the National Academy of Sciences USA*, 57, 2893–2897.

Sheldon, I. M. (2002). Regulation of biotechnology: Will we ever 'freely' trade GMOs? *European Review of Agricultural Economics*, 29(1), 155–176.

Shetty, M. J., Chandan, K., Krishna, H. C., & Aparna, G. S. (2018). Genetically modified crops: An overview. *Journal of Pharmacognosy and Phytochemistry*, 7(1), 2405–2410.

Smyth, S. J. (2017). Genetically modified crops, regulatory delays, and international trade. *Food and Energy Security*, 6(2), 78–86.

Srivastava, S. P. (2022). GM crops, farmers and consumers concerns: Role of advocacy. *World Journal of Advanced Research and Reviews*, 13(2), 50–56.

Stenberg, J. A. (2017). A conceptual framework for integrated pest management. *Trends in Plant Science*, 22(9), 759–769.

Stout, M., & Davis, J. (2009). Keys to the increased use of host plant resistance in Integrated Pest Management. In *Integrated Pest Management: Innovation-Development Process*, edited by R. Peshin & A. K. Dhawan (Dordrecht: Springer), 163–181.

Tepfer, M., Gaubert, S., Leroux-Coyau, M., Prince, S., & Houdebine, L. M. (2004). Transient expression in mammalian cells of transgenes transcribed from the Cauliflower mosaic virus 35S promoter. *Environmental Biosafety Research*, 3, 91–97.

Turnbull, C., Lillemo, M., & Hvoslef-Eide, T. A. (2021). Global regulation of genetically modified crops amid the gene edited crop boom – a review. *Frontiers in Plant Science*, 12, 630396.

Varzakas, T. H., Arvanitoyannis, I. S., & Baltas, H. (2007). The politics and science behind GMO acceptance. *Critical Reviews in Food Science and Nutrition*, 47, 335–361.

Way, M. J., & Van Emden, H. F. (2000). Integrated pest management in practice – pathways towards successful application. *Crop Protection*, 19(2), 81–103.

Weale, A. (2010). Ethical arguments relevant to the use of GM crops. *New Biotechnology*, 27(5), 582–587.

Zalabák, D., Pospíšilová, H., Šmehilová, M., Mrízová, K., Frébort, I., & Galuszka, P. (2013). Genetic engineering of cytokinin metabolism: Prospective way to improve agricultural traits of crop plants. *Biotechnology Advances*, 31(1), 97–117.

16 Socio-economic impact of genetically modified crops

Anu Sharma and Vishal Sharma

16.1 Introduction

Genetically modified organisms (GMOs) are the product of a novel technology that involves the insertion of DNA into the genome of an organism. Production of GM crops requires the transfer of new DNA into plant cells, after which the cells are grown using tissue culture methods. The cells finally develop into plants, the seeds of which will inherit the new DNA and thus will exhibit new/different characteristics and traits. The new trait could be anything from increasing the shelf life of the fruit, to making it resistant to a particular disease/pest or to improving a crop's stress tolerance etc. The newly inserted DNA becomes part of the GM plant's genome. Hence, GM plants are also known as transgenic or genetically engineered plants because they are generated using transgenic methods.

Transgenic methods include molecular methods that enable the transfer of a gene or a group of genes from an individual of one species to an individual or individuals of a different species. Presently two common approaches are used for introducing purified genes into plant cells: one includes the use of the Ti plasmid of *Agrobacterium tumefaciens* to transfer the gene and the second, known as the gene gun or electroporation, involves the use of a metal particle/fiber or an electric pulse to puncture the cell wall and transfer the gene into the nucleus. Transgenic crops are developed with the objective to improve various aspects like crop yield, nutrient composition, resistance to insects/pests, and to lessen environmental pollution.

The history of GM crops is not very old. GM *Flavr Savr* tomato was first launched in the United States in 1994. This tomato was genetically altered to increase its shelf life by delaying its ripening and softening. It was introduced onto the market for consumers only after detailed evaluation was carried out by federal agencies who declared it fit for consumption.

Since then, GM crops have become enormously popular. In 2003, The World Health Organisation (WHO) and Food and Agricultural Organization (FAO) of United States have developed certain recommendations and standards for ascertaining the safety of these GM crops. In 2005, other GM

DOI: 10.4324/9781003278566-16

crops like alfalfa and sugar beets were introduced into US markets. Later, in 2015, the US government approved an application for the development of a genetically engineered salmon for human consumption and in 2017 GM apples were made available.

Since the introduction of this novel technology to farmers in 1996, the area cultivated with GM crops had increased to 180 million hectares by 2015. Farmers from all over the world very rapidly accepted this technology, and assessment of its socio-economic impacts (SEIs) have considerably contributed to the "GM debate" which is a hot topic worldwide. SEIs cannot be strictly defined. They can be both positive and negative, and they often influence different groups of stakeholders at the same time in indistinct ways. The specific impact depends on the context in which they are being used. In this chapter we will focus on the socio-economic impacts of these GM crops from various viewpoints.

16.2 Socio-economic impacts of GM crops: what do they mean?

Historically, every technology that has been developed has had certain economic, political, social, cultural, or ethical impacts on different spheres of society. No change can take place in a social vacuum. Thus, throughout the history of humankind, innovations both technological and scientific have always influenced socio-economic relations, sometimes in an indistinct manner and sometimes with highly visible effects. This is also true with the cultivation and adoption of GM crops.

The socio-economic and cultural impacts of using GM crops have received less attention as compared to their technological aspects. In part this is because it takes years to explore the socio-economic impacts of any technology and, when the impact does become clear, it has already become widespread and deeply institutionalized. For instance, the Green Revolution led to the creation of a new class of agricultural labourers, altering gender relations by shifting the burden of farming onto the shoulders of women. By the time social scientists understood this process, it had been ingrained in the social setup. Also, even when the use of a technology is discontinued, its socio-economic impacts will continue to exist and might affect the society, history, and its people in the ages to come.

The use of GM crops may bring about many changes that might cause irreparable damage. To use or not use GM crops/organisms, therefore, has always been a controversial topic. Many ethical and social aspects of GMOs emanate from the manipulation of life forms and processes. GM crops have the capability to enter the gene pool of wild populations and contaminate conventional crops and this potential contamination by GMOs is one threat that needs to be looked into. This harsh reality, that socio-economic impacts could linger long after a technology has become obsolete, emphasizes the critical importance of studying the potential socio-economic impacts of GM crops before they are introduced into the public domain.

16.3 How are socio-economic considerations assessed?

Socio-economic assessments can be carried out by applying two kinds of approaches – ex post or ex ante. In the case of ex-post assessment, the impact of technology is assessed *after* its introduction, based on data that have been collected after studying the consequences of the technology within a specific country/region. Information is collected in the form of surveys done in the areas that have already been directly affected by the application of the technology. One example of this approach was a study of Bt cotton in South Africa that focussed on the role of institutions in the type and level of benefits that farmers received after the introduction of the technology (Gouse et al., 2005, 2009). From the studies it was shown that adoption of Bt cotton by smallholder cotton farmers on the Makhathini Flats in South Africa was being prevented because of institutional failure.

Ex-ante assessments, however, study the effects of a technology *before* its introduction on the ground. These studies help authorities to arrive at a conclusion as to whether a technology should or should not be authorized for introduction. In these kinds of assessments, no data are available, so data have to be identified on the basis of assumptions or from identical or comparable cases. One example of this approach was a study carried out on the introduction of GM banana in Uganda by Kikulwe and colleagues (Kikulwe et al., 2008). Here, it was predicted that due to people's negative impressions about GM technology, there would not be wide acceptance and adoption of the crop; a potential risk that was addressed by establishing communication with the people and clearing their doubts about GM crops.

16.4 Why should we study the socio-economic impacts of GM crops?

Assessment of potential socio-economic impacts of GM crops is important for several reasons and these are described in the sections below (Dano, 2007).

16.4.1 Moral obligation towards society

There is always a moral, ethical, and social responsibility on the shoulders of the scientists who introduce any kind of new technology into society. The potential socio-economic impacts of technologies beyond the controlled confines of laboratories and greenhouses need to be assessed before their introduction into the world. Yet, the role of scientists and technologists does not end once a technology leaves the laboratory and enters the society; in fact, it becomes even more important.

16.4.2 Inter-generational responsibility

A technology that is introduced into the world has far reaching consequences. It has implications on future generations as well. An inter-generational responsibility rests on the shoulders of technology developers and government agencies involved in its regulation. Assessing the socio-economic impacts of GM crops becomes important as it will guarantee that the adverse effects are reduced or avoided. This will safeguard the interests and necessities of present generations as well as of those to come.

16.4.3 Social acceptance

By carrying out a detailed study on the potential socio-economic impacts of GMOs, developers and regulators can get an idea of the extent of acceptance of the technology or the product by the common people.

16.4.4 Reducing long-term costs

When socio-economic impacts are studied, a major worry always remains the costs. Costs are related to the mechanism of participation of various stakeholders and the time involved in going through these processes. But one must understand that investing in studies on socio-economic impacts now will save them the possible additional expenses that might be incurred due to adverse impacts of the technology. Hence, additional costs and inconveniences may be avoided or minimized if socio-economic considerations are taken into account in the decision-making process. This will also be beneficial in the long term because, once technology is introduced into society, the impacts that it will produce will be irreversible changes that might not be under anyone's control. Thus, a thorough and honest investigation of socio-economic impacts is essential. Developers of technology need to show a strong sense of moral and ethical responsibility.

16.5 Socio-economic impacts of GM crops

There are many socio-economic impacts, negative as well as positive, of growing and using GM crops. Socio-economic impacts from various viewpoints are discussed below.

16.5.1 Farm-level impacts

Farmers had various socio-economic expectations concerning GM crops, of which the most important were the potential increases in yield and income and a reduced expenditure on fertilisers and pesticides. To understand the farm-level impacts of all kinds of GM crops and the underlying mechanisms of socio-economic effects, detailed analysis is required that takes into account the typology of farms, farm characteristics, and the rates at which GM crops have been adopted (Kathage et al., 2015). Results have shown that GM crops fulfilled the expectations of adopting farmers by providing benefits such as increased yield, increased profit, a reduced need for pesticides, and enhanced nutrient composition. In fact, biofortified crops which have enhanced nutrient composition are expected to increase the nutrient status of the farmers thus leading to enhanced farmer wellbeing. There have been numerous success stories – a vitamin A rich orange sweet potato and the *Flavr Savr* tomato are two such examples. These biofortified crops provide sufficient levels of micronutrients to the targeted populations.

However, not every farmer will gain the same advantage from GM crops, as the amount of benefit depends on a number of factors concerning both the farm and the farmer (Hall et al., 2013). Farm typology is related to farm and farmer characteristics. Farm characteristics include a number of factors like location of the country/region, size, income by crop and livestock type, ownership and organic certification; farmer characteristics include gender, access to education and training, income from agriculture and from sources other than the agriculture, and time devoted to farming. The location of adoption of GM crops can directly influence their socio-economic impact. For example, their impact on food security in developed countries who are already self-sufficient is likely to be negligible, whereas in developing countries, such as in Africa, introduction of an insect-resistant GM crop may positively impact food security because in such countries 30–50 percent of crops are lost due to insect/pest attacks (Deloitte & Touche, 2015).

16.5.2 Income-related aspects

Income-related aspects include the expenditure incurred while growing the crop and income earned after harvesting the crop. Farmers have to spend on a number of things like the purchase of seeds, fertiliser, pesticides/herbicides, irrigation of the fields, machinery/equipment, and the workforce involved. The output of the whole process is the yield, which will be sold by the farmer in the market. A farmer's income will largely depend on the crop quality and local market characteristics.

The potential increases in yield and economic return depend on a variety of factors. The more heterogeneous these factors are, the more variable will be the resulting benefits and costs. A change or improvement in one factor may be offset by other factors. For instance, let's take an example of an insect-resistant GM crop. Cultivation may lead to substandard yield if other factors like the kind of GM crop being used or the level of pest infestation in the field are limiting. Climatic and soil conditions also play a role. Market characteristics are also involved. For example: Have GM crops been already introduced in the market? If yes, then what was their demand and level of societal acceptance?

Adoption of GM crops by some growers may also have an effect on those who choose not to cultivate them. The presence of GM crops on the market can affect the availability of non-GM seeds and output prices. Non-adopting farmers will have to bear additional costs of segregation measures or damage that can occur in cases of cross pollination. However, they may benefit too, perhaps in cases where insect-resistant GM crops have caused a reduction in local pest populations.

As an example, in India, despite the controversies surrounding the introduction of Bt cotton, its adoption has uplifted the status of cotton growers and enhanced the agricultural economy. As per reports, Bt-cotton growers gained profits to the tune of Rs. 1877 per acre (US$38) and 126 kg/acre of yield increase has been registered. A net increase of 18 percent (Rs. 15,841/US$321) in yearly household expenditures has been recorded in Bt-cotton grower families, which points towards the economic prosperity of cotton growers (Kathage & Qaim, 2012). Adoption of Bt cotton increased the income of India's agri-biotech industry by 22 times. Owing to these benefits, more than 80 percent of the Indian farmers have now adopted Bt cotton. Bt-cotton adoption has greatly contributed to making India a world leading grower and exporter of cotton; the period 2002–11 is often termed as the white gold era for the Indian cotton industry (James, 2011).

16.5.3 Health related aspects

Health aspects include factors such as the wellbeing and fitness of the farmers/farm workers as well as the local population residing there. For example, decreased pesticide usage leads to healthier, longer, and more productive life spans (Bennet et al., 2006; Krishna & Qaim, 2012). In studies conducted on Bt-cotton cultivation in India between 2002 and 2008, it was found that in spite of secondary pest outbreaks and increased pesticide usage against those secondary pests, total pesticide use had declined significantly. Moreover, farmers who had not cultivated Bt cotton also benefitted from a reduction in pesticide applications (Krishna & Qaim, 2011).

GM crops with enhanced nutrient content directly benefit consumer health by giving them needed micronutrients that are otherwise negligible in their day to day diet. The reduced need for physical labour, lower pesticide usage, and better income from the increased yield have had indirect positive effects on the people's health.

Bawa and Anilakumar (2013) have also reviewed the risks and public concerns related to GM crops. The process of gene transfer involves insertion of DNA at new sites. These genomic changes can lead to many unpredictable changes in the metabolism of the organism arising from the formation of new kinds of proteins. Some of the commonly raised concerns (Malarkey, 2003) about the possible effects of GM crops on human wellbeing are:

1 Toxicity caused by novel genes and gene products.
2 Threat of expression of novel antigenic proteins or alterations in levels of protein allergens in the body.

3 Risk of the side effects arising due to changes in host metabolic pathways or overexpression of toxic or pharmacologically active substances.
4 Probability of changes in nutrient composition due to introduction of new traits that make it totally different from conventional counterparts.

So, to summarize, the health risks associated with consumption of GM crops are resistance against antibiotics, enhanced sensitivity towards allergens, unnatural changes in nutrient composition of food, and toxicity. The bacterium *Bacillus thuringiensis* (Bt) is widely used for producing GM crops. It produces Bt toxins that have insecticidal properties and has been used by farmers as an insecticide spray against many lepidopteran and coleopteran insects. Bt toxins belong to the Cry (crystal) protein family and Cry toxins are produced in crops that contain Bt genes. One such protein is Cry9c, which was incorporated into a corn variety with the trade name StarLink (Aventis Crop Science, Research Triangle Park, NC). The effects of StarLink corn were studied by US regulatory agencies and due to the probability of allergens produced in humans from its consumption, in 1998 the US Environmental Protection Agency discontinued its use for human consumption yet allowed its continued use as animal feed.

Malnutrition is a menace in the developing world. As per the data provided by the Food and Agriculture Organisation of the United Nations, it is estimated that about 811 million people faced hunger in 2020. Globally, about 149 million children under the age of 5 years were stunted due to malnutrition (UNICEF/WHO/WB Joint Malnutrition estimates, 2021). Malnutrition is caused by a lack of food and where children aren't consuming enough calories. Moreover, the food that they do eat lacks the vitamins and minerals required for their proper growth. Half of the world's population has deficiencies in iron, iodine, zinc, folic acid, and vitamin A. Currently, biofortification – the process of developing a crop with increased micronutrient content in its edible parts – has emerged as a cost-effective and promising technology to make micronutrients available to the deficient populations on a global level.

Such crops are developed by several biotechnological methods (Garg et al., 2018). Rice is an example where many attempts have been made to increase the nutrient content of the grain. It is considered a favourable crop for biofortification because of its low allergenic properties (Suzuki et al., 2003). Researchers successfully expressed the human gene lactoferrin in dehusked rice which led to an astonishing 120 percent in iron content. However, even this enhancement is not sufficient to fulfil the daily need of adults (Suzuki et al., 2001). To solve this problem, ferritin gene was made to express in rice because lactoferrin gene expression did not lead to desired result. As compared to lactoferrin, ferritin is readily absorbed in human intestine. Similarly, Masuda et al. (2013) showed that introduction of ferritin gene and mugineic acid (ferric iron chelator) biosynthetic genes effectively enhanced the seed iron level in rice thus improving its iron content.

16.5.4 Social aspects

Social impact is defined as the aftereffects that human populations have to bear due to any public or private actions; and these actions may alter many aspects of people's lives. Social aspects include all aspects related to life, work, and organization as members of society. Cultural impacts that include changes in values and belief systems are also studied under social aspects (IOCPG, 2003).

Thus, in order to study the social impact of GM crops we have to consider components that affect the working environment such as working hours and overtime, remunerations and health coverage/allowances, skill enhancement and training, and the machinery and safety equipment available. These factors directly influence the life of workers at a farm. Moreover, social interactions between adopters and non-adopters of GM crops can also change and there can be a shift/change in the buyer–supplier relationship.

Ethical and cultural effects such as degrading moral values (concerning good agricultural practices and handling of natural resources), using local and native knowledge and cultural practices related to farming or the provision of justice (regarding the availability of technology and any kind of discrimination between GM crop users and non-users) are also taken into consideration. There is always a probability of pollination and mingling of products of GM crops and non-GM crops and hence there is an urgent need to establish a system that will ensure their co-existence. Coexistence here refers to the ability to grow and sell the produce from both traditional crops and GM crops simultaneously within the same agricultural system. Coexistence is more of an economic concern than a safety issue.

16.5.5 Gender-related aspects

A few studies have focussed on the impact of GM crops on gender (Chambers et al., 2014). Participation of women in food production has grown massively since the 1970s. Food and Agriculture Organisation Report in 2011 stated that women form 20–50 percent of the agricultural labour force in developing countries (FAO, 2011). In developed countries like the United States, Australia, and New Zealand, the percentage of women in farming increased between 1980 and 2010 and as per the United States Department of Agriculture 2017 census, women comprised of about 36 percent of the total agriculturists in the USA.

A detailed study on gendered agricultural production systems needs to be undertaken to understand that the roles undertaken by women and men vary according to locations, and these differences need to be taken into account before doing any agricultural research (Bock, 2006). Thus, the issue of gender-suitable technologies becomes relevant to GM crops as well. In many regions, agricultural technologies like large and/or heavy machinery, such as tractors, threshers etc., are seen to be linked with masculinity and part of male dominion (Brandth, 2006).

Globally, women probably farm on a smaller scale as compared to men. Despite the fact that GM crops are considered a female-appropriate or gender-neutral technology, it is expected that women are less likely to adopt GM crops readily. Prior studies conducted on the relationship between gender and agriculture have revealed varied results. On the one hand, they found it doubtful to believe that sex of the farmer had any role in the adoption, production, and marketing of GM crops in both developed and developing countries (Feldman & Welsh, 1995; Schafer, 2002; Sundari & Gowri, 2002; Prugl, 2004). Yet, they also show it is a fact that females often face restrictions in their work which restrain them from improving their wages and ability of household to obtain proper food and nutrition. Many other factors such as lack of education, money, extension services, and land are also reported (Ransom & Bain, 2011; Quisumbing et al., 2014). In both developed and developing countries, the roles of women farmers might be different but the constraints that they face during their work are somewhat similar.

A few studies have pointed to the fact that commercialization of GM crops have different effects on men and women. For instance, if we consider India, adoption of Bt cotton has benefitted the female labourers who pick the cotton because increased yields has meant an increase in their working hours and, in turn, increased income (Subramanian & Qaim, 2010). On the other hand, male workers who were previously employed in chemical spraying experienced a fall in their income because less chemicals were needed to be sprayed, thus decreasing their labour time. This aspect can also affect women too. For example, in Colombia, the introduction of HR cotton led to the recruitment of fewer women labourers for weeding, which hampered their income (Zambrano et al., 2013).

One more aspect that has attracted attention of researchers is the role played by women in the decision-making process in agricultural households. In Colombia and the Philippines, it was found that women had contributed equally with men in the decision-making and supervision of Bt cotton and Bt maize, respectively (Yorobe & Smale, 2012; Zambrano et al., 2013).

16.5.6 Impact along the supply chain

The supply chain involves all the stakeholders – developers/producers, intermediaries, and consumers. Socio-economic impacts arising as a result of introduction of GM crops can influence the working lives and profits of all the stakeholders of the supply chain. Effects on the supply chain can be seen in many directions.

16.5.6.1 Bidirectional effects

These include GMO regulations and coexistence rules at international and national levels. Regulations regarding GMO certification and the intellectual property rights in the form of patents, licences, etc. are also included.

16.5.6.2 Upstream effects

At the top of the supply chain, we have GM seed producing companies and industrialists involved with the manufacture of products required in farming such as herbicides/insecticides or fertilisers. They gain their profits from the farmers who are adopting GM crops and buying their seeds. Conversely, producers of non-GM seeds and other herbicides might experience a downfall in sales. Similarly, companies producing insecticides might also experience low sales due to the introduction of insect-resistant GM crops. GM crop adoption affects research and development too, with the acceptance of GM crops leading to increased investment in their development.

16.5.6.3 Downstream effects

At the bottom of the supply chain we have the intermediaries between the consumers and the farm. GM crops can influence many aspects like market access and trade interests, logistics, and governance mechanisms. The scale of these effects will depend upon what kind of producer the country is, whether it is a large-scale or small-scale producer, whether the country is involved in the trade of the crop at the international level, and the type and size of the supply shift due to the adoption of GM crops.

Depending upon the purpose that has to be served, GM crop adoption might affect the supply chain differently. For example, the feed industry will earn profits because of the lower rates for raw materials if increased GM crop cultivation leads to higher yields with lower prices. However, as more and more GM crops are grown there will be less feed available that is GMO-free and this will have several possible effects on the organic farming industry. Organic arable farmers would be able to charge more as non-GM feed is scarcer and organic livestock producers would be able to charge more as livestock fed on non-GM feed will become rarer. Yet, conversely, the non-GM feed they need will become much more expensive. The introduction of new GM products under the various types of the coexistence rules enforced by different regulating authorities, labelling schemes, and intellectual property rights can therefore affect both the structure and performance of the supply chain. This, in turn, can cause both costs and benefits for the different actors along the supply chain as noted above.

16.5.7 Food security

According to the United Nations' Committee on World Food Security, food security is defined as a situation in which all the population has access to safe, enough, and nutritious food at all times. In addition, the food available should fulfil their food preferences and should provide enough calories required for a safe and healthy life. In other words, food security is the measure of food availability and the ability of an individual to access it.

Thus, it's really unfortunate that, even today, food security is a distant dream for a large chunk of the population across the globe. As per the data provided by Food and Agricultural Organization of United States in 2020 (see www.fao.org), about 9.9 percent of the world's population is undernourished. Eradicating hunger is one of the United Nations' Millennium Development Goals (United Nations, 2012).

It pointed out four facets of food security:

1 Availability of good quality and sufficient quantities of food that is supplied domestically.
2 Individuals should have enough resources for obtaining foods for a well-balanced and nutrient rich diet;
3 Provision for a healthy diet, safe drinking water, sanitation, and health facilities. All these facilities are an important part of an individual's well-being.
4 Food should be available in sufficient quantities throughout the year regardless of any economic or climatic crisis or seasonal food scarcity.

Thus, food security is a multi-dimensional concept and all the four dimensions must be fulfilled simultaneously to achieve it (Ruane & Sonnino, 2011). GM crops are sometimes seen as a way to achieve this goal. Some researchers view the adoption and growing of GM crops as a way of improving food security (Juma, 2011; Borlaug, 2007), whereas others think of this technology as a threat to food security (Shiva et al., 2011).

The role of GM crops in ensuring food security has always remained a hot topic for debate. GM crops increase yield and ensure food availability. They may also enhance food quality and nutrient composition. Finally, GM crop cultivation influences farmers' income and their access to food. In countries with lower agriculture production and limited access to resources, GM crops can improve food availability (Qaim & Kouser, 2013). Yet, even though GM crops can be an important tool for improving food security (Dibden et al., 2009) – by incorporating certain traits such as insect and/or herbicide resistance, drought/salinity tolerance, increasing shelf life of fruits etc., and improving food access by increasing the income of farmers and by enhancing the nutrient composition of the crops – they cannot alone improve the grim food situation all over the world.

Farmers as well as consumers therefore need to choose whether or not they want to cultivate and consume GM crops. Consumers take into account several factors such as acceptance, availability, and the price comparison between GM and non-GM crops. Today, several countries have launched GM-related labels (GMO or GMO-free) through which a buyer can distinguish between GM and non-GM food products. Socio-economic impacts at the consumer level are mostly related to the costs involved in labelling the products and their willingness to spend extra amounts to acquire or avoid specific products. Thus, GM crops alone can't lower the hunger index but they can be an important component in a broader food security strategy.

16.5.8 Impacts on farmers' rights to save seeds

Farmers all over the world follow the traditional practice of saving seeds from the produce for sowing the next season. The saved seeds are shared, reused, exchanged, and sometimes sold. Farmers belonging to developing countries widely practise traditional seed saving and free exchange of planting materials, whereas in developed countries such practices are rarely seen because of the industrial nature of the agriculture there. These traditional seed saving practices by farmers are the basis of the foundation of the immense genetic diversity in agriculture today. However, certain regulations, like the IPR (intellectual property rights) system on seeds, are viewed as a potential threat to the long-term food security of rural communities. The International Treaty on Plant Genetic Resources for Food and Agriculture (ITPGRFA) under the Food and Agriculture Organization (FAO) protects the right of farmers to save and exchange seeds. Also, governments are entrusted to pass legislations to safeguard farmers' rights. However, countries under international trade agreements, such as the Agreement on Trade-Related Aspects of Intellectual Property Rights (TRIPS) of the WTO, are bound to pass laws for protecting the IPRs of seed companies and they find it hard to preserve the rights of the farmers. In spite of the flexibilities given under the TRIPS Agreement, only a few countries have shown the courage and interest to protect farmers' rights to seeds.

The Green Revolution has exhibited how expensive technologies promote the dependence of poor farmers, who lack the capital required for adopting a new technology, on credit providers. In a similar way, GMOs too have the potential to further establish rural inequalities and stigmatize poor farmers, leading to long-term impacts on the capability of farmers to take decisions on how to plant their own farms.

16.5.9 Environmental impacts

Currently, the global environmental situation is quite grim. Issues such as increasing population, climate change, and biodiversity loss are having a tremendous impact on our environment. A decrease in forest cover combined with the fumes emitted by automobiles and factories have led to a rise in carbon dioxide levels. These factors accelerate global warming which leads to climate change. Scientists predict that global temperature will rise by 1.4–5.8°C by 2100. Climate change causes disturbance in rainfall patterns and thus agricultural practices too are disturbed. Soaring human population growth has accelerated forest destruction, increased water pollution, and degraded water quality. The loss of habitat has resulted in loss of biodiversity. Thus, to retain forest cover, it is essential to make sure that no more forest lands are brought under cultivation and that the food production needs for the growing populace are met from the area already under cultivation. GM crops with their capability to enhance yield could thus ensure that sufficient food grains are produced that can meet the demands of the growing population, without putting more land under cultivation. In this way, GM crops can positively impact environment. It has been estimated that GM crops have increased crop production by 22 percent and reduced pesticide use by 37 percent (Klümper & Qaim, 2014; Kovak et al., 2022). In order to achieve similar yield increase with non-GM crops, a further 300 million acres of land would have to be put under cultivation which would make the Earth's environmental status even worse (Zhang et al., 2016).

Brookes and Barfoot (2020) found that use of GM insect-resistant and herbicide-tolerant varieties lowered the environmental degradation caused by herbicides. There was a reduction in pesticide spraying by about 8.3 percent. In China, it is estimated that Bt cotton usage reduced pesticide use by 78,000 tons in 2001 (Pray et al., 2002). Additionally, from 1996 to 1998, the University of California carried out a study on Bt cotton fields and found that the average cost reduction in pesticide usage was between 25 and 65 dollars per acre and a 5 percent increase in the yield was recorded during the same period. In addition to these benefits, Bt cotton required fewer foliar sprays against other cotton pests and thus less money was spent on insecticide usage, ultimately benefitting the farmer.

Unguarded use of pesticides can aggravate the threat of poisoning in field workers (Damalas & Eleftherohorinos, 2011). Insect-resistant GM crops have the ability to produce a specific kind of protein that works as a pesticide. This is the reason that the GM crops require less pesticide usage as compared to conventional crops. Thus, insect-resistant GM crops can have a direct role in environment protection by lowering a crop's pesticide footprint. Moreover, it has been shown that with GM crops the number of cases of pesticide poisoning in field workers has also reduced (Kouser & Qaim, 2011).

Herbicide-tolerant GM crops can also alter the crop management system, since no herbicide is required and there is a reduction in soil preparation. Such changes benefit the environment by reducing the greenhouse gas emissions emitted by the machines used in the field. Kovak et al. (2022) found that GM crops also reduced greenhouse gas emissions in Europe. There was a 7.5 percent reduction in total agricultural greenhouse gas emissions by using GM crops.

Virus-resistant GM crops benefit the environment by reducing local viral loads. Local crops which are sensitive to the same virus are also cross-protected by the use of these virus-resistant GM crops. Similarly, drought-resistant and salinity-tolerant GM crops can reduce the usage of water resources in the area.

GM crops are thoroughly evaluated and studied for environmental effects before they are made commercially available in the market. Firstly, the developers of the crop assess their impacts and then regulatory bodies of the countries where these crops are to be introduced study the pros and cons of their adoption in detail. In the United States, these regulatory bodies include agencies like the US Food and Drug Administration (FDA), the US Environmental Protection Agency (EPA), and the US Department of Agriculture (USDA). These agencies make a detailed risk assessment of the interactions and effects of GM crops on the environment.

And yet, in spite of all these safety measures, the usage of GM crops has ignited a debate regarding the impact they have on the environment. Researchers have indicated that GM crops have actually led to a net increase in herbicide usage and have enhanced the threat of emergence of herbicide-resistant weeds (Brooks & Barfoot, 2020). In fact, in a study carried out in Brazil, pesticide usage increased 1.6-fold between 2000 and 2012 and pesticide usage for soybean tripled. This study indicated that introduction of GM crops in Brazil led to an overall increase in pesticide use which in turn increased the risks of environmental and human exposure. Similar results have been observed in studies carried out in many other countries of the world (Peshin & Zhang, 2014; Benbrook, 2012; Pengue, 2005; Landrigan & Benbrook, 2015). These studies strongly suggest that repercussions of GM crop usage are unprecedented and therefore detailed studies over a long period of time are required before giving them the nod of approval.

Incidences of glyphosate-resistant ryegrass in Australia in 1996 created an alarming situation, showing that resistant varieties may emerge, resulting in more pesticide use. In addition, there have been 22 more insect species resistant to glyphosate that have evolved (Powles et al., 1998). This evolution of 'Superweeds' and 'Superpests' is a potential risk to the environment. Superweeds can emerge when transgenes enter the weedy relatives of the GM crop (Hails & Morley, 2005; Halfhill et al., 2002). For example, a weed that has acquired a transgene that provides drought resistance will experience enhanced fitness to water stress and it will spread in the field quickly, often suppressing the growth of the other beneficial plants. Also, these Superweeds can cause serious problems for farmers in that they are resistant to most of the available herbicides and their presence will accelerate this resistance in the wild population.

Evolution of resistant varieties is a phenomenon that has already been seen in some pest populations – Superpests – and farmers and scientists can't ignore this aspect. Current GM approaches are not very efficient in countering this problem and therefore a long-term approach will be important in determining how long this technology will benefit humankind (Mannion & Morse, 2012).

Sometimes, GM crops affect the diversity present in various ecosystems in a way that is beyond imagination. One such incident was the Monarch butterfly controversy (Losey et al., 1999) in which an experiment was designed to study effect N4640-*Bt* maize on the Monarch butterfly (*Danaus Plexippus*) larval feeding cycle. Monarch larvae's main nutrition is derived from milkweed (*Asclepias curassavica*), which commonly occurs in and around the edges of corn fields. As maize pollen shedding coincided with the Monarch butterfly's larval feeding cycles, researchers decided to study the effect that N4640-*Bt* maize pollen had on the growth of Monarch butterfly larvae. A comparative study was made between milkweed dusted with N4640-*Bt* maize pollen and milkweed dusted with traditional corn pollen.

They found that larvae fed on N4640-*Bt* maize pollen showed poor growth and higher mortality and thus came to the conclusion that N4640-*Bt* maize significantly impacted Monarch butterfly populations. This experiment generated a lot of controversy and the whole experimental design was questioned. That being so, such studies, however, point towards the potential risks that GM crops can indirectly pose on our biodiversity.

In sum, there are a lot of concerns that the use of GM crops degrades agricultural ecosystems and those studies which predict negative outcomes of GM crops are very worrying. Yet all these research studies producing varying results makes it quite impossible to predict the pros and cons of GM crops. If we assess the situation on a global scale, studies indicate that GM crops offer more positives than negatives (Meftaul et al., 2020) and that, in fact, GM crops have contributed towards a healthier and cleaner environment.

16.6 Conclusions

Farming is an age-old practice and immense diversity is found in the types of crops grown in the world and the conditions under which this is carried out. Introduction of genetic engineering technology into the field has generated a cascade of both social and economic effects.

On the basis of the studies conducted, GM crops have benefitted both small- and large-scale farmers of cotton, soybean, maize, and canola. However, benefits to small-scale farmers depend on many other factors like access to credit facilities, extension services, demand in the market, and assistance from the authorities.

To improve the food security of global populations, GM crops will have an important role to play. Traits like increase in yield, pest/insect resistance, protection of the crop from biotic and abiotic stresses, and enhanced nutritional quality will have to be developed. But the developers shall have to keep long-term goals in mind which will be helpful in attaining the benefits on a large scale and for longer periods. Even if the above-mentioned traits are achieved, hunger alleviation by GM crops will not be easily achieved. Thus, socio-economic impacts posed by the progressive adoption of GM crops are endless and a lot more research over a longer period of time is required to ascertain exact effect.

References

Bawa, A., & Anilakumar, K. (2013). Genetically modified foods: Safety, risks and public concerns – a review. *Journal of Food Science and Technology*, 50(6), 1035–1046.

Benbrook, C. M. (2012). Impacts of genetically engineered crops on pesticide use in the U.S. – the first sixteen years. *Environmental Sciences Europe*, 24, 24.

Bennet, R., Morse, S., & Ismael, Y. (2006). The economic impact of genetically modified cotton on South African smallholders: Yield, profit and health effects. *Journal of Development Studies*, 42(4), 662–677.

Bock, B. B. (2006). Rurality and gender identity: An overview. In *Rural Gender Relations: Issues and Case Studies*, edited by B. B. Bock & S. Shortall (Wallingford, UK: CABI Publishing), 155–164.

Borlaug, N. (2007). Feeding a hungry world. *Science*, 318, 359.

Brandth, B. (2006). Agricultural body-building: Incorporations of gender, body and work. *Journal of Rural Studies*, 22, 17–27.

Brookes, G., & Barfoot, P. (2018). *GM Crops: Global Socio-economic and Environmental Impacts 1996–2016* (Dorchester, UK: PG Economics), 9.

Brookes, G., & Barfoot, P. (2020). Environmental impacts of genetically modified (GM) crop use 1996–2018: Impacts on pesticide use and carbon emissions. *GM Crops & Food*, 11(4), 215–241.

Chambers, J. A., Zambrano, P., Falck-Zepeda, J., Gruère, G., Sengupta, D., & Hokanson, K. (2014). *Agricultural Technologies for Africa: A State of Affairs* (Washington, DC: IFPRI).

Damalas, C. J., & Eleftherohorinos, C. I. (2011). Pesticide exposure, safety issues, and risk assessment indicators. *International Journal of Environmental Research and Public Health*, 8(5), 1402–1419.

Dano, E. C. (2007). Potential socio-economic, cultural and ethical impacts of GMOs: Prospects for socio Economic Impact Assessment. In *Biosafety First*, edited by T. Traavik & L. C. Lim (Trondheim: Tapir Academic Publishers).

Deloitte & Touche (2015). *Reducing Food Loss Along African Agricultural Value Chains*. Member of Deloitte Touche Tohmatsu Limited, UK.

Dibden, J., Potter, C., & Cocklin, C. (2009). Contesting the neoliberal project for agriculture: Productivist and multifunctional trajectories in the European Union and Australia. *Journal of Rural Studies*, 25(3), 299–308.

FAO (2011). *The State of Food and Agriculture. Women in Agriculture: Closing the Gender Gap for Development* (Rome: FAO). https://www.fao.org/publications/sofa/2010-11/en/

Feldman, S., & Welsh, R. (1995). Feminist knowledge claims, local knowledge, and gender divisions of agricultural labor-constructing a successor science. *Rural Sociology*, 60, 23–43.

Garg, M., Sharma, N., Sharma, S., Kapoor, P., Kumar, A., Chunduri, V., & Arora, P. (2018). Biofortified crops generated by breeding, agronomy, and transgenic approaches are improving lives of millions of people around the world. *Frontiers in Nutrition*, 5, 12. doi:doi:10.3389/fnut.2018.00012.

Goto, F., & Yoshihara, T. (2001). Improvement of micronutrient contents by genetic engineering development of high iron content crops. *Plant Biotechnology*, 18, 7–15.

Gouse, M., Piesse, J., Thirtle, C. G., & Poulton, C. (2009). Assessing the performance of GM Maize amongst small holders in Kwazulu-Natal, South Africa. *AgBioForum*, 12(1), 78–89.

Gouse, M., Pray, C. E., Kirsten, J. E., & Schimmelpfennig, D. (2005). A GM subsistence crop in Africa: The case of Bt white maize in S Africa. *International Journal of Biotechnology*, 7(1–3).

Hails, R. S., & Morley, K. (2005). Genes invading new populations: A risk assessment perspective. *Trends in Ecology and Evolution*, 20(5), 245–251.

Halfhill, M. D., Millwood, R. J., Raymer, P. L., & Stewart, Jr., C. N. (2002). *Bt-transgenic oilseed rape hybridization with its weedy relative, Brassica rapa. Environmental Biosafety Research*, 1(1), 19–28.

Hall, C., Knight, B., Ringrose, S., & Knox, O. (2013). *What Have Been the Farm-Level Economic Impacts of the Global Cultivation of GM Crops? Systematic Review*. CEE Review 11–002, Collaboration for Environmental Evidence. https://www.environmentalevidence.org/wp-content/uploads/2014/07/CEE11-002.pdf

IOCPG (2003). Principles and guidelines for social impact assessment in the USA (by Interorganizational Committee on Principles and Guidelines for Social Impact Assessment). *Impact Assessment and Project Appraisal*, 21, 231–250.

James, C. (2011). *Global Status of Commercialized Biotech/GM Crops*. ISAAA Brief No. 43 (Ithaca, NY: ISAAA).

Juma, C. (2011). Preventing hunger: Biotechnology is key. *Nature*, 479, 471–472.

Kathage, J., & Qaim, M. (2012). Economic impacts and impact dynamics of Bt (Bacillus thuringiensis) cotton in India. *Proceedings of the National Academy of Sciences*, 109(29), 11652–11656.

Kathage, J., Gómez-Barbero, M., & Rodríguez-Cerezo, E. (2015). *Framework for the Socio-Economic Analysis of the Cultivation of Genetically Modified Crops*. European GMO Socio-Economics Bureau, 1st Reference Document. Science and Policy Reports, Institute for Prospective Technological Studies, Joint Research Centre, European Commission, Seville, Spain.

Kikulwe, E., Wesseler, J., & Falck-Zepeda, J. (2008). *Introducing a Genetically Modified Banana in Uganda: Social Benefits, Costs, and Consumer Perceptions*. IFPRI discussion papers. International Food Policy Research Institute (IFPRI).

Klümper, W., & Qaim, M. A. (2014). Meta-analysis of the impacts of genetically modified crops. *PLoS One*, 9(11), e111629. doi:10.1371/journal.pone.0111629.

Kouser, S., & Qaim, M. (2011). Impact of Bt cotton on pesticide poisoning in smallholder agriculture: A panel data analysis. *Ecological Economics*, 70(11), 2105–2113.

Kovak, E., Blaustein-Rejito, D., & Qaim, M. (2022). Genetically modified crops support climate change mitigation. *Trends in Plant Science*, 27(7), 627–629.

Krishna, V. V., & Qaim, M. (2011). Pesticide reduction sustainability of Bt technology in India. International Congress, August 30–September 2, Switzerland.

Krishna, V. V., & Qaim, M. (2012). *Bt* cotton and sustainability of pesticide reductions in India. *Agricultural Systems*, 107, 47–55.

Landrigan, P. J., & Benbrook, C. (2015). GMOs, herbicides, and public health . *New England Journal of Medicine*, 373(8), 693–695.

Lonnerdal, B. (2003). Genetically modified plants for improved trace element nutrition. *Journal of Nutrition*, 133, 1490–1493.

Losey, J., Raynor, L., & Carter, M. E. (1999). Transgenic pollen harms Monarch larvae. *Nature*, 399, 214. doi:10.1038/20338.

Malarkey, T. (2003). Human health concerns with GM crops. *Mutation Research*, 544 (2/3), 217–221.

Mannion, A. M., & Morse, S. (2012). Biotechnology in agriculture: Agronomic and environmental considerations and reflections based on 15 years of GM crops. *Progress in Physical Geography: Earth and Environment*, 36(6), 747–763.

Masuda, H., Kobayashi, T., Ishimaru, Y., Takahashi, M., Aung, M. S., Nakanishi, H., … Nishizawa, N. K. (2013). Iron-biofortification in rice by the introduction of three barley genes participated in mugineic acid with soybean ferritin gene. *Frontiers in Plant Science*, 4, 132. doi:10.3389/fpls.2013.00132.

Meftaul, I. M., Venkateswarlu, K., Dharmarajan, R., Annamalai, P., Asaduzzaman, M., Parven, A., & Megharaj, M. (2020). Controversies over human health and ecological impacts of glyphosate: Is it to be banned in modern agriculture? *Environmental Pollution*, 263(Pt A), 114372.

Pengue, W. (2005). Transgenic crops in Argentina: The ecological and social debt. *Bulletin of Science, Technology & Society*, 25(4), 314–322.

Peshin, R., & Zhang, W. (2014). Integrated Pest Management and pesticide use. In *Integrated Pest Management*, edited by D. Pimentel & R. Peshin (Amsterdam: Springer), 1–46.

Powles, S. B., Lorraine-Colwill, D. F., Dellow, J. J., & Preston, C. (1998). Evolved resistance to glyphosate in rigid rye grass (Lolium rigidum) in Australia. *Weed Science*, 46(5), 604–607.

Pray, C. E., Huang, J., Hu, R., & Rozelle, S. (2002). Five years of Bt cotton in China – the benefits continue. *The Plant Journal*, 31(4), 423–430.

Prugl, E. (2004). Gender orders in German agriculture: From the patriarchal welfare state to liberal environmentalism. *Sociologia Ruralis*, 44, 349–372.

Qaim, M., & Kouser, S. (2013). Genetically modified crops and food security. *PLoS One*, 8(6), e64879.

Quisumbing, A. R., Meinzen-Dick, R., Raney, T. L., Croppenstedt, A., Behrman, J. A., & Peterman, A. (2014). Closing the knowledge gap on gender in agriculture. In *Gender in Agriculture: Closing the Knowledge Gap*, edited by A. R. Quisumbing, R. Meinzen-Dick, T. L. Raney, A. Croppenstedt, J. A. Behrman, & A. Peterman (Dordrecht: Springer), 3–27.

Ransom, E., & Bain, C. (2011). Gendering agricultural aid: An analysis of whether international development assistance targets women and gender. *Gender & Society*, 25, 48–74.

Ruane, J., & Sonnino, A. (2011). Agricultural biotechnologies in developing countries and their possible contribution to food security. *Journal of Biotechnology*, 156 (4), 356–363.

Schafer, R. (2002). Transformations of Ovambo society and changes in agricultural systems in northern Namibia – gender relations and tradition farming and ecological knowledge. *Anthropos*, 97, 73–87.

Shiva, V., Barker, D., & Lockhart, C. (2011). *The GMO Emperor has No Clothes* (New Delhi: Navdanya International).

Subramanian, A., & Qaim, M. (2010). The impact of Bt cotton on poor households in rural India. *Journal of Development Studies*, 46, 295–311.

Sundari, S., & Gowri, V. (2002). New agricultural technology: A gender analysis. *Indian Journal of Social Work*, 63, 517–539.

Suzuki, Y. A., Shin, K., & Lönnerdal, B. (2001). Molecular cloning and functional expression of a human intestinal lactoferrin receptor. *Biochemistry*, 40, 15771–15779.

Suzuki, Y. A., Kelleher, S. L., Yalda, D., Wu, L., Huang, J., Huang, N., & Lönnerdal, B. (2003). Expression, characterization, and biologic activity of recombinant human lactoferrin in rice. *Journal of Pediatric Gastroenterology and Nutrition*, 36, 190–199.

Theil, E., Burton, J., & Beard, J. (1997). A sustainable solution for dietary iron deficiency through plant biotechnology and breeding to increase seed ferritin control. *European Journal of Clinical Nutrition*, 51, S28–S31.

Theil, E. C. (2011). Iron homeostasis and nutritional iron deficiency. *Journal of Nutrition*, 141, 724–728.

United Nations (2012). *The Millennium Development Goals Report 2012* (New York: United Nations).

Yorobe, Jr., J. M., & Smale, M. (2012). Impacts of Bt maize on smallholder income in the Phillipines. *AgBioForum*, 15, 152–162.

Zambrano, P., Lobnibe, I., Cabanilla, D. B., Maldonado, J. H., & Falck-Zepeda, J. (2013). *Hiding in the plain sight: Women and GM crop adoption*. Paper presented at the 17th ICABR ConferenceInnovation and Policy for the Bioeconomy, June 18–21. Ravello, Italy.

Zhang, C., Wohlhueter, R., & Zhang, H. (2016). Genetically modified foods: A critical review of their promise and problems. *Food Science and Human Wellness*, 5(3), 116–123.

17 GM crops – friends or foes?

Summarizing the argument

Kanchan Sharma, Kawaljeet Kour, Nancy Bhura, Jasmeet Kour and Bhaskar Jyoti

17.1 Introduction

The term "genetically modified" (GM) is considered controversial; on the one hand, it is beneficial both for the producers and consumers of a particular kind of food and yet, on the other hand, it is associated with potential biomedical hazards and adverse environmental effects. Its history can be traced back to the mid-1990s with the advent of a technique to transfer genes across species (Avery et al., 1994). GM plants, also known as transgenic plants, are designed to incorporate desirable characteristics such as insect resistance, disease resistance, delayed ripening, tolerance against abiotic stress, herbicide tolerance, increased nutritional composition, increased yield, high ornamental value, production of vaccines that are edible for humans, production of biodegradable plastics, their use as bioreactors for producing nutraceuticals, therapeutic agents, antigens etc. (Zhang et al., 2016).

The foremost plants that were genetically modified included antibiotic-resistant tobacco and petunias. In the USA, some species of tomato were genetically modified to delay ripening as far back as the 1940s and these were approved by the Food and Drug Administration (FDA). After this, several other transgenic crops also received approval from the FDA such as canola with a modified oil composition and herbicide-resistant cotton and soybeans, etc. Also, GM crops that have the potential to tolerate abiotic stress have been shown to have better growth than non-GM crops. Nowadays a variety of GM foods are available in the market such as potatoes, rice, cotton, maize, strawberries, sugar beet, carrots, papaya, brinjal and many more (Bawa & Anilakumar, 2013) and genetic modification has been used across the globe for producing pharmaceuticals, for crop modification and improvement, and to create industrial chemicals, etc. Thus, genetically engineered plants have the potential to affect modern society in many ways. Yet, there still remains the risk, and fear, that the use of GM crops as food for humans and animals may cause health problems (Verma, 2013).

DOI: 10.4324/9781003278566-17

17.2 Genetically modified crops as friends

17.2.1 Economic benefits

Between 2006–2012, there was a global increase in farm income generated from GM food reaching up to $116 billion, which was much more than in the previous 10 years (Brookes & Barfoot, 2014). James and Brookes estimated that maximum economic gains (42 percent) were reported from enhanced yield due to advanced genetics as well as from pest and weed resistance; with the decreased costs of production due to less usage of pesticides and insecticides contributing to the remaining 58 percent (Zhang et al., 2016). Genetically engineered crops can prove to be a great substitution for agricultural practices which can be disastrous to the environment (Garcia-Li et al., 2014). Brookes and Barfoot (2014) estimated a reduction in the use of pesticides on the GM crop area since 1996 of 448 million kg and the environmental impact associated with herbicide and insecticide use representing a reduction of 9 percent.

17.2.2 Products for therapeutic purposes

With advancements in plant biotechnology and genetic engineering it has become possible to alter the gene expression of plants as desired. Hence, one can now enable the expression of a viral or bacterial antigen in the edible part of plant cells. These foods can then be used as vaccines and are able to stimulate the immune system for the production of antibiotics. GM crops such as rice, maize, soybean, and potatoes are considered to be effective bearers of edible vaccines to fight against various infections including those from *Escherichia coli*, the rabies virus, *Helicobacter pylori*, and the hepatitis virus (Nicolia et al., 2014; Ellstrand & Hancock, 1999; Hare & Chua, 2002; Aggarwal, 2012; Reichman et al., 2006). With the help of genetic engineering, the genes responsible for expressing viral or bacterial antigens can be directly transferred to the plant cell nuclei or chloroplasts. These antigens act as a stimulator for a whole-body response, leading to the production of antibodies and providing immunity against selected pathological microflora (Ubalua, 2009). To achieve target effects, the presence of active substances in the edible part of the fruits and vegetables is crucial. Only then they can stimulate mucosal immunity due to the processes of digestion and absorption in the alimentary canal of organisms. The lyophilizate of these plants, a freeze-dried form, is the commercially recommended form of these resistance stimulating plants, providing lowered risk of contamination and more favourable storage conditions (Kramkowska et al., 2013).

17.2.3 Agronomic benefits

In the late 1990s a significant increase in yield of food crops was observed in the USA which was attributed to the introduction of GM crops (Brookes &

Barfoot, 2014). Brookes and Barfoot also reported that the rise in global production of millions of tons of soybean, corn, cotton, and canola was as a result of advancements in plant biotechnology (Cong et al., 2013). The yield effect of GM crops as compared with traditional varieties depends on many factors: the current and previous crop grown, their specific trait characteristics, agricultural practices, rate of pest infestation, market characteristics, cost of seed used, farming practices used, etc. (Krishna & Qaim, 2012; Mutuc et al., 2012). The use of bioengineering in agriculture has the potential to improve the quality of food and provide solutions to problems like food availability, poverty, and environmental conservation. GM crops offer great way to satisfy the basic needs of humans by increasing the global food supply (Bakshi, 2011). With the introduction of newly developed bio-fortified GM crops, farmers' nutritional status is expected to increase which will ultimately contribute to their wellbeing; however, the adoption of these crops could have different impacts on wealthier and poorer farmers (Lipton, 2007).

The principal GM crops grown commercially are soybeans, corn, cotton, and canola with insecticide- and herbicide-resistant properties, sweet potato which is virus resistant, rice with enhanced iron and vitamin content, and many others that are able to withstand extreme weather conditions. There are also GM bananas that are used to produce vaccines to treat hepatitis B, trees of several fruits and nuts that are able to yield years earlier than traditional cultivars, and plants that are used for the production of biodegradable plastics (Bawa & Anilakumar, 2013). No doubt that the rise in incorporation of GM crops in agriculture has created many benefits but it also poses a threat to biodiversity since genetic engineering may favour monocultures – plantings of a single kind – of crops best suitable for certain conditions or to produce a single product (James et al., 1998; Sweet & Shepperson, 1997). This ecological imbalance has also attributed to more natural ecosystems being transformed into agricultural lands.

In the UK the most common GM crops that have been approved for use as food are slow-ripening tomatoes, soya tolerant to wide-range of herbicide called glyphosate and maize having insect-resistant properties. These tomatoes are frequently used for the production of tomato paste, reducing waste production and lowering processing costs. It is beneficial to the consumer as it is cheaper than its non-GM counterparts and also has a thicker consistency which is desirable. It was reported that the use of glyphosate-tolerant crops by the farmers reduces the use of more expensive, persistent and poisonous herbicides and favours the use of a single, safe and rapidly-degrading herbicide (Dewar et al., 2000). Herbicide-tolerant soya was most commonly grown in USA during late 1990s and was also popular there.

GM maize, also approved in the UK to use as food, contains the Bt gene which is obtained from *Bacillus thuringiensis*. This gene is responsible for the production of a particular kind of protein which is poisonous to some insects; more specifically, caterpillars. Decades previously the bacteria themselves were also used as an insecticide in organic farming. Variants of the Bt gene have been incorporated into several food crops in the USA which include cotton, sugar beet, maize, and potato. However, among these crops the most significant effect was observed in cotton, leading to a reduction in the need for insecticide as compared with its conventional counterparts (De Maagd et al., 1999). In fact, a significant reduction in the use of harsh pesticides such as organophosphates was observed after the introduction of GM crops. Along with these effects, it was reported that Bt protein had no adverse effect on bees or other insects, mammals, birds, or fish. In addition to insect-resistant varieties, other GM crops include traits such as virus resistance, increased nutritional value including levels of vitamins and micro-nutrients (iron, iodine, folate, etc.), enhanced colour, and better flavour.

Research has been done to genetically modify crops to obtain modified oils, either tailored to meet particular needs of processors or with pharmaceutical or industrial value, like for the production of biodegradable plastics, for producing starch for industrial purposes, the production of fragrances and pigments, and for creating edible vaccines which will be safe and cheap. For the latter, human testing had been carried out for diseases caused by *Escherichia coli* and hepatitis B and will benefit those living in areas where drugs, clean needles or syringes are not frequently available (Napier et al., 1999; Sayanova et al., 1997; Barro et al., 1997; Mor et al., 1998).

17.2.4 Production of genetically modified plants as vaccines

A vaccination is a drug that boosts the ability of the body's defense to fight illnesses. Because of the multiple benefits that plant-based vaccine production may provide, it has received a lot of attention. Plants can be utilized as bioreactors in the manufacturing of vaccines against animal and human diseases. The use of plant-derived vaccines against contagious diseases, autoimmune conditions, and tumours may be able to overcome some of the key issues that standard immunisation has. Vaccine prices, the necessity for storage and distribution chains from the manufacturer to the point of use, and the reliance on injections are all issues in underdeveloped countries. Vaccines made from plants do not have these problems. Many plant species, including tomato, banana, corn, lupine, lettuce, potato, tobacco, and others, are used in the production of vaccines (Takeyama et al., 2015; Das et al., 2021).

A vaccine antigen in plants was first demonstrated in 1990. This was done by Curtiss and Cardineau by expressing antigen A (allergen of protein at surface) of *Streptococcus mutans* in tobacco (Curtiss & Cardineau, 1990). This was followed by plant interpretation of hepatitis B (allergen) and heat altered enterotoxin by *E. coli* for diarrhoea (Fahad et al., 2015). All these were expressions of a vaccine antigen within plants. Various livestock have also been protected by feeding them with GM plants like alfalfa that express antigens to safeguard against various infectious viruses, such as foot and mouth disease virus (FMDV) (Ruiz et al., 2015).

17.2.5 Plant-based vaccine development process

Most important steps in the development process of plant-based vaccines are: the appropriate combination between one or more immunological defensive antigens, the layout of genetic material, and enhancers that will increase the expression of the immune response against toxicity in desired plants and genetic transformation (Meric et al., 2021). The developmental process is shown in Figure 17.1.

17.2.6 Applications in veterinary and human medicines

A vaccine against infectious bronchitis virus (IBV) in chickens was created by using palatable diced potato tubers that exhibit S1 glycoprotein. Chickens

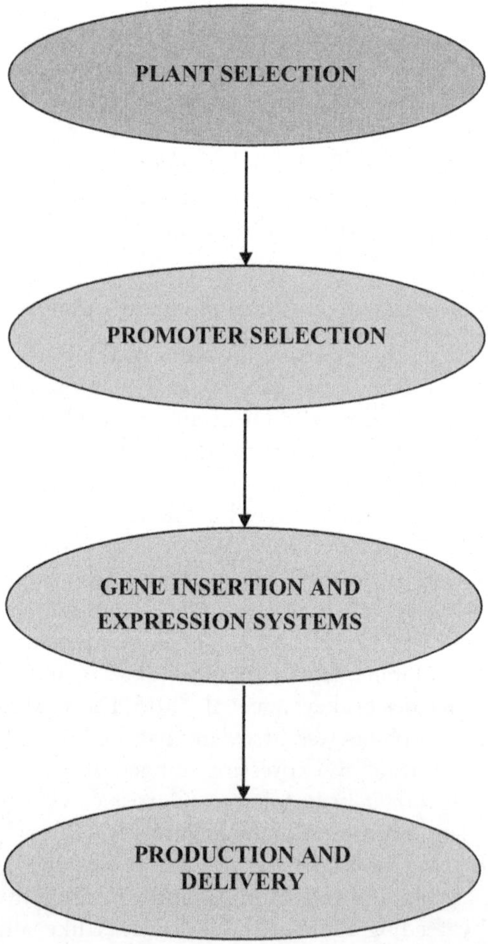

Figure 17.1 Steps involved in development of plant-based vaccine.
Source: Authors.

that were given the oral vaccine exhibited a virus-specific immune reaction and were resistant to IBV (Bhuiyan et al., 2021). With plant-made VP2 protein, an antibiotic against the infectious bursal disease virus (IBDV) was developed. The chickens were given preparations obtained from the leaves of Arabidopsis or recombinant seed. They were shielded to an extent comparable to that of a commercial injectable vaccination (Gómez et al., 2013). Maize has been utilized as a consumable carrier system targeting TGEV (transmissible gastroenteritis coronavirus). Piglets were given enveloped spike protein as an antigen to produce neutralising antibodies (Egelkrout et al., 2020). Post-weaning diarrhoea (PWD) in pigs caused by *enterotoxin E. coli* (ETEC) was also a subject for developing a vaccination system from plants that exhibited immunogenicity and partial protection after being orally delivered to piglets by generating ETEC F4 fimbriae protein in tobacco alfalfa leaves and barley seeds (Kolotilin et al., 2012; Girma et al., 2019).

Different research teams in various parts of the globe have developed vaccines based on immune response such as the immunogenicity of tobacco mosaic virus (TMV) against bovine herpes virus (BHV) (Dummer & Leite et al., 2014), the structural protein E2 of bovine viral diarrhoea virus (BVDV) in transgenic alfalfa (Aguirreburualde et al., 2013), and *B. anthracis* in GM tobacco, potato, and tomato plants.

The subunit B of cholera toxin (CTB) of *V. cholerae* is a vaccine developed from plants that induces mucosal and serum protection. *Agrobacterium tumefaciens* harboring an allele that encodes cholera toxin B altered with a pattern mainly for a localized signal in endoplasmic reticulum, known as SEKDEL, underneath regulation of the CaMV35S, was used to carry out nuclear transformation of tobacco plants (Girma et al., 2019). The L1 main capsid protein coding sequence of the human papillomavirus HPV type 11 (HPV11) was injected into potato plants to create transgenic potato plants (Esparza & Monroy, 2014). An Ebola virus subunit vaccine has been created by using a geminiviral replicon system in *Nicotiana benthamiana*. This was created by connecting the Ebola glycoprotein (GP1) to the C terminus of a humanized IgG antibody (Diamos & Mason, 2019).

Plant-based edible vaccines have a number of advantages over injectable vaccinations, including cost- and time-effectiveness. Because plant-based procedures do not require cold-chain storage and or soil-rich land, it is often substantially less expensive to produce plants with the capability to generate medicinal proteins in huge amounts (Su et al., 2021). Hence, if demand for the vaccine rises, plant-based pharmaceutical manufacturing can more easily scale up to meet increased demand. Furthermore, edible vaccines mean minimal prices for surgical supplies, no need for health practitioners, and a reduced defile hazard (Kurup & Thomas, 2020), competence by multiple components (allowing multiple antigens to approach microfold cells at the same time), and good efficiency in eliciting a mucosal immune response (Bhatia et al., 2015; Jan et al., 2016).

Yet, there are some drawbacks: i.e. low levels of heterologous protein expression, dosage issues, immunological resistance, cross pollination of modified plants with other species etc. In addition, gene silencing, as well as its effects on insects and soil microbes, are key roadblocks to advancement (Arpaia et al., 2020).

17.3 Genetically modified crops as foes

17.3.1 Potential adverse health effects

Consuming GM foods can sometimes lead to allergenicity and prove toxic, which is attributed to the use of protective proteins to obtain pest resistance. They may also cause introduction of allergens into the food chain. GM soybean containing a methionine-rich 2S albumin storage protein was reported to be a potent allergen. However, its genetic modification was carried out to raise its methionine content, which indicates a higher nutritional profile, since it was for use as animal feed (Nordlee et al., 1996). Consequently, GM crops could prove as safe as their conventional counterparts. These allergens are commonly the glycoproteins (Kaeppler, 2000) and are present naturally in a few sources such as peanuts, shellfish, tree nuts, and some legumes which predicts that only few foods could lead to allergenic reactions. Also, the allergenicity occurs when these allergens are present in concentrated amounts in the food and the individuals are sensitized to them over the time. Thus, there are few chances that allergenicity could occur from GM foods (Verma, 2013). The detrimental health effects of GM foods include their toxicity, allergenicity, and transfer of antibiotic resistance. The transfer of an allergen may occur from a GM crop into a non-allergenic target crop which ultimately leads to the formation of a neo-allergen (Bawa & Anilakumar, 2013). However, it is easy to test the allergenicity of GM food using in vitro tests such as RAST or immunoblotting when the gene is from a crop of known allergenicity (Nordlee et al., 1996; Noteborn et al., 1995).

In the late 1990s, it was reported that GM potatoes are toxic to mammals. These potatoes were genetically modified to produce Galanthus nivalis agglutinin (GNA) – a kind of natural insecticide usually found in snowdrops. The potatoes were found to cause damage to vital organs, the immune system, and the stomach lining (Pusztai & Ewen, 1999). However, it was reported in later research that the damage to the vital organs was not because of the genes but as a result of a particular component of the genetic engineering process. It was shown by the study that all these issues were caused by the 35S cauliflower-mosaic-virus (CaMv) promoter which was introduced in almost all GM foods and crops and which might be due to its linkage to a non-target chromosome which led to the switching on of wrong genes.

17.3.3 Effect of removal or inactivation of genes

The altered properties of GM foods are due to alteration in their genes which may involve incorporation of new genes. Moreover, the removal of genes from plants or organisms might result in the production of desirable or undesirable traits. Thus it must be considered that removal or inactivation of a gene may have detrimental effects on food quality. For example, coffee can be genetically engineered to form decaffeinated coffee. In such coffee plants, the genes responsible for caffeine synthesis are turned off or deleted but this may have an undesirable side effect. Inhibition of aflatoxin synthesis is a characteristic of caffeine. The aflatoxin is a potent toxin and also a carcinogen produced by certain kind of molds. Thus, the decaffeinated coffee beans become more potent to contamination by molds producing aflatoxin. The aflatoxin may remain active through the food preparation process and lead to toxicity.

Many environmentalists are also of the opinion that the pesticidal gene product of GM crops may prove toxic to non-target consumers. For example, Bt genes were incorporated with the aim to give better pest control. However, its endotoxins adversely effected the health of non-target species (Betz et al., 2000). Depending on the strain of Bt genes, its range of toxicity is, however, narrow and was limited to specific groups of insects, lepidoptera or coleoptera. The random gene insertion, transgene stability, and genomic disruption are speculated to be due to gene transfer that may result from unpredictable gene expression. This gene flow causes losses in weed diversity (Radosevich et al., 1992).

7.3.4 Toxicity testing of the whole plant

GM foods must be tested for their probable effects before allowing them to enter the food chain. It has also been argued by concerned experts that GM foods should be subjected to the same testing and approval procedures as for medicines – i.e. clinical trials (Halford & Shewry, 2000). Testing the toxicity of GM foods may help in the assessment of health risks. Food contains a wide number of chemicals. The toxicity of the components of GM foods such as Bt cotton can be assessed in experimental animals at high doses. One can assess the extent of safety of a whole GM food by comparing it with the non-GM food or the constituent from which it is derived (Betz et al., 2000). But, it might be not be possible to test whole foods at high doses. Additionally, the overexpression of an existing protein or toxicologically active constituent may cause ill health effects leading to greater exposure to that particular constituent.

17.3.5 Disruption of the food web

One of the biggest worries about developing pest-resistant GM plants is the possibility that the transgenic features of these crops could be transmitted to the other weed varieties which will result in the resistance, and thus amount, of weeds (Raman, 2017). It's also likely that if more pests die due to insect-resistant varieties, competitive rivalry will decrease, and tiny pests will become severe issues. Their use may also lead to the migration of pests towards another communities of plants which were not affected previously. Propagation of these effects to higher levels in the food web hierarchy is claimed to impact the animals and plants that humans eat (Bawa & Anilakumar, 2013). More research studies on toxicity are also required to see whether herbicide or plant residues (pest-resistant) have a significant impact on groups of microbe species found in the soil, including fungi, bacteria, nematodes, and other microbes (Zhang et al., 2016).

17.3.6 Western agriculture's dominance, transnational corporations' monopolies, and misuse of the poor

People residing in developing countries consider that transnational companies (TNCs) in the advanced economies control most of the biotech industry and that their sole intention is to make money (Ramaswami et al., 2012; Singh, 2021). An example of this is that many herbicide-resistant crops grown through gene modification fall under the category of essential crops in European agribusiness. Technology of allele termination by TNCs has also been condemned because it was seen as a step towards establishing 'ownership' over GM seed manufacturing (Jasanoff, 2016). The unintentional spread of this technology to other variants and similar genera of a certain crop via cross-pollination, leading to massive infertility, was a concern associated with its deployment. Pollen from terminator trait-carrying crops has also been claimed to contaminate the farms of producers who have refused or cannot afford the technique. Furthermore, levying exorbitant costs for the use of seeds will result in local farmers losing control of cooperatives. It is also expected that more privatizations will raise legal and economic hurdles to the usage of varieties (Partelow et al., 2019).

Activists see Golden Rice as more of a publicity stunt for the biotech industry rather than a gift to the world's poor population. According to Charles Margulis of the Greenpeace Genetic Engineering Campaign, the sector has deliberately utilized Golden Rice in an endeavour by industrialized countries to assuage rising suspicion of its innovative foods. Poor farmers in developing countries could be disadvantaged by TNCs, for example, over ownership of the GM seeds TNCs create, which would be just one social issue highlighted by Third World countries over the introduction of GM plants (Philips, 2008). The situation is complicated further by the fact that most of the genetic resources and, as a result, the diversity on which genome modification is based are found in developing countries, and the developing world's farmers are paying for the products that originated from their nation's own resources. In order to support the Third World, additional plants should have been included in the targets, and all these countries should indeed be assisted in moving away from costly and complicated crop cultivation and toward moderate, sustainable techniques (Schulman, 2020).

Bringing GM foods to the marketplace is a very time-consuming and costly procedure. Consumers are worried that copying these innovative species of plants will drive up the seed prices to the point that small farm owners as well as Third World countries will not be able to afford them. Farmers would have to purchase new seeds every year and, as a result, they would suffer financially (Hamburger, 2018). Another issue regarding the adoption of GM species of plants is the reduction of export markets because the goods manufactured by them will be substituted by goods generated via genetic modification in industrialized nations (Taheri et al., 2017). In poor countries, this is projected to result in unemployment and a reduction in international revenues.

The issues of poverty and biodiversity loss in food crops have also been discussed, which are exacerbated by the transnational domination of agribusiness dependent on GM foods (Taheri et al., 2017). It has been claimed that to maintain food and nutritional security, relatively low cost, more readily available, and safer alternatives to transgenic rice, such as green leafy vegetable, moringa, amaranth, sweet potato, and so on, should be employed.

17.3.7 Toxicity to reproductive and developmental systems

There are several studies on this topic but research is limited. M13 DNA given to pregnant rats as food was subsequently found in the body parts of foetuses and infants, implying a potential mother-to-child mode of transmission through the placenta (Shen et al., 2022). Vehicle DNA consumed by the mother could pose a risk to the developing foetus. It was found that the birth rates in swine fed with GM corn plunged by up to 80 percent because of the presence of high levels of Fusarium molds, even though it is stated that corn developed from *Bacillus thuringiensis*, i.e., Bt corn encoding crystal (Cry) peptides, is less infected with aflatoxins (Yang et al., 2021; Ojiambo et al., 2018). In further research, the offspring of rats fed GM soya died at a high rate (56 percent), and the surviving progeny had stunted growth (Lohmiller et al., 2020).

GM rice expressing Xa21 caused increased body weight in pregnant rats and also an increase in the body mass, total height, and length of tail in the foetus, while transgenic rice generating trypsin inhibitor in cowpea increased body length in male rats and red blood cell density, haemoglobin, and leukocyte amount in female rats (Gaharwar et al., 2021). Bar genes added into GM potatoes in reproductive and developmental research showed no detrimental impacts, which could be related to the reduced amount of GM potatoes in foodstuffs that in turn disguises any negative implications (Seek Rhee et al., 2005; Wang et al., 2014). First and foremost, transgenic foods must be evaluated for unanticipated health impacts on vulnerable populations such as children, as their ingestion is unavoidable beyond the first year.

17.3.8 Gene transfer studies in human volunteers

In relation to the dangers of GM foods, there are many uncertainties. "Foreign proteins that have never been in the human food chain will soon be consumed in large proportions," one critic predicted (Bawa & Anilakumar, 2013). "But now we're being expected to think that everything is fine with GM foods since we haven't seen any deceased folks yet," says another (Butler, Reichhardt, Abbott, Dickson, & Saegusa, 1999).

As of January 2009 there had only been one human trial that examined the impact of GM food consumption. Seven human volunteers, who had had their large intestines removed for medical reasons, took part in the study. These participants consumed GM soy to determine if its DNA was transferred to the bacteria that dwell normally in the human gut. Before the start of the trial, scientists discovered that three out of seven participants already had transgenes in the bacteria living in their gut transferred from GM soya. The scientists determined that the gene transfer had not occurred throughout the experimentation since the low-frequency transfer didn't rise after the eating of GM soy. They also reported that the transgene would not survive transit through the intact gastrointestinal system in individuals with full digestive tracts (Netherwood et al., 2004). Since then, West et al. (2019) conducted a randomized double-blind crossover design for acute consumption of two single breakfasts, containing either seed oil generated from transgenic *Camelina sativa* plants or commercially blended fish oil. Allocation of 36 healthy people was done into two groups and venous blood samples were collected after the postprandially. There was no adverse health effects reported excepting few minor illnesses such as minor upper respiratory tract infections, minor nose bleed, and headaches.

According to two studies, there were no substantial differences in the quality and nutritional value of animal feeds that included material derived from GM plants (Flachowsky, Chesson, & Aulrich, 2005; Beagle et al., 2006). And, other studies showed that no recombinant DNA or novel protein residues were discovered in any organ or tissue samples tested from animals fed GM plants (Nordlee, Taylor, Townsend, Thomas, & Bush, 1996; Streit, Beach, Register, Jung, & Fehr, 2001).

17.3.9 Reduced efficacy of antibiotic therapy

The general community is equally worried about the risks of gene transfer from plants to microorganisms. It is hypothesized that human and animals feeding on foods that are genetically modified with antibiotic resistance marker genes (e.g., Npt II gene possessing neomycin phosphotransferase for resistance to kanamycin and neomycin, or Ampr gene possessing β-lactamase for resistance to ampicillin) may result in gene transmission from the GM food to the microflora in the human gut and animals or pathogens in the surroundings (Kaeppler, 2000).

However, since no proof of allergenicity or toxicity associated with the Npt II (neomycin phosphotransferase II) gene (which is utilized to grow GM plants currently on the market) has been found, it is considered safe to use. Furthermore, humans are vulnerable to eating kanamycin-resistant microorganisms that are naturally present in the environment. According to reports, the human gut has 10^{12} bacteria that are kanamycin-resistant, and eating a tomato with the Npt II gene increases the frequency of bacteria that are kanamycin-resistant in the stomach by only 10^{-6} percent. In addition, the enzyme encoding for Npt II is inactivated or degraded by acidic conditions in the stomach or rumen. ATP is also required for the function of Npt II and this is found only in extremely low amounts in the gut. It has been shown that when the Ampr gene is used to select bacterial recombinants, it is not transmitted to plants. Furthermore, because the Ampr gene does not code for any plant products, it is considered safe. The Npt II and Ampr genes have therefore been approved for use in GM plants (Flavell, Dart, Fuchs, & Fraley, 1992; Read, 2000; Jefferson, 1987).

Yet, the general public is hesitant to accept this fact. In response to public opinion, scientists have also generated selectable marker genes based upon non-resistance such as green fluorescent protein (GFP) encoding gene and glucuronidase gene (Uid A) (Jefferson, 1987; Ebinuma, Sugita, Matsunaga, & Yamakado, 1997; H. Kaeppler, Menon, Skadsen, Nuutila, & Carlson, 2000). A study committee from the National Academies of Science, Engineering, and Medicine concluded that there were no sufficient data available which showed any difference in risks to human health when conventionally bred crops and currently commercialized genetically engineered crops are compared as per (Bayer, 2020).

Removal of antibiotic resistance genes includes many strategies (Jaiwal, Sahoo, Singh, & Singh, 2002). These include:

1st strategy: Cloning a gene (selectable marker) and a transgene on two unique transfer DNA (T-DNA) molecules in single or multiple *Agrobacterium tumefaciens* strains for plant transformation. The transgenic and selectable marker genes are thus placed at loci that are expected to recombine frequently in the next generation, allowing the transgene to be isolated from the selectable marker gene (De Block & Debrouwer, 1991; Daley, Knauf, Summerfelt, & Turner, 1998).

2nd strategy: For removing the selectable marker gene from the genome of the plant, surround it with direct repeats of recognition sites for a site-specific recombinase, allowing it to be easily removed via recombinase-mediated site-specific recombination. The R/Rs system of *Zygosaccharomyces rouxii*, the Cre/lox recombination system of bacteriophage P1, the Flp/frt recombination system of yeast 2m plasmid, and the Cre/lox recombination system of bacteriophage P1 are all examples of this type. The first round of transformation creates transgenic plants with the selection marker between two directly orientated recognition sites for the respective recombinase, which is a frequent feature of these systems. The recombinase reaction is triggered after expression of recombinase, which can be achieved by crossing in plants that already express the enzyme, transient expression by the second transformation, or the use of an inducible promoter (Dale & Ow, 1991; Russell, Hoopes, & Odell, 1992; Kilby, Davies, Snaith, & Murray, 1995; Onouchi, Nishihama, Kudo, Machida, & Machida, 1995; Lyznik, Rao, & Hodges, 1996; Vergunst, Jansen, & Hooykaas, 1998). It is also possible to delete a marker gene by putting it on a transposable element, which gets removed after transposition (Ebinuma et al., 1997; Yoder & Goldsbrough, 1994). The transgene may be mobile on its own, and transposase activation permits the desired transgene to be relocated to a different chromosomal site. The two transgenes may be separated through genetic crosses and/or segregation (Goldsbrough, Lastrella, & Yoder, 1993).

3rd strategy: The deletion of DNA based on intrachromosomal homologous recombination between two homologous sequences, such as integrating the att sequence of bacteriophage (Zubko, Scutt, & Meyer, 2000), can be used to create marker-free GM plants.

17.3.10 Possible transfer of antibiotic-resistant genes to bacteria in the gastrointestinal tract

Fears are increasing about the possibility of transgene transfer from GM crops to bacteria living in the gastrointestinal tract (Uzogara, 2000). The fact that antibiotic resistance coding genes have been inserted into the genome of several GM crops alongside the GM trait transgene is a source of particular worry. These antibiotic resistance genes were introduced to make it easier to choose GM positive transformed plants.

One source of concern is the prospect that markers being used as antibiotic resistance genes in transgenic crops could be horizontally transmitted to pathogenic bacterial gut, lowering antimicrobial therapy effectiveness. Although this chance is modest (Halford & Shewry, 2000), alternative marker genes have been developed such as the jellyfish green fluorescent protein (GFP) gene. Richards et al. (2003) determined that GFP has a low allergenicity risk after testing its toxicity and allergenicity in male rats for 26 days.

Single bacterial species can pick up DNA fragments from GM plants. *Acinetobacter* sp. has been demonstrated to efficiently ingest transgene DNA from GM sugar beet (Gebhard & Smalla, 1998) and GM potatoes (de Vries, Meier, & Wackernagel, 2001). These tests were, however, carried out under standardized laboratory conditions that are unlikely to be met in natural settings. To the best of the authors' knowledge, there has been no in vivo investigation that has shown that DNA fragments from GM crops are reliably integrated into the mammalian gut microbiota.

Even though the intestinal plasmid was able to modify bacteria in vitro, research into the integration of ampicillin resistance coding plasmid DNA by the gut microbiome of mono-associated rats found no uptake in vivo (Wilcks, van Hoek, Joosten, Jacobsen, & Aarts, 2004). There are, however, a few technological constraints that impede the detection of exceptional integration events, so they need to be considered. For example, a bacterial transformant can take years to proliferate to the point where it can be detected using existing analytical methods (Nielsen et al., 2014). In the USA, penicillin-resistant S. *pneumoniae*, for example, took decades to reach 21.5 percent of isolates (Doern et al., 2001).

Although the amalgamation of DNA fragments from GM plants into bacteria inside the gut microbiome is a possibility, only a single human investigation has found evidence of DNA transfer from gene of GM soya into the microbiota of the subject. In this investigation, the CP4-EPSPS (CP4 variant of 5-enolpyruvyl-shikimate-3-phosphate synthase) transgene was discovered in the gut microbiota of three out of seven ileostomists, but this did not escalate after eating a GM soya meal (Netherwood et al., 2004). If the amount of antibiotics or herbicides consumed is significant, only genes imparting resistance to antibiotics or herbicides such as glyphosate with antibacterial capabilities would be capable of being chosen. It's also worth noting that most research on the establishment of antibiotic resistance exclude the notion of random transformation with linear DNA fragments as a contributing factor (Gay & Gillespie, 2005). In short, considering the important consequences for the health of humans and animals that could result from the creation of antibiotic-resistant pathogenic bacteria via horizontal gene transfer, it is reasonable that the next generation of GM crops is developed without antibiotic resistance genes (Nawaz et al., 2019).

17.4 Concluding remarks

Globally, several countries have reported technological strategies to enhance their food production based on biotechnology. GM foods are the outcome of these strategies and can be classified into three generations. Foods from the first generation possess high herbicide tolerance, insect resistance, and tolerance to environmental stress; foods from the second generation possess high nutritional profile; whereas third-generation foods produce pharmaceuticals. One of the biggest constraints related to these GM foods are food-related allergies regardless of the fact that this technology is able to provide sustainable agricultural productivity. It is the US FDA and other regulatory bodies worldwide which have the authority to ensure the safety of these foods. This debatable topic regarding the safety of GM foods does not seem to come to a halt even in the face of various scientific backings which have undoubtedly supported their safe use. Nevertheless, these foods are beneficial in many aspects, and they should be adopted under specified conditions averting potential risks.

References

Aggarwal, S. R. (2012). What's fueling the biotech engine – 2011 to 2012. *Nature Biotechnology*, 30(12), 1191–1197.

Aguirreburualde, M. S. P., Gómez, M. C., Ostachuk, A., Wolman, F., Albanesi, G., Pecora, A., & Wigdorovitz, A. (2013). Efficacy of a BVDV subunit vaccine produced in alfalfa transgenic plants. *Veterinary Immunology and Immunopathology*, 151(3–4),315–324.

Arpaia, S., Christiaens, O., Giddings, K., Jones, H., Mezzetti, B., Moronta-Barrios, F., … Dietz-Pfeilstetter, A. (2020). Biosafety of GM crop plants expressing dsRNA: Data requirements and EU regulatory considerations. *Frontiers in Plant Science*, 11, 940.

Avery, O. T., Macleod, C. M., & McCarty, M. (1994). Studies on the chemical nature of the substance inducing transformation of pneumococcal types: Induction of transformation by a deoxyribonucleic acid fraction isolated from pneumococcus type III. *Journal of Experimental Medicine*, 79(2), 137–158.

Bakshi, A. (2011). Potential adverse health effects of Genetically Modified crops. *Journal of Toxicology and Environmental Health*, 6(3), 211–225.

Barro, F., Rooke, L., & Bekes, F. (1997). Transformation of wheat with high molecular weight subunit genes results in improved functional properties. *Biotechnology*, 15, 1295–1299.

Bawa, A. S., & Anilakumar, K. R. (2013). Genetically modified foods: Safety risks and public concerns – a review. *Journal of Food Science and Technology*, 50(6), 1035–1046.

Bayer (2020). Unlocking hidden – potential biotechnology and GMOs. Retrieved January 17, 2021, from https://www.bayer.com/en/crop-science/crop-science-innovations-biotech-gmos

Beagle, J., Apgar, G., Jones, K., Griswold, K., Radcliffe, J., Qiu, X., … Iqbal, M. (2006). The digestive fate of *Escherichia coli* glutamate dehydrogenase deoxyribonucleic acid from transgenic corn in diets fed to weanling pigs. *Journal of Animal Science*, 84(3), 597–607.

Betz, F. S., Hammond, B. G., & Fuchs, R. L. (2000). Safety and advantages of *Bacillus thuringiensis*-protected plants to control insect pests. *Regulatory Toxicology and Pharmacology*, 32, 156–173.

Bhatia, S., Sharma, K., Dahiya, R., & Bera, T. (2015). *Modern Applications of Plant Biotechnology in Pharmaceutical Sciences* (New York: Academic Press).

Bhuiyan, M. S. A., Amin, Z., Rodrigues, K. F., Saallah, S., Shaarani, S. M., Sarker, S., & Siddiquee, S. (2021). Infectious bronchitis virus (gammacoronavirus) in poultry farming: Vaccination, immune response and measures for mitigation. *Veterinary Sciences*, 8(11), 273.

Brookes, G., & Barfoot, P. (2014). Economic impact of GM crops: The global income and production effects 1996–2012. *GM Crops & Food*, 5(1), 65–75.

Butler, D., Reichhardt, T., Abbott, A., Dickson, D., & Saegusa, A. (1999). Long-term effect of GM crops serves up food for thought. *Nature*, 398(6729), 651–656.

Cong, L., Ran, F. A., Cox, D., Lin, S., Barretto, R., Habib, N., … Marraffini, L. A. (2013). Multiplex genome engineering using CRISPR/Cas systems. *Science*, 339 (6121), 819–882.

Curtiss, R. I., & Cardineau, C. A. (1990). Oral immunization by transgenic plants. World Patent Application, WO 90/02484. Patent record available from the World Intellectual Property Organization (WIPO).

Dale, E. C., & Ow, D. W. (1991). Gene transfer with subsequent removal of the selection gene from the host genome. *Proceedings of the National Academy of Sciences*, 88(23), 10558–10562.

Daley, M., Knauf, V., Summerfelt, K., & Turner, J. (1998). Co-transformation with one *Agrobacterium tumefaciens* strain containing two binary plasmids as a method for producing marker-free transgenic plants. *Plant Cell Reports*, 17(6), 489–496.

Das, T., Samantarai, R., & Panda, P. (2021). Prospects of plant based edible vaccines in combating COVID-19 and other viral pandemics: A review. *e-planet*, 19(1), 1–18.

De Block, M., & Debrouwer, D. (1991). Two T-DNA's co-transformed into *Brassica napus* by a double *Agrobacterium tumefaciens* infection are mainly integrated at the same locus. *Theoretical and Applied Genetics*, 82(3), 257–263.

de Vries, J., Meier, P., & Wackernagel, W. (2001). The natural transformation of the soil bacteria *Pseudomonas stutzeri* and *Acinetobacter* sp. by transgenic plant DNA strictly depends on homologous sequences in the recipient cells. *FEMS Microbiology Letters*, 195(2), 211–215.

De Maagd, R. A., Bosch, D., & Stiekema, W. (1999). *Bacillus* thuringiensis toxin-mediated insect resistance in plants. *Trends in Plant Science*, 4, 9–13.

Dewar, A. M., Haylock, L. A., Bean, K. M., & May, M. J. (2000). Delayed control of weeds in glyphosate-tolerant sugar beet and the consequences on aphid infestation and yield. *Pest Management Science*, 56, 345–350.

Diamos, A. G., & Mason, H. S. (2019). Modifying the replication of geminiviral vectors reduces cell death and enhances expression of biopharmaceutical proteins in *Nicotiana benthamiana* leaves. *Frontiers in Plant Science*, 9, 1974.

Doern, G. V., Heilmann, K. P., Huynh, H. K., Rhomberg, P. R., Coffman, S. L., & Brueggemann, A. B. (2001). Antimicrobial resistance among clinical isolates of *Streptococcus pneumoniae* in the United States during 1999–2000, including a comparison of resistance rates since 1994–1995. *Antimicrobial Agents and Chemotherapy*, 45(6), 1721–1729.

Dummer, L. A., & Leite, F. P. L. (2014). Bovine herpes virus glycoprotein D: A review of its structural characteristics and applications in vaccinology. *Veterinary Research*, 45(1).

Ebinuma, H., Sugita, K., Matsunaga, E., & Yamakado, M. (1997). Selection of marker-free transgenic plants using the isopentenyl transferase gene. *Proceedings of the National Academy of Sciences*, 94(6), 2117–2121.

Egelkrout, E., Hayden, C., Fake, G., Keener, T., Arruda, P., Saltzman, R., ... Howard, J. (2020). Oral delivery of maize-produced porcine epidemic diarrhea virus spike protein elicits neutralizing antibodies in pigs. *Plant Cell, Tissue and Organ Culture (PCTOC)*, 142(1), 79–86.

Ellstrand, N. P. H., & Hancock, J. F. (1999). Gene flow and introgression from domesticated plants into their wild relatives. *Annual Review of Ecology, Evolution, and Systematics*, 30, 539–563.

Esparza , J. F. C., & Monroy, A. (2014). Expression of the capsid protein of human papillomavirus in plants as an alternative for the production of vaccines. In *Plant-derived Pharmaceuticals: Principles and Applications for Developing Countries*, edited by K. L. Hefferon (Wallingford, UK: CAB International), 91.

Fahad, S., Khan, F. A., Pandupuspitasari, N. S., Ahmed, M. M., Liao, Y. C., Waheed, M. T., ... Huang, J. (2015). Recent developments in therapeutic protein expression technologies in plants. *Biotechnology Letters*, 37(2), 265–279.

Flachowsky, G., Chesson, A., & Aulrich, K. (2005). Animal nutrition with feeds from genetically modified plants. *Archives of Animal Nutrition*, 59(1), 1–40.

Flavell, R. B., Dart, E., Fuchs, R. L., & Fraley, R. T. (1992). Selectable marker genes: Safe for plants? *Bio/technology*, 10(2), 141–144.

Gaharwar, U. S., Verma, A., Singh, R., Prasad, T., & Rajamani, P. (2021). Health risks and environmental concerns of GM crop adoption. In *Policy Issues in*

Genetically Modified Crops, edited by P. Singh, A. Borthakur, A. A. Singh, K. Kumar, & K. K. Singh (New York: Academic Press), 371–400.

Garcia-Yi, J., Lapikanonth, T., Vionita, H., Vu, H., Yang, S., Zhong, Y., ... Wesseler, J. (2014). What are the socio-economic impacts of genetically modified crops worldwide? A systematic map protocol. *Environmental Evidence*, 3(1), 1–17.

Gay, P. B., & Gillespie, S. H. (2005). Antibiotic resistance markers in genetically modified plants: A risk to human health? *The Lancet Infectious Diseases*, 5(10), 637–646.

Gebhard, F., & Smalla, K. (1998). Transformation of *Acinetobacter* sp. strain BD413 by transgenic sugar beet DNA. *Applied and Environmental Microbiology*, 64(4), 1550–1554.

Girma, B., Shegu, D., Muluneh, A., & Woldemariyam, F. T. (2019). Vaccine production in transgenic plants for animal and human diseases. *Archives of Veterinary and Animal Sciences*, 1(1), 2.

Goldsbrough, A. P., Lastrella, C. N., & Yoder, J. I. (1993). Transposition mediated re–positioning and subsequent elimination of marker genes from transgenic tomato. *Bio/technology*, 11(11), 1286–1292.

Gómez, E., Lucero, M. S., Zoth, S. C., Carballeda, J. M., Gravisaco, M. J., & Berinstein, A. (2013). Transient expression of VP2 in *Nicotiana benthamiana* and its use as a plant-based vaccine against infectious bursal disease virus. *Vaccine*, 31(23), 2623–2627.

Halford, N. G., & Shewry, P. R. (2000). Genetically modified crops: Methodology, benefits, regulation and public concerns. *British Medical Bulletin*, 56(1), 62–73.

Hamburger, D. J. (2018). Normative criteria and their inclusion in a regulatory framework for new plant varieties derived from genome editing. *Frontiers in Bioengineering and Biotechnology*, 6, 176.

Hare, P. D., & Chua, N. H. (2002). Excision of selectable marker genes from transgenic plants. *Nature Biotechnology*, 20(6), 575–580.

Jaiwal, P. K., Sahoo, L., Singh, N. D., & Singh, R. P. (2002). Strategies to deal with the concern about marker genes in transgenic plants: Some environment-friendly approaches. *Current Science*, 83(2), 128–136.

James, R. R., Difazio, S. P., Brunner, A. M., & Strauss, S. H. (1998). Environmental effects of genetically engineered woody biomass crops. *Biomass and Bioenergy*, 14(4), 403–414.

Jan, N., Shafi, F., Hameed, O., Muzaffar, K., Dar, S., Majid, I., & Nayik, G. A. (2016). An overview on edible vaccines and immunization. *Austin Journal of Nutrition and Food Sciences*, 4(2), 1078.

Jasanoff, S. (2016). *The Ethics of Invention: Technology and the Human Future* (New York: WW Norton & Company).

Jefferson, A. R. (1987). Assaying chimeric genes in plants: The GUS gene fusion system. *Plant Molecular Biology Reporter*, 5, 387–405.

Kaeppler, H. F. (2000). Food safety assessment of genetically modified crops. *Agronomy Journal*, 92(4), 793–797.

Kaeppler, H., Menon, G., Skadsen, R., Nuutila, A.-M., & Carlson, A. (2000). Transgenic oat plants via visual selection of cells expressing green fluorescent protein. *Plant Cell Reports*, 19(7), 661–666.

Kilby, N. J., Davies, G. J., Snaith, M. R., & Murray, J. A. (1995). FLP recombinase in transgenic plants: Constitutive activity in stably transformed tobacco and generation of marked cell clones in Arabidopsis. *The Plant Journal*, 8(5), 637–652.

Kolotilin, I., Kaldis, A., Devriendt, B., Joensuu, J., Cox, E., & Menassa, R. (2012). Production of a subunit vaccine candidate against porcine post-weaning diarrhea in high-biomass transplastomic tobacco. *PLoS One*, 7(8), e42405.

Kramkowska, M., Grzelak, T., & Czyzewska, K. (2013). Benefits and risks associated with genetically modified food products. *Annals of Agricultural and Environmental Medicine (AAEM)*, 20(3), 413–419.

Krishna, V., & Qaim, V. (2012). Bt cotton and sustainability of pesticide reductions in India. *Agricultural Systems*, 107, 47–55.

Kurup, V. M., & Thomas, J. (2020). Edible vaccines: promises and challenges. *Molecular Biotechnology*, 62(2), 79–90.

Lipton, M. (2007). Plant breeding and poverty: Can transgenic seeds replicate the 'Green Revolution' as a source of gains for the poor? *Journal of Development Studies*, 43, 31–62.

Lohmiller, J. J., Swing, S. P., & Hanson, M. M. (2020). In Reproduction and breeding. In *The Laboratory Rat*, edited by G. J. Krinke (New York: Academic Press), 157–179.

Lyznik, L. A., Rao, K., & Hodges, T. K. (1996). FLP-mediated recombination of FRT sites in the maize genome. *Nucleic Acids Research*, 24(19), 3784–3789.

Meriç, S., Gümüş, T., & Ayan, A. (2021). Plant-based vaccines: The future of preventive healthcare? In *Botany: Recent Advances and Applications*, edited by B. K. Ghimire (London: IntechOpen).

Mor, T. S., Gomez-Tim, M. A., & Palmer, K. E. (1998). Perspective, edible vaccines – a concept coming of age. *Trends in Microbiology*, 6, 449–453.

Mutuc, M., Rejesus, R., Pan, S., & Yorobe, J. (2012). Impact assessment of Bt corn adoption in the Philippines. *Journal of Agricultural & Applied Economics*, 44, 117–135.

Napier, J. A., Michaelson, L. V., & Stobart, A. K. (1999). Plant desaturases: Harvesting the fat of the land. *Current Opinion in Plant Biology*, 2, 123–127.

Nawaz, M. A., Mesnage, R., Tsatsakis, A. M., Golokhvast, K. S., Yang, S. H., Antoniou, M. N., & Chung, G. (2019). Addressing concerns over the fate of DNA derived from genetically modified food in the human body: A review. *Food and Chemical Toxicology*, 124, 423–430.

Netherwood, T., Martín-Orúe, S. M., O'Donnell, A. G., Gockling, S., Graham, J., Mathers, J. C., & Gilbert, H. J. (2004). Assessing the survival of transgenic plant DNA in the human gastrointestinal tract. *Nature Biotechnology*, 22(2), 204–209.

Nicolia, A., Manzo, A., Veronesi, F., & Rosellini, D. (2014). An overview of the last 10 years of genetically engineered crop safety research. *Critical Reviews in Biotechnology*, 34(1), 77–88.

Nielsen, K. M., Bøhn, T., & Townsend, J. P. (2014). Detecting rare gene transfer events in bacterial populations. *Frontiers in Microbiology*, 4, 415.

Nordlee, J. A., Taylor, S. L., Townsend, J. A., Thomas, L. A., & Bush, R. K. (1996). Identification of a Brazil-nut allergen in transgenic soybeans. *New England Journal of Medicine*, 334(11), 688–692.

Noteborn, H. P. J. M., Bienenmann-Ploum, M. E., Van den Berg, J. H. J., Alink, G. M., Zolla, L., Reynaerts, A., ... Kuiper, H. A. (1995, January). Safety assessment of the *Bacillus thuringiensis* insecticidal crystal protein CRYIA (b) expressed in transgenic tomatoes. In *ACS Symposium Series*, Vol. 605 (Washington, DC: American Chemical Society), 134–147.

Ojiambo, P. S., Battilani, P., Cary, J. W., Blum, B. H., & Carbone, I. (2018). Cultural and genetic approaches to manage aflatoxin contamination: Recent insights provide opportunities for improved control. *Phytopathology*, 108(9), 1024–1037.

Onouchi, H., Nishihama, R., Kudo, M., Machida, Y., & Machida, C. (1995). Visualization of site-specific recombination catalyzed by a recombinase from *Zygosaccharomyces rouxii* in *Arabidopsis thaliana*. *Molecular and General Genetics* MGG, 247(6), 653–660.

Partelow, S., Abson, D., Schlüter, A., Fernández-Giménez, M., von Wehrden, H., & Collier, N. (2019). Privatizing the commons: New approaches need broader evaluative criteria for sustainability. *International Journal of the Commons*, 13(1).

Phillips, T. (2008). Genetically modified organisms (GMOs): Transgenic crops and recombinant DNA technology. *Nature Education*, 1(1), 213.

Pusztai, A., & Ewen, S. W. B. (1999). Effect of diets containing genetically modified potatoes expressing *Galanthus nivalis* lectin on rat small intestine. *Lancet*, 354, 1353–1354.

Radosevich, S. R., Ghersa, C. M., & Comstock, G. (1992). Concerns a weed scientist might have about herbicide-resistant crops. *Weed Technology*, 6(3), 635–642.

Raman, R. (2017). The impact of Genetically Modified (GM) crops in modern agriculture: A review. *GM Crops & Food*, 8(4), 195–208.

Ramaswami, B., Pray, C. E., & Lalitha, N. (2012). The spread of illegal transgenic cotton varieties in India: Biosafety regulation, monopoly, and enforcement. *World Development*, 40(1), 177–188.

Read, D. (2000). *Use of Antibiotic Resistance Marker Genes in Genetically Modified Organisms*. Environmental Risk Management Authority, Wellington, New Zealand.

Reichman, J. R., Watrud, L. S., Lee, L. S., Burdick, C. A., Bollman, M. A., Storm, M. J., ... Mallory-Smith, C. (2006). Establishment of transgenic herbicide-resistant creeping bentgrass (*Agrostis stolonifera* L.) in nonagronomic habitats. *Molecular Ecology*, 15(13), 4243–4255.

Richards, H. A., Han, C.-T., Hopkins, R. G., Failla, M. L., Ward, W. W., & Stewart Jr, C. N. (2003). Safety assessment of recombinant green fluorescent protein orally administered to weaned rats. *The Journal of Nutrition*, 133(6), 1909–1912.

Ruiz, V., Mozgovoj, M. V., Dus Santos, M. J., & Wigdorovitz, A. (2015). Plant-produced viral bovine vaccines: What happened during the last 10 years? *Plant Biotechnology Journal*, 13(8), 1071–1077.

Russell, S. H., Hoopes, J. L., & Odell, J. T. (1992). Directed excision of a transgene from the plant genome. *Molecular and General Genetics* MGG, 234(1), 49–59.

Sayanova, O., Smith, M. A., & Lapinskas, P. (1997). Expression of a borage desaturase cDNA containing an N-terminal cytochrome b(5) domain results in the accumulation of high levels of delta(6)- desaturated fatty acids in transgenic tobacco. *The Proceedings of the National Academy of Sciences*, USA, 94, 4211–4216.

Schulman, A. H. (2020). The impact of GM crops on agriculture. In *Genetically Modified and Irradiated Food*, edited by D. V. Andersen (New York: Academic Press), 195–213.

Seek Rhee, G., Hyun Cho, D., Hyuck Won, Y., Hyun Seok, J., Sun Kim, S., Jun Kwack, S., ... & Sik Choi, K. (2005). Multigeneration reproductive and developmental toxicity study of bar gene inserted into genetically modified potato on rats. *Journal of Toxicology and Environmental Health, Part A*, 68(23–24),2263–2276.

Shen, C., Yin, X. C., Jiao, B. Y., Li, J., Jia, P., Zhang, X. W., ... Liu, J. P. (2022). Evaluation of adverse effects/events of genetically modified food consumption: A systematic review of animal and human studies. *Environmental Sciences Europe*, 34(1), 1–33.

Singh, P. (2021). Genetically modified crops in India: Politics, policies, and political economy. In *Policy Issues in Genetically Modified Crops*, edited by P. Singh, A. Borthakur, A. A. Singh, K. Kumar, & K. K. Singh (New York: Academic Press), 75–96.

Streit, L. G., Beach, L. R., Register, J. C., Jung, R., & Fehr, W. R. (2001). Association of the Brazil nut protein gene and Kunitz trypsin inhibitor alleles with soybean protease inhibitor activity and agronomic traits. *Crop Science*, 41(6), 1757–1760.

Su, H., Yakovlev, I. A., Van Eerde, A., Su, J., & Clarke, J. L. (2021). Plant-produced vaccines: Future applications in aquaculture. *Frontiers in Plant Science*, 12, 718775.

Sweet, J. B., & Shepperson, R. (1997). The impact of genetically modified herbicide tolerant oilseed rape in UK. *Acta Horticulturae*, 459, 225–234.

Taheri, F., Azadi, H., & D'Haese, M. (2017). A world without hunger: Organic or GM crops? *Sustainability*, 9(4), 580.

Takeyama, N., Kiyono, H., & Yuki, Y. (2015). Plant-based vaccines for animals and humans: Recent advances in technology and clinical trials. *Therapeutic Advances in Vaccines*, 3(5–6), 139–154.

Ubalua, A. (2009). Transgenic plants: Successes and controversies. *Biotechnology and Molecular Biology Reviews*, 4(6), 118–127.

Uzogara, S. G. (2000). The impact of genetic modification of human foods in the 21st century: A review. *Biotechnology Advances*, 18(3), 179–206.

Vergunst, A. C., Jansen, L. E., & Hooykaas, P. J. (1998). Site-specific integration of Agrobacterium T-DNA in *Arabidopsis thaliana* mediated by Cre recombinase. *Nucleic Acids Research*, 26(11), 2729–2734.

Verma, S. R. (2013). Genetically modified plants: Public and scientific perceptions. *ISRN Biotechnology*, 7, 1–11.

Wang, E. H., Yu, Z., Hu, J., Jia, X. D., & Xu, H. B. (2014). A two-generation reproduction study with transgenic Bt rice TT51 in Wistar rats. *Food and Chemical Toxicology*, 65, 312–320.

West, A. L., Miles, E. A., Lillycrop, K. A., Han, L., Sayanova, O., Napier, J. A., … Burdge, G. C. (2019). Postprandial incorporation of EPA and DHA from transgenic *Camelina sativa* oil into blood lipids is equivalent to that from fish oil in healthy humans. *British Journal of Nutrition*, 121(11), 1235–1246.

Wilcks, A., van Hoek, A. H., Joosten, R. G., Jacobsen, B. B., & Aarts, H. J. (2004). Persistence of DNA studied in different ex vivo and in vivo rat models simulating the human gut situation. *Food and Chemical Toxicology*, 42(3), 493–502.

Yang, Y., Deng, M., Chen, J., Zhao, X., Xiao, K., He, W., … Tan, C. (2021). Starch supplementation improves the reproductive performance of sows in different glucose tolerance status. *Animal Nutrition*, 7(4), 1231–1241.

Yoder, J. I., & Goldsbrough, A. P. (1994). Transformation systems for generating marker–free transgenic plants. *Bio/technology*, 12(3), 263–267.

Zhang, C., Wohlhueter, R, & Zhang, H. (2016). Genetically modified foods: A critical review of their promise and problems. *Food Science and Human Wellness*, 5, 116–123. doi:10.1016/j.fshw.2016.04.002.

Zubko, E., Scutt, C., & Meyer, P. (2000). Intrachromosomal recombination between attP regions as a tool to remove selectable marker genes from tobacco transgenes. *Nature Biotechnology*, 18(4), 442–445.

Index

Taylor & Francis eBooks

www.taylorfrancis.com

A single destination for eBooks from Taylor & Francis with increased functionality and an improved user experience to meet the needs of our customers.

90,000+ eBooks of award-winning academic content in Humanities, Social Science, Science, Technology, Engineering, and Medical written by a global network of editors and authors.

TAYLOR & FRANCIS EBOOKS OFFERS:

- A streamlined experience for our library customers
- A single point of discovery for all of our eBook content
- Improved search and discovery of content at both book and chapter level

REQUEST A FREE TRIAL

support@taylorfrancis.com

Routledge Taylor & Francis Group
CRC Press Taylor & Francis Group